11th International Scientific Conference of Civil and Environmental Engineering for PhD. Students and Young Scientists (YS2019)

IOP Conference Series: Materials Science and Engineering
Volume 566

Tatranska Lomnica, Slovakia
25 - 26 April 2019

Editors:

Dusan Katunsky
Peter Mesaros
Vlasta Ondrejka Harbul'akova

ISBN: 978-1-5108-9392-4
ISSN: 1757-8981

Printed from e-media with permission by:

Curran Associates, Inc.
57 Morehouse Lane
Red Hook, NY 12571

Some format issues inherent in the e-media version may also appear in this print version.

Institute of Physics
Dirac House, Temple Back
Bristol BS1 6BE UK

Phone: 44 1 17 929 7481
Fax: 44 1 17 920 0979

techtracking@iop.org

Additional copies of this publication are available from:

Curran Associates, Inc.
57 Morehouse Lane
Red Hook, NY 12571 USA
Phone: 845-758-0400
Fax: 845-758-2633
Email: curran@proceedings.com
Web: www.proceedings.com

Table of contents

Volume 566

11th International Scientific Conference of Civil and Environmental Engineering for PhD. Students and Young Scientists (YS2019)

25–26 April 2019, Tatranská Lomnica, Slovakia

Published online: 9 August 2019

Papers

PREFACE

The 11th International Scientific Conference of Civil and Environmental Engineering for the PhD. students and young scientists below 35 years under auspices of the dean of the Civil Engineering Faculty of the Technical University of Košice assoc. prof. Peter Mésároš.
The event was held in Tatranské Matliare, the High Tatras, Slovakia on 25th - 26th of April 2019. The traditional meeting of PhD. students, their supervisors and young scientists working in the field of civil engineering and environmental science was organised by the Faculty of Civil Engineering of the Technical University of Košice, Slovakia.
The Young Scientist 2019 offered an international platform for the dissemination of original research results; provide a pleasant environment to present new, great ideas of young researchers and discovered advances in the field of civil engineering and related interdisciplinary topics. The conference provided an opportunity for PhD. students and young researchers to discussed scientific topics and research results of topical issues in civil engineering, to share their knowledge and experiences, to collaborate in innovative designs and proposals. In addition, the valuable presentations were extended by a social program including the interesting field trip in the High Tatras.

The Conference assembled more than 50 delegates, from the Slovakia (Technical University of Košice, Slovak University of Technology in Bratislava), the Czech Republic (Czech Technical University in Prague; Brno University of Technology, Mendel University in Brno, VSB-Technical University of Ostrava), from the Croatia (Josip Juraj Strossmayer University of Osijek) and from the Poland (Cracow University of Technology of Cracow) including PhD. students and other representatives from the universities. The scientific contributions involved a wide range of research fields: Building Physics and Services, Sustainable Architecture and Energy Efficiency, Statics, Dynamics and Modeling, Structural Engineering and Bridges, Transport and Geotechnical Engineering, Construction Technology, Organization and Management, Construction Economics, Marketing and Management, Building Information Modeling, Innovations in Construction Design and Technology, Environmental Engineering, Healthy Indoor Environment, Material Engineering and Recycling as well as Sustainable Water Management.
Forty contributions have been selected to be published in the volume of IOP Conference Series: Materials Science and Engineering. The organizing committee appreciates the enormous work of the both the editorial board and reviewers in the preparation of this volume. Many thanks also to all authors for their flexibility and meeting deadlines.

prof. Dušan Katunský
assoc. prof. Peter Mésároš
assis. prof. Vlasta Ondrejka Harbuľáková

ORGANISING COMMITTEE

assis. prof. Vlasta ONDREJKA HARBUĽÁKOVÁ, PhD.
Ing. arch. Viktor KARĽA
Ing. Jaroslav KOŠIČAN
Ing. Martin LAVKO
Ing. Jana SMETANKOVÁ
Ing. Lenka ŠTULEROVÁ

REVIEW STATEMENT

All papers published in this volume of *IOP Conference Series: Materials Science and Engineering* have been peer reviewed per two reviewers through processes administered by the Editors. Reviews were conducted by expert referees to the professional and scientific standards expected of a proceedings journal published by IOP Publishing.

LIST OF REVIEWERS

prof. Stanislav KMEŤ
prof. Dušan KATUNSKÝ
prof. Vincent KVOČÁK
prof. Jindřich MELCHER
prof. Jiří HIRŠ
prof. Zuzana VRANAYOVÁ
prof. Adriana EŠTOKOVÁ
prof. Zbyšek PAVLÍK
prof. Mária KOZLOVSKÁ
prof. Magdaléna BÁLINTOVÁ
assoc. prof. Peter MÉSÁROŠ
assoc. prof. Martina ZELEŇÁKOVÁ
assoc. prof. Andrej ŠKRINÁR
assoc. prof. Anna SEDLÁKOVÁ
assoc. prof. Brigita SALAIOVÁ
assoc. prof. Daniela KÁPOSZTÁSOVÁ
assoc. prof. Peter KAPALO
assoc. prof. Silvia VILČEKOVÁ
assoc. prof. Stanislav SEITL
assoc. prof. Viktor BORZOVIČ
assoc. prof. Martin LIDMILA
assoc. prof. Marián ROVŇÁK
assoc. prof. Ján KANÓCZ
assoc. prof. Miloslav BAGOŇA
assoc. prof. Alena SIČÁKOVÁ
assoc. prof. Martin LOPUŠNIAK
assoc. prof. Viktor BORZOVIČ
assoc. prof. Marián VERTAĽ
assoc. prof. Magdaléna ŠTUJBEROVÁ
assoc. prof. Ján MANDULA
assoc. prof. Kamila KOTRASOVÁ
assoc. prof. Marián VERTAĽ
assis. prof. Eva TERPÁKOVÁ,
assis. prof. Filip PACHLA
assis. prof. Zuzana POÓROVÁ
assis. prof. Jozef JUNÁK
assis. prof. Martin KOVÁČ
assis. prof. Martin MOHAPL
assis. prof. Miloš SLIVANSKÝ
assis. prof. Mohamad AL ALI
assis. prof. Erika DOLNÍKOVÁ
assis. prof. Peter PLATKO
assis. prof. Richard BALÁŽ
assis. prof. Peter SABOL
assis. prof. Slávka HARABINOVÁ
assis. prof. Štefan DEMČÁK
assis. prof. Štefan KUŠNÍR
assis. prof. Tomáš MANDIČÁK
assis. prof. Viktória BAJZECEROVÁ
assis. prof. Vlasta ONDREJKA HARBUĽÁKOVÁ
assis. prof. Marcela ONDOVÁ
assis. prof. Marek ZOZULÁK
Ing. arch. Viktor KARĽA

IOP Conf. Series: Materials Science and Engineering **566** (2019) 011003 doi:10.1088/1757-899X/566/1/011003

Peer review statement

All papers published in this volume of *IOP Conference Series: Materials Science and Engineering* have been peer reviewed through processes administered by the proceedings Editors. Reviews were conducted by expert referees to the professional and scientific standards expected of a proceedings journal published by IOP Publishing.

IOP Conf. Series: Materials Science and Engineering **566** (2019) 012001 doi:10.1088/1757-899X/566/1/012001

Use of heat mirrors on field of renovation of window structures

S Bizonova[1]

[1] Institute of Architectural Engineering, Faculty of Civil Engineering, Technical University of Kosice, Vysokoskolska 4, 040 23 Kosice, Slovakia

Abstract. The paper presents the results of the analysis of several variants of glass systems using low emissivity heat mirror films. The subject of the paper is the determination of the thermal-optical parameters of the structure as well as the analysis of surface temperatures of the considered glass systems, which were obtained by calculation. The result is optimization of the structure which will be the subject of further research.

1. Introduction

For a longer time, the current topic of renovation of existing buildings may vary in scope. It may be a complex design or, conversely, only minimal changes, depending on investor´s requirements and source options. Building renovation also addresses the issue of replacement of window structures.

After the millennium, the first three-chamber plastic windows (frames) with insulating double glass unit experienced a great boom. Year-to-year, progress has been made and improved frames and glass systems have been introduced to the market. Improvement happened in case of thermal insulating and optical properties of glass systems. A huge trend was the replacement of the original windows as the household heat consumption was observed as well as the users noticed a significant outflow of heat through the windows, up to 47% [1,2] approximately 10% of total energy consumption [3]. Conventional double glass systems will no longer be able to meet the standards, the use of a triple insulating unit will therefore be necessary. The main drawbacks of this solution are increased system weight and impaired light transmission [4]. The solution could be a glass system consisting of two 4mm thick glass panes and additional foils.

However, the normative system of the Slovak Republic in relation to energy saving is still intensifying. The current requirement (recommended value) for U value through windows up to year 2020 is $U_{W,r1}$= 1,0 W/(m^2.K), the normalized value being $U_{W,N}$= 1,4 W/(m^2.K). Since 2020, the target U value has been approved so far $U_{W,r2}$= 0,6 W/(m^2.K). The requirements laid down by the standard are binding on new buildings, but if they are functionally, technically and financially feasible, renewed buildings should also meet them [5].

However, the values obtained by the calculation declared by the window manufacturers may not coincide with the real values obtained by in-situ measurements [6]. Any errors or deviations in production and installation into existing perimeter structure may not be taken into account. For this reason, certain reserves are in place and therefore improved thermal insulation properties should be considered in the design.

In case of renovation of existing buildings, their partial designs are considered as a cost-effective solution, the replacement of the glass system in the existing frame. This alternative is based on the fact that the windows (frames) installed after 2000 have at least the adequate thermal insulation features and, if they do not show any defects, their replacement would be unprofitable. It is also important to note that the ratio of the frame surface is significantly lower in relation to the area of the glass system, which logically implies that it is necessary to draw attention mainly to the thermal-optical parameters of the glass system.

2. Progressive glass system - heat mirror film

The heat mirror films work on the principle of reflection of long-range infrared radiation emitted by surrounding surface while being highly transparent [7]. The glass system with the use of the heat mirror is significantly lower in comparison to a conventional triple glass system of the same thickness as the intermediate glass sheet is replaced by one or more heat mirror films. The original gap between in the glass system is thus divided into two or more parts that can be filled with different gases.

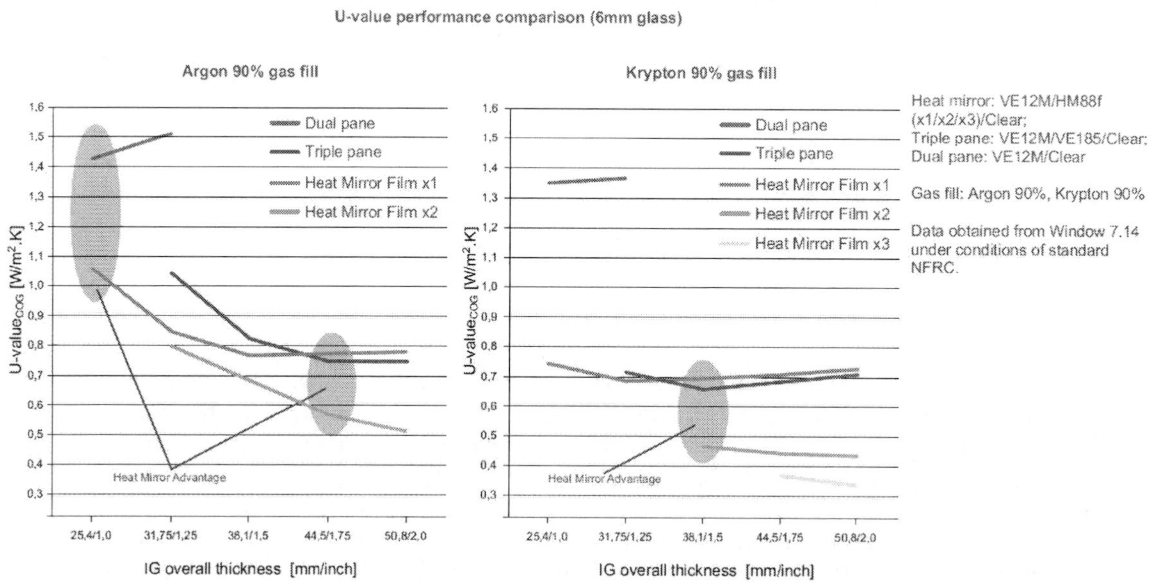

Figure 1. U-value performance comparison (6mm glass.) [10]

Low emissivity heat mirror films were in 1958 discovered as a material applicable to the glass system, have since been used primarily as space thermal insulators [7]. Designed for the patent, 1976 John C. C. Fan and Frank Bachner of the institute of Technology, 1982 [8] under number 4337990 [9]. At the same time, since 1978, Stephen E. Selkowitz and colleagues [7,10] form the University of California, Berkley, also studied the possibilities of using heat mirrors. Films can be applied to glass or plastic element and, depending on application, can reduce thermal losses by 25-57% [7]. Effectively eliminate UV radiation (x<99%). In the form of monolayer coatings, In_2O_3: Sn, doped SnO_2, Cd_2SnO_4; as the two layers are applied: ZnS/metal/ZnS, Bi_2O_3 and TiO_2/Ag/TiO_2 [11].

Table 1. Thermal-optical parameters of heat mirror films. [12]

Name	Emissivity ε [%]	Solar heat gain coefficient g [%]	T_{sol} [%]	R_{sol} [%]	T_{vis} [%]	R_{vis} [%]	T_{UV} [%]
Heat Mirror 88	12,2	69	65	23	88	6	1,6
Heat Mirror TC 88	10,9	56	52	25	80	2	2,0
Heat Mirror 77	7,0	51	48	41	79	13	1,6
Heat Mirror SC 75	5,5	40	37	46	76	11	1,4
Heat Mirror 66	4,3	38	35	55	65	26	1,3
Heat Mirror 55	3,4	32	29	62	54	36	1,1
Heat Mirror 44	3,1	25	23	70	44	48	0,8

Heat Mirror 33	1,7	18	16	77	32	59	0,6

In the introduction there is an outline of the possibility of replacing the glass system as an alternative to replacing the entire window. Three variations were chores for the glass system of the original thickness of 4mm; each of which has several alternatives. They differ in number of heat mirror layers used (2-6pcs/44mm) and at the same time in the thickness of chamber between the gas filled glasses (Krypton, Argon); 12-5mm/44mm. The most effective gas layer thickness for Krypton is 8, in the case of Argon 12 mm [13]. The different variations differ in the type of heat mirror used, which significantly influences the resulting thermal-optical properties of the glass system. In the table 2 there are compositions of original (commonly available) and newly designed glass systems.

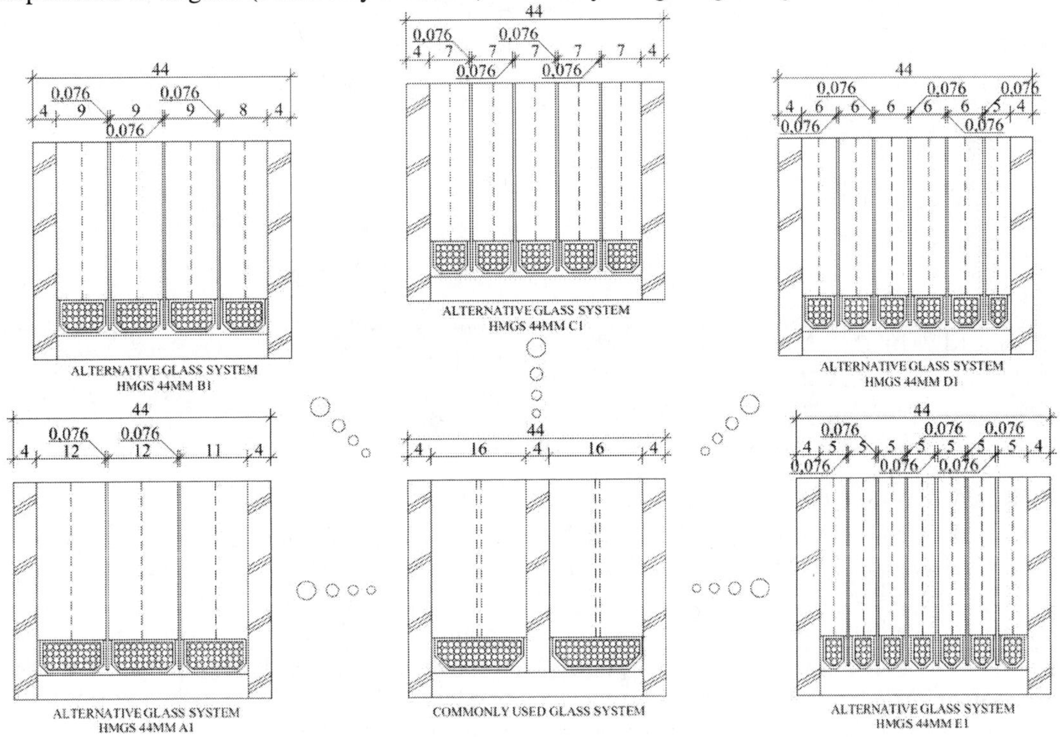

Figure 2. Original and substitutional glass systems; thickness 44mm.

The boundary conditions of the calculation model were chosen according to [5,14,15]. The required normalized value of the internal surface temperature to avoid condensation is determined by considering the average outdoor temperature of the coldest month in the year (January) for the building location according to STN EN ISO 13790/NA [15]. Kosice is for the month of January in region III. and therefore has a regression formula for the monthly average outside temperature:

$$y = -0{,}004x - 2{,}5 \qquad (1)$$

where y is average monthly outdoor temperature in °C
x height above sea level of town in m [15].

$$y = -0{,}004.230 - 2{,}5$$

$$y = -3{,}42°C\ (Kosice, region\ III.)$$

Internal design temperature is $\theta_i = +20°C$. The results are obtained using the Window 7.4.14 [16] software according to the NFRC standard with external design temperature changed. For more accurate results, it would be useful to use dynamic simulation model and the thermal reference year [17,18].

Table 2. Input parameters- compositions of selected glass systems; thickness 44mm.

HM 88+HM SC 75; 44MM

HMGS 44MM A1	
PLANITHERM ULTRA N SGG	4MM
Air 5%, Krypton 95%	12MM
HM88 Suspended film	0.076MM
Air 5%, Krypton 95%	12MM
HM Solar control 75 suspended film	0.076MM
Air 5%, Krypton 95%	11MM
PLANICLEAR SGG	4MM

HMGS 44MM B1	
PLANITHERM ULTRA N SGG	4MM
Air 5%, Krypton 95%	9MM
HM88 Suspended film	0.076MM
Air 5%, Krypton 95%	9MM
HM88 Suspended film	0.076MM
Air 5%, Krypton 95%	9MM
HM Solar control 75 suspended film	0.076MM
Air 5%, Krypton 95%	8MM
PLANICLEAR SGG	4MM

HMGS 44MM C1	
PLANITHERM ULTRA N SGG	4MM
Air 5%, Krypton 95%	7MM
HM88 Suspended film	0.076MM
Air 5%, Krypton 95%	7MM
HM88 Suspended film	0.076MM
Air 5%, Krypton 95%	7MM
HM88 Suspended film	0.076MM
Air 5%, Krypton 95%	7MM
HM Solar control 75 suspended film	0.076MM
Air 5%, Krypton 95%	7MM
PLANICLEAR SGG	4MM

HMGS 44MM D1	
PLANITHERM ULTRA N SGG	4MM
Air 5%, Krypton 95%	6MM
HM88 Suspended film	0.076MM
Air 5%, Krypton 95%	6MM
HM88 Suspended film	0.076MM
Air 5%, Krypton 95%	6MM
HM88 Suspended film	0.076MM
Air 5%, Krypton 95%	6MM
HM Solar control 75 suspended film	0.076MM
Air 5%, Krypton 95%	5MM
PLANICLEAR SGG	4MM

HMGS 44MM E1	
PLANITHERM ULTRA N SGG	4MM
Air 5%, Krypton 95%	5MM
HM88 Suspended film	0.076MM
Air 5%, Krypton 95%	5MM
HM88 Suspended film	0.076MM
Air 5%, Krypton 95%	5MM
HM88 Suspended film	0.076MM
Air 5%, Krypton 95%	5MM
HM88 Suspended film	0.076MM
Air 5%, Krypton 95%	5MM
HM Solar control 75 suspended film	0.076MM
Air 5%, Krypton 95%	5MM
PLANICLEAR SGG	4MM

HM 77; 44MM

HMGS 44MM A2	
PLANITHERM ULTRA N SGG	4MM
Air 5%, Krypton 95%	12MM
HM77 Suspended film	0.076MM
Air 5%, Krypton 95%	12MM
HM77 Suspended film	0.076MM
Air 5%, Krypton 95%	11MM
PLANICLEAR SGG	4MM

HMGS 44MM B2	
PLANITHERM ULTRA N SGG	4MM
Air 5%, Krypton 95%	9MM
HM77 Suspended film	0.076MM
Air 5%, Krypton 95%	9MM
HM77 Suspended film	0.076MM
Air 5%, Krypton 95%	9MM
HM77 Suspended film	0.076MM
Air 5%, Krypton 95%	8MM
PLANICLEAR SGG	4MM

HMGS 44MM C2	
PLANITHERM ULTRA N SGG	4MM
Air 5%, Krypton 95%	7MM
HM77 Suspended film	0.076MM
Air 5%, Krypton 95%	7MM
HM77 Suspended film	0.076MM
Air 5%, Krypton 95%	7MM
HM77 Suspended film	0.076MM
Air 5%, Krypton 95%	7MM
HM77 Suspended film	0.076MM
Air 5%, Krypton 95%	7MM
PLANICLEAR SGG	4MM

HMGS 44MM D2	
PLANITHERM ULTRA N SGG	4MM
Air 5%, Krypton 95%	6MM
HM77 Suspended film	0.076MM
Air 5%, Krypton 95%	6MM
HM77 Suspended film	0.076MM
Air 5%, Krypton 95%	6MM
HM77 Suspended film	0.076MM
Air 5%, Krypton 95%	6MM
HM77 Suspended film	0.076MM
Air 5%, Krypton 95%	6MM
HM77 Suspended film	0.076MM
Air 5%, Krypton 95%	5MM
PLANICLEAR SGG	4MM

HMGS 44MM E2	
PLANITHERM ULTRA N SGG	4MM
Air 5%, Krypton 95%	5MM
HM77 Suspended film	0.076MM
Air 5%, Krypton 95%	5MM
HM77 Suspended film	0.076MM
Air 5%, Krypton 95%	5MM
HM77 Suspended film	0.076MM
Air 5%, Krypton 95%	5MM
HM77 Suspended film	0.076MM
Air 5%, Krypton 95%	5MM
HM77 Suspended film	0.076MM
Air 5%, Krypton 95%	5MM
HM77 Suspended film	0.076MM
Air 5%, Krypton 95%	5MM
PLANICLEAR SGG	4MM

HM SC 75; 44MM

HMGS 44MM A3	
PLANITHERM ULTRA N SGG	4MM
Air 5%, Krypton 95%	12MM
HM Solar control 75 suspended film	0.076MM
Air 5%, Krypton 95%	12MM
HM Solar control 75 suspended film	0.076MM
Air 5%, Krypton 95%	11MM
PLANICLEAR SGG	4MM

HMGS 44MM B3	
PLANITHERM ULTRA N SGG	4MM
Air 5%, Krypton 95%	9MM
HM Solar control 75 suspended film	0.076MM
Air 5%, Krypton 95%	9MM
HM Solar control 75 suspended film	0.076MM
Air 5%, Krypton 95%	9MM
HM Solar control 75 suspended film	0.076MM
Air 5%, Krypton 95%	8MM
PLANICLEAR SGG	4MM

HMGS 44MM C3	
PLANITHERM ULTRA N SGG	4MM
Air 5%, Krypton 95%	7MM
HM Solar control 75 suspended film	0.076MM
Air 5%, Krypton 95%	7MM
HM Solar control 75 suspended film	0.076MM
Air 5%, Krypton 95%	7MM
HM Solar control 75 suspended film	0.076MM
Air 5%, Krypton 95%	7MM
HM Solar control 75 suspended film	0.076MM
Air 5%, Krypton 95%	7MM
PLANICLEAR SGG	4MM

HMGS 44MM D3	
PLANITHERM ULTRA N SGG	4MM
Air 5%, Krypton 95%	6MM
HM Solar control 75 suspended film	0.076MM
Air 5%, Krypton 95%	6MM
HM Solar control 75 suspended film	0.076MM
Air 5%, Krypton 95%	6MM
HM Solar control 75 suspended film	0.076MM
Air 5%, Krypton 95%	6MM
HM Solar control 75 suspended film	0.076MM
Air 5%, Krypton 95%	6MM
HM Solar control 75 suspended film	0.076MM
Air 5%, Krypton 95%	5MM
PLANICLEAR SGG	4MM

HMGS 44MM E3	
PLANITHERM ULTRA N SGG	4MM
Air 5%, Krypton 95%	5MM
HM Solar control 75 suspended film	0.076MM
Air 5%, Krypton 95%	5MM
HM Solar control 75 suspended film	0.076MM
Air 5%, Krypton 95%	5MM
HM Solar control 75 suspended film	0.076MM
Air 5%, Krypton 95%	5MM
HM Solar control 75 suspended film	0.076MM
Air 5%, Krypton 95%	5MM
HM Solar control 75 suspended film	0.076MM
Air 5%, Krypton 95%	5MM
HM Solar control 75 suspended film	0.076MM
Air 5%, Krypton 95%	5MM
PLANICLEAR SGG	4MM

COMMONLY USED GLASS SYSTEMS, 44MM

GS 44MM X1	
PLANITHERM ULTRA N SGG	
Air 5%, Krypton 95%	4MM
PLANICLEAR SGG	16MM
Air 5%, Krypton 95%	4MM
PLANICLEAR SGG	16MM
	4MM

GS 44MM X2	
PLANICLEAR SGG	
Air 5%, Krypton 95%	4MM
PLANICLEAR SGG	16MM
Air 5%, Krypton 95%	4MM
PLANICLEAR SGG	16MM
	4MM

GS 44MM X3	
PLANICLEAR SGG	
Air 5%, Argon 95%	4MM
PLANICLEAR SGG	16MM
Air 5%, Argon 95%	4MM
PLANICLEAR SGG	16MM
	4MM

3. Results and discussion

The resulting values of the evaluated parameters for glass systems of thickness 44 mm. In the case of commonly used glass systems (see table 2) of 44 mm thickness, after taking into account the boundary conditions, the surface temperatures on the external glass pane surface range form -2.4°C ~ -3.5°C; on the inner surface in the range of +14.7 ~ +17.3°C. A summary of the results of the comparison glass systems can be found in table 3.From the point of view of the thermal insulation ability of the original glass system (X1, X2, X3), the most advantageous system is X1 with U_{COG}=0,889 W/(m^2.K) with acceptable solar factor g=51.3% and visible transmittance τ_{vis}=73.5%, but more than 37% of the ultraviolet radiation is released into the interior. System with the most unacceptable U value is system X3 with U_{COG}=1.619 W/(m^2.K), resulting in a solar factor higher by 21.7%, the transmittance of the visible part of spectrum by this glass system is at approximately the same level (2.1% difference) and the ultraviolet radiation transmittance is nearly 53%. In the case of glass system X1, it can be said that its parameters, and in particular the calculated temperatures of the inner surface of the glass, meet the actual requirements, and there is no risk of condensation on the surface in the middle of the glass system surface under normal outdoor temperatures. If there are an increased number of people in the considered room where the glass system is located and hence increased temperature and relative humidity there is a risk of condensation on the inner glass. If the outdoor temperature drops below -10°C (in Kosice there are several such days during the winter), the relative humidity of indoor air is approx. 70% and the indoor temperature is 20°C, the dew point temperature under the given conditions would was $\theta_{dp;20°C;70\%}$=14.36°C. With these parameters, the condensation on the inner surface of the glass would be more than likely for the glass systems X2 a X3 [19]. The temperature over the circumference of the glass system is further aggravated by the influence of the spacer and therefore its linear thermal transmittance need to be taken into account. However, the risk of condensation (and icing) is also threatened on the outside of the glass systems where U_{COG} is in the range of 0.4 ~ 1.0 W/(m^2.K) [20]. An adverse phenomenon can be avoided by using a low emissivity coating with a maximum emissivity 0.2 at position #1.

IOP Conf. Series: Materials Science and Engineering **566** (2019) 012001 doi:10.1088/1757-899X/566/1/012001

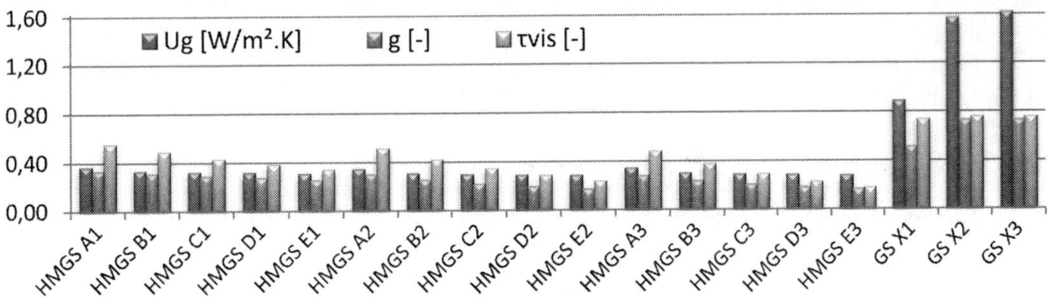

Figure 3. Graph of comparison of key parameters of selected glass systems, thickness 44mm.

The proposed progressive glass systems are designed using three types of heat mirror films, each of which includes five alternatives that vary in the number of heat mirror films used. For HMGS 1 ~ 3, that U-value is in the range of 0.282 ~ 0,368 W/(m^2.K). Interestingly, due to the use of the heat mirror, the ultraviolet transmittance coefficient value dropped to zero, this effect can be observed for all 15 chosen glass systems. The temperature in the middle of the inner surface of the glass pane is in the range of +19.3 ~ +19.6°C. These values are significantly closer to indoor air temperature (+20°C), which not only eliminates the possibility of condensation on the glass surface but eliminates the thermal discomfort of occupants due to the unpleasant coldness radiation from the indoor glass pane.

Table 3. Resulting parameters of evaluated glass systems, thickness 44mm.

HM 88+HM SC 75; 44MM										
Name	Boundary conditions	Dimensions of glass system [mm]	Number of layers	Thickness [mm]	U [W/(m².K)]	g[-]	τ_{vis}[-]	τ_{uv}[-]	Outer glass surface teperature [°C]*	Inner glass surface temperature [°C]*
HMGS 44MM A1	-3,42;20°C (STN 73 0540-2,3)	1240x1250	7	43,15	0,368	0,334	0,549	0,000	-4,0	+19,3
HMGS 44MM B1			9	43,23	0,336	0,313	0,486	0,000	-4,0	+19,4
HMGS 44MM C1			11	43,30	0,323	0,291	0,431	0,000	-4,0	+19,4
HMGS 44MM D1			13	43,38	0,317	0,274	0,383	0,000	-4,0	+19,5
HMGS 44MM E1			15	43,46	0,309	0,253	0,340	0,000	-4,0	+19,5
HM 77; 44MM										
HMGS 44MM A2	-3,42;20°C (STN 73 0540-2,3)	1240x1250	7	43,15	0,348	0,305	0,514	0,000	-4,0	+19,4
HMGS 44MM B2			9	43,23	0,310	0,253	0,419	0,000	-4,0	+19,5
HMGS 44MM C2			11	43,30	0,298	0,216	0,345	0,000	-4,0	+19,6
HMGS 44MM D2			13	43,38	0,291	0,191	0,285	0,000	-4,0	+19,6
HMGS 44MM E2			15	43,46	0,286	0,168	0,237	0,000	-4,0	+19,6
HM SC 75; 44MM										
HMGS 44MM A3	-3,42;20°C (STN 73 0540-2,3)	1240x1250	7	43,15	0,342	0,281	0,482	0,000	-4,0	+19,4
HMGS 44MM B3			9	43,23	0,304	0,238	0,375	0,000	-4,0	+19,5
HMGS 44MM C3			11	43,30	0,293	0,206	0,293	0,000	-4,0	+19,6
HMGS 44MM D3			13	43,38	0,287	0,184	0,229	0,000	-4,0	+19,6
HMGS 44MM E3			15	43,46	0,282	0,163	0,179	0,000	-4,0	+19,6

*Surface temperatures are related to the center of the glass system.

COMMONLY USED GLASS SYSTEMS; 44MM										
Ozn.	Boundary conditions	Dimensions of glass system [mm]	Number of layers	Thickness [mm]	U [W/(m².K)]	g[-]	τ_{vis}[-]	τ_{uv}[-]	Outer glass surface teperature [°C]*	Inner glass surface temperature [°C]*
GS 44MM X1	-3,42;20°C (STN 73 0540-2,3)	1240x1250	5	44,0	0,889	0,513	0,735	0,3721	-3,5	+17,3
GS 44MM X2			5	44,0	1,575	0,730	0,756	0,5275	-2,9	+14,9
GS 44MM X3			5	44,0	1,619	0,730	0,756	0,5275	-2,4	+14,7

*Surface temperatures are related to the center of the glass system.

4. Conclusion

As a result of the increasing heat protection of buildings, replacement of windows is an often solution to a high heat leakage through them. Since the glass system forms the majority of the area of the structure, it is worth considering, exchanging only the glass system itself. The selected progressive glass systems were evaluated in terms of thermal-optical parameters and their surface temperature calculations. The results indicate the possibility of profitable replacement of the original glass system as progressive, which could be put into the original e.g. plastic frame. The heat insulating properties of heat mirror system is 60 ~ 75% better than the original glass systems. In the vast majority of cases, they exclude the risk of condensation of water vapour on the inner surface of the glass, which is the most observable problem in the winter period; during summer period effectively prevent undesired overheating of interior. The choice of the most appropriate system from analyzed depends on the conditions in which heat mirror insulating unit will be installed.

5. References

[1] J Magyar and S Palkova 2014 *Stare okna treba vymenit ako prve* (Bratislava: Jaga)
[2] B Mempouo et al. 2010 *Novel window technologies and the code for sustainable homes in UK (International Journal of Low-Carbon Technologies* vol 5*)* Issue 4 pp 167-174
[3] R Hart et al. 2019 *Thermal performance and potential annual energy impact of retrofit thin-glass triple-pane glazing in US residential buildings (Building Simulation* vol 12*)* Issue 1 (Springer Professional) pp 79-86
[4] P Holzinger 2011 *Thin glass technology for insulating glass production (International conference on architectural and automotive glass)* pp 556-578
[5] STN 73 0540-2/Z1 2016 *Thermal protection of buildings. Thermal performance of buildings and components. Part 2: Functional requirements* (Bratislava: SUTN)
[6] P Jurica et al. 2016 *Results of long term experimental testing of windows for low energy houses (4th international conference of applied technology)* (Belgium)
[7] S E Selkowitz 1978 *Transparent heat mirrors for passive solar heating applications (Proceedings of Passive Systems* vol 78*)* pp 124-132
[8] Patent no. US 4337990 1982 *https://patentscope.wipo.int/search/en/detail.jsf?docId=US37309178&recNum=4&office=&queryStrque=ALLNAMES%3A%28John+Fan+Frank+Bachner%29&prevFilter=&sortOption=Pub+Date+Desc&mamaxR=4*
[9] J C C Fan and F J Bachner 1976 *Transparent heat mirrors for solar energy applications (Applied Optics* vol 15*)* Issue 4 pp 1012-1017
[10] D Aratesh et al. 1989 *The design and testing of hightly insulating glazing system for use with conventional window system (Journal of solar energy engineering* vol 111*)*
[11] C M Lampert 1981 *Heat mirror coatings for energy conserving windows (Solar energy materials)* pp 1-41
[12] Katalog heat mirror- izolacni skla Interm *http://www.izolacniskla.cz/pdf/heat_mirror_odborny_katalog.pdf*
[13] Vitro 2016 *Glass technical document TD 101. Gas space convention effects on U-values in insulating glass units*
[14] STN 73 0540-3 2012 *Thermal protection of buildings. Thermal performance of buildings and components. Part 3: Properties of environments and building products* (Bratislava: SUTN)
[15] STN EN ISO 13790/NA/Z1 2012 *Energy performance of buildings. Calculation of energy use for space heating and cooling* (Bratislava: SUTN)
[16] Lawrence Berkley Laboratory 2017 *Window 7.4.14 nfrc Simulation manual*
[17] H Hens 2011 *Applied building physics, boundary conditions, building performance and material properties* (Berlin: Wilhelm Ernst & Sohn) ISBN 978-3-433-02962-6
[18] M Zozulak et al. 2016 *Hygrothermal initial conditions. Case studies. (Applied mechanics and materials)* (Switzerland) pp 273-278
[19] S Bizonova 2019 *Pouzitie nizkoemisivnych vrstiev heat mirror pri obnove okennych konstrukcii (Seminar doktorandov 2019)* (Kosice: Technical university of Kosice)
[20] H Glaser ang J Szyska 2011 *Condensation on the outside surface of window glazing-what are the key parameters and how to avoid with low-E coatings (Proceedings of glass performance days)* (Finland) pp 212-217

Acknowledgements
This paper was created and realized with support of research project VEGA 1/0389/17 "Transparent construction in interaction of the contemporary architecture and on the human health".

IOP Conf. Series: Materials Science and Engineering **566** (2019) 012002 doi:10.1088/1757-899X/566/1/012002

Application of micro wire on the sample made from asphalt mixture with using bitumen

J Bokomlaško[1] **and J Mandula**[1]

[1] Department of Geotechnics and traffic Engineering, Technical University of Košice, Vysokoškolská 4, Košice, Slovakia

Abstract. The micro wire are new possible sensors, when serves for the contactless measurement of the deformations, which arise in the sample. We can have correct measurement, with right length, diameter and application of the micro wire on the sample. The micro wire was used to two samples, which were made of asphalt mixture AC 11. Bitumen was used in order to glue the micro wire to the test sample. The signal was sufficiently strong for both test specimens. Measurement was made on a sample in the form of a beam. The reaction of the micro wire to the load applied was evaluated. Measurement was done in cycles. In total, three measurement cycles were performed. Measurement results were evaluated.

1. Introduction

At present, development and improved of materials is on high level. Therefore it is important to develop newer and smaller measurement sensors. The sensors should be able to record the changes which have been caused with various types of mechanical stress. On the bases of these sensors we are able to determine material properties. Amorphous magnetic micro wire covered with glass belongs to category of new sensors. These are sensors which consist of metallic core. Iron core is protected with glass packaging. The change of magnetic field is excited by sensor which is placed without any contact above the micro wire. The time at which the micro wire is pre-magnetized is measured. In the article it is marked as H signal and physical unit is microsecond [μs]. If is micro wire tensile, it is lengthened and time pre-magnetized is longer. The micro wire can be used to measure temperature and mechanical stress simultaneously. Moreover, it can be used to monitor material expansion. In contrast with other sensors they can scan contact parameters because micro wire diameter enables to measure the inside of the structure of materials. In order to measure correctly, it is important to select the right micro wire. They differ in diameter core, material core, diameter of glass packaging and expansion of micro wire. The aim of research is to find the right diameter micro wire, optimal length and application on the test samples. The application of micro wire to the measurement test sample must be substantially in order not to get wrong measurements. The choice of incorrect way of micro wire application will influence the whole measurement. Appropriate number of measurement is required to find optimal application, length and diameter of micro wire. The purpose of the article is to describe the way how apply micro wire on the sample using bitumen. Afterwards, the force of micro wire signal and reaction of micro wire to the loading were measured.

IOP Conf. Series: Materials Science and Engineering **566** (2019) 012002 doi:10.1088/1757-899X/566/1/012002

2. Application of micro wire

The micro wire was applied directly to the sample while using bituminous. The length of the micro wire was 4 cm. The micro wire was inserted into the prepared groove on the sample. In order that, micro wire is straight and measurement accuracy is achieved. Subsequently, the bituminous type 50/70 is heated to 120 ° C. The micro wire is covered with bitumen. This guarantees good interaction between the test sample and the micro wire. In this way, micro wires were applied to two test samples.

The first test sample has the shape of a beam of 400 x 50 x 15 mm (Figure 1.). It is a mixture of AC 11 and 50/70 bitumen.

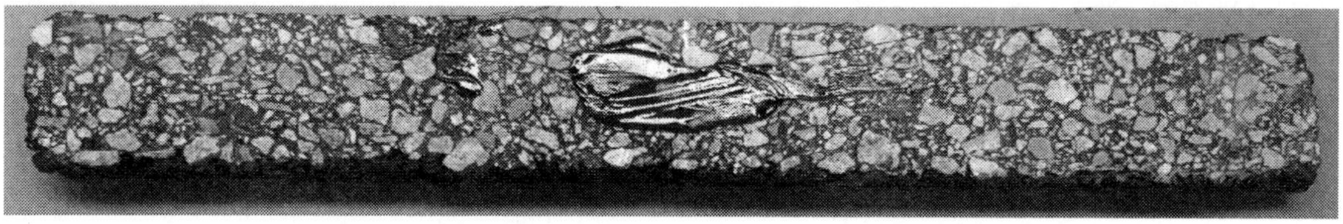

Figure 1. Direct micro wire application using bitumen on the beam (1st test sample).

After application of the micro wire on test sample, the H signal was detected without any deformation on the sample. The micro wire response was correct and the beam was loaded with a bending force. The Marshal test sample has a cylindrical shape (Figure 2.) with diameter d = 100 mm and high h = 75 mm. The Marshal test sample was made of asphalt mixture type AC 11 with binder type bitumen 50/70.When the micro wire had capped bitumen, H signal was detected on the test sample without any deformation. The micro wire response was correct.

Figure 2. Direct micro wire application using bitumen on the 2nd test sample.

3. Measure with micro wire

Measurements were made on a beam test sample (Figure 1).The test sample thickness was set at 15 mm (to achieve deformation even at small force and therefore the micro wire can respond to the change). The resistance of the asphalt mixture was not assessed. The signal strength of the micro wire, its response to the deformation induced, and the method of applying the micro wire to the test sample were evaluated. The test sample was tested for a three-point bend so that bending was induced (Figure 3). The micro wire was applied to the upper surface of the specimen where there is the tensile force. Moreover, an indicator watch was used to measure the deformation.

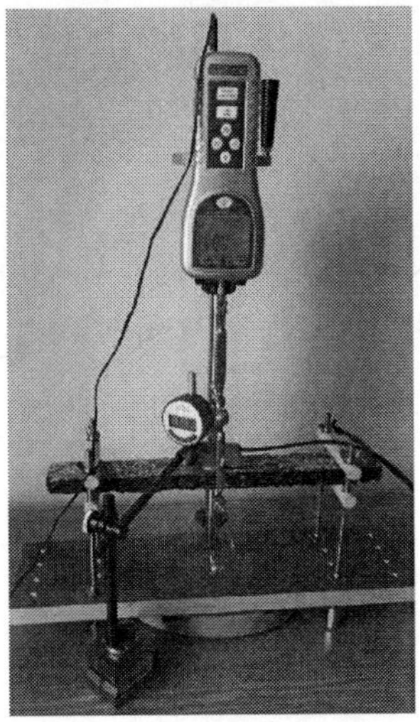

Figure 3. three-point bend with 1st test sample.

Beam was tested in three load cycles where each cycle started with an initial force of 0 N. Subsequently, the test sample was loaded to 10 N and deformation was recorded. Afterwards, the sample was gradually loaded to 20, 30 and 40 N. For each load, the deformation of the test sample was also recorded. After 40 N, the force was reduced in steps of 30 N, 20 N and 10 N until the test sample returned to 0 N. Altogether, the test was repeated three times and the resulting values are shown in Table 1.

Table 1. The course of the measurement beam in the three-point bend device.

Load	Measurement					
	1st Cycles		2nd Cycles		3rd Cycles	
Force [N]	H signal [μs]	Deformation [mm]	H signal [μs]	Deformation [mm]	H signal [μs]	Deformation [mm]
0	785.69	0.000	783.49	0.000	782.90	0.000
10	783.78	0.159	779.73	0.202	779.00	0.195
20	778.66	0.495	775.50	0.513	774.90	0.505
30	773.74	0.924	768.22	1.012	768.00	1.005
40	761.22	1.481	757.68	1.512	757.00	1.530
30	767.70	0.924	762.61	1.012	761.90	1.005
20	776.61	0.450	771.38	0.513	770.50	0.505
10	780.68	0.165	777.93	0.202	777.00	0.195
0	784.49	0.000	780.78	0.000	780.05	0.000

Measured values were plotted. The graph in Figure 4 shows the dependence between H signal and force. In the first measurement, the higher values of the H signal were measured. In the second and third measurements the sample was stabilized and almost identical values were measured (the curves in the graph are overlap).The second graph shows the dependence between H signal and deformation (Figure 5.). There was also stabilization of the test sample as we can see on the 2nd and 3rd load cycles (curves overlap).

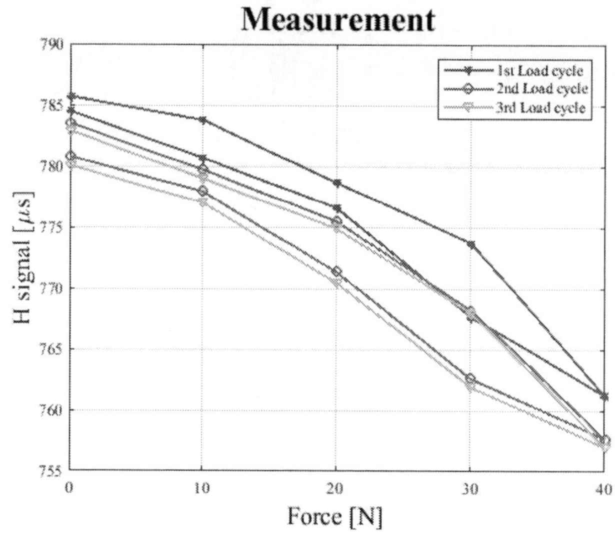

Figure 4. Direct micro wire application using bitumen on the 1st test sample, Force + H signal.

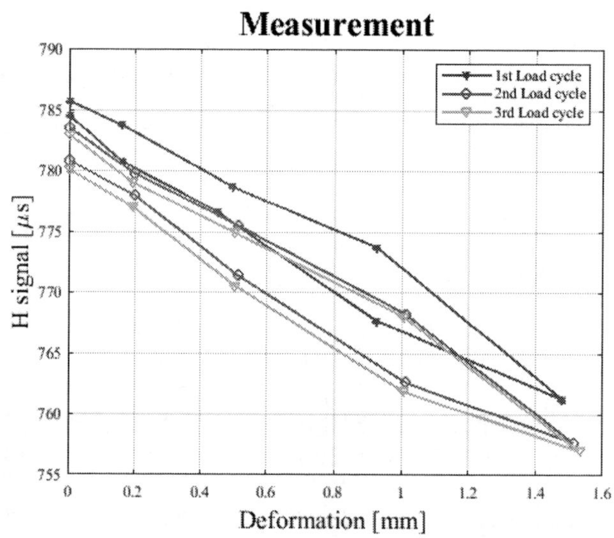

Figure 5. Direct micro wire application using bitumen on the 1st test sample, Deformation + H signal.

4. Conclusion

The application of micro wire using bitumen shows its self as the appropriate way of attachment to the test sample. Due to the fact that the same material is used in asphalt mixture the effect of non-interaction between micro wire and test sample is minimalised. Attachment of micro wire on test sample is complicated because it is necessary to heat bitumen to 110 °C. This is disadvantage. During the test correct reaction of micro wire in unloaded state was detected. The result was positive because of this reason the loading test was performed. The graph show positive results of measurement. The first test of loading cycle shows the fact, that there was consolidation of the test sample and micro wire. The second and third loading cycles represent repeatability and it is expected that the following cycles will be repeated accordingly.

IOP Conf. Series: Materials Science and Engineering **566** (2019) 012002 doi:10.1088/1757-899X/566/1/012002

5. References

[1] Praslicka D, Blazek J, Smelko M, Hudak J, Cverha A, Mikita I, Varga R and Zhukova A 2013 IEEE Transactions on Magnetics *Possibilities of measuring stress and health monitoring in materials using contact-less sensor based on magnetic micro wires* **49** IEEE pp 128 - 131

[2] Sabol R, Rovnak M, Klein P, Vazquez M and Varga R 2015 IEEE Transactions on Magnetics *Mechanical Stress Dependence of the Switching Field in Amorphous Micro wires* **51** IEEE

[3] Antonov A S, Borisov V T, Borisov O V, Prokoshin A F and Usov N A 2000 Journal of Physics D: Apploed Physics *Residual quenching stresses in glass-coated amorphous ferromagnetic micro wires* **33** pp 1161-1168

[4] Varga R 2012 UPJŠ Prírodovedecká fakulta ústav Fyzikálnych Vied *Sklom potiahnuté magnetické mikrodrôty* Univerzita Pavla Jozefa Šafárika v Košiciach 76

[5] Mandula J and Bokomlasko J 2017 XXVI. vedecké sympózium s medzinárodnou účasťou *Meranie deformácií asfaltových zmesí s využitím mikro drôtov* Hrádok pp 89 - 93

[6] Zhang D, Hou S, Bian J and He L 2016 Engineering Fracture Mechanics *Investigation of the micro-cracking behaviour of asphalt mixtures in the indirect tensile test* **163** pp 416 - 425

[7] Hajikarimi P, Moghadas Nejad F, Khodaii A, Fini A H 2018 Construction and Materials *Introducing a stress-dependent fractional nonlinear viscoelastic model for modified asphalt binders* **183** pp 102 - 113

Acknowledgments

This work was supported by the Slovak Research and Development Agency under the contract No. APVV-15-0777.

This research has been carried out within the project NFP 26220220051 Development of progressive technologies for utilization of selected waste materials in road construction engineering, supported by the European Union Structural funds.

The Importance of green roofs and Sustainable development

L Bříza[1]

[1] Brno University of Technology, Faculty of Civil Engineering, Department of Technology, Mechanization and Construction Management, Czech Republic

Abstract. The sustainable development is the observance to the principles of respect for nature, environmental protection and respensible approach to social and economic standards of urban development. Green roofs are an important and inherent aspect of such development. For current and especially future climate conditions they are also represent a signification factor of water management and are one of the ways to keep cities habitable for future generations as well. Green roofs can efficiently retain water and gradually release it back into the environmental in a place where was a driving rain without water affected site literally just flowed. The desired side effect of the process may be overall cooling of the environmental. Simultaneously, the vegetation constructions contribute to the elimination of the so-called heat island effect, and thus contribute to the natural change of water.

1. Sustainable development

Sustainable development is such a way of developing a human society that naturally combines the full conservation of the environment with social and economic progress. It is wrong to interpret sustainable development only as environmental protection. The right concept of Sustainable development is based on three pillars that support the system equally. These are Social, Economic and Environment dimensions (Figure 1).

1.1. Social dimension

The Social dimension focus on balancing inequalities between social groups or individuals. The aim of the social dimension is to eliminate poverty locally and globally, to create access to basic medical help and to ensure hygienic conditions. An important part is also the suppression of discrimination, racism, xenophobia, or religious intolerance. The goal is to achieve a higher level of life in conscious modesty.

1.2. Economic dimension

The Economic dimension consists of the economic activities of human society, the interaction of human society with each other and interaction with the environment. In terms of sustainable development, this is the most problematic dimension. It is very difficult to grasp and is a great challenge for sustainable development to deal with it.

1.3. Environment dimension

The Environmental dimension recognizes the importance of ecosystems, which they appropriately appreciate, either materially or spiritually. One of the theses is that the limited system does not allow unlimited growth. Thus, the environmental dimension has a significant impact on both the Social dimension and the Economic dimension. Initial interventions into other dimensions have led to sustainable development. The overriding principle of the Environment dimension is the protection of biodiversity in all its forms, while: "the highest level is to be understood as a cultural diversity that is of no less importance for the maintenance of the dynamic balance and stability of human society than biological diversity for the preservation of ecosystems" (Rynda).

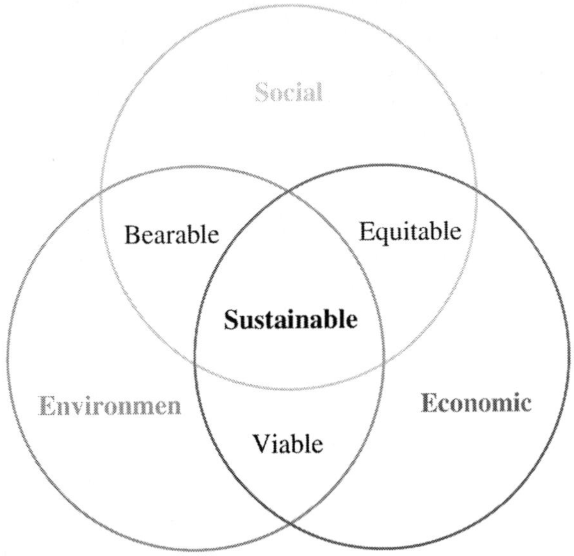

Figure 1. Dimensions of Sustainable development.

1.4. Objectives of Sustainable development

The most well-known definition of the sustainable development objectives comes from the World Commission on Environment and Development (WCED): "Sustainable development is development that meets the needs of the present without compromising the ability of future generations to meet their own needs." In addition, one of comprehensible slogans of sustainable construction is quote by Antoine de Saint-Exupéry: "We do not inherit the Earth from our ancestors, we borrow it from our children."

2. Sustainable construction

Sustainable construction responds to the requirements of sustainable development under the principle of a new way of designing, building and using buildings. They respect the principles of all three pillars, social, economic and environment.

The construction industry, including all building activities (eg buildings, silos, roads, bridges, dams, etc.), is the main material and energy consumer of natural resources. At the same time, the construction industry is among the major environmental polluters and the land user.

The traditional, still predominant, approach to designing buildings is based on three basic requirements; the quality of the design solution, the cost and time it takes to build the building. Always looking for the ideal compromise of these three requirements from the point of view of the builder. However, this approach does not include broader aspects from the point of view of the environmental impact and the social and cultural quality of the functions of the realized constructions. The new concept must be more comprehensive and include sets of criteria that can be divided into three areas:

- quality of the environment
- economic efficiency and limitations
- social and cultural contexts

These areas precisely replicate the basic dimensions of sustainable development from which the basic criteria of sustainable construction result for the construction sector. While the traditional approach is based on the principle of maximal economic effect without any more significant impact on the environment, the new concept of sustainable construction emphasizes the importance of reducing the negative environmental effects of buildings. Therefore, the concept of sustainability is not only about the inclusion of environmental aspects related to global climate change.

IOP Conf. Series: Materials Science and Engineering **566** (2019) 012003 doi:10.1088/1757-899X/566/1/012003

2.1. Main tasks of sustainable construction

Professor Ing. Petr Hájek, CSc. presents these tasks (magazine Stavebnictví, issue 11-12/07):

Environmental aspects

- Energy; the aim is to increase the energy efficiency of construction and using of buildings.
- Materials; the aim is to make more efficient use of material resources.
- Emissions / Waste; the aim is to reduce emissions and waste.
- Water; the aim is to reduce the consumption of quality water.
- Soil; the aim is to contribute to the sustainable development of settlements.

Economic aspects

- Costs of implementation; the aim is to optimize the costs of building implementation while ensuring maximum functional quality and minimal environmental waste throughout the life cycle.
- Operating costs; the aim is to optimize operating costs throughout the life cycle.
- Durability; the aim is to ensure long-term durability of the building.
- Local economy; the aim is to support the development of the local economy and employment.

Socio-cultural aspects

- Quality; the aim is to improve the quality and functionality of the indoor and outdoor environments of buildings.
- Safety; the aim is to ensure the safety of the indoor environment and surrounding buildings.
- Society; the aim is to positively influence the local social climate and employment.
- Culture; the aim is to protect and maintain the cultural heritage.

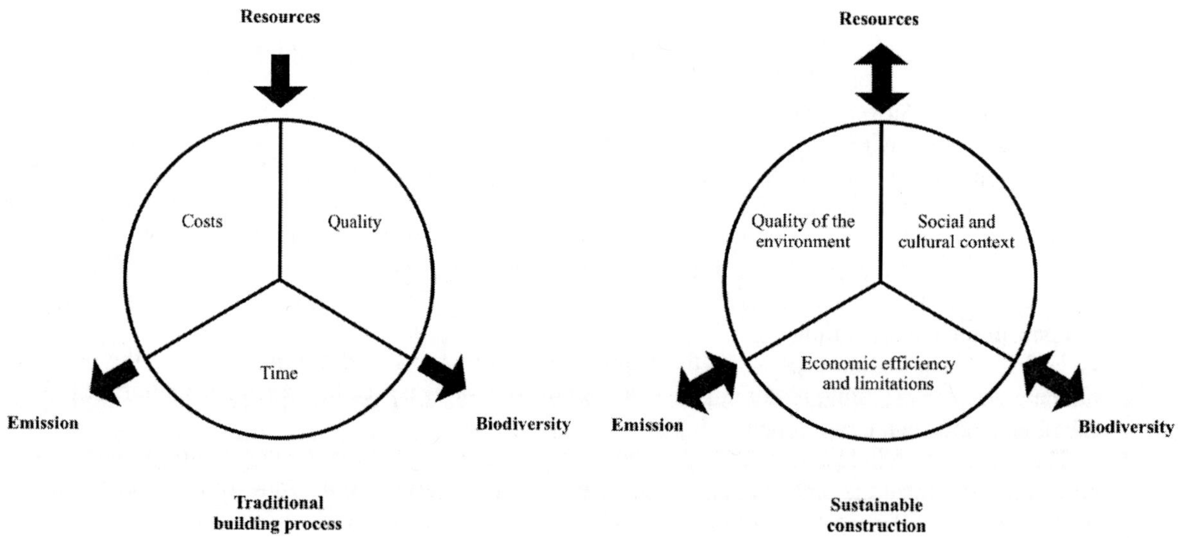

Figure 2. Traditional building process and Sustainable construction.

The road to sustainable construction is not a search for one or several universal technical solutions, but the application of new design principles, new materials and technologies for their development, new construction technologies, including its organization, new methods of assessment and evaluation of buildings, etc., while preserving the architectural and structural variations and variability in design of buildings. Such an approach, however, requires accepting certain changes in the concept of design architecture, but also in the concept of a design solution.

3. Water, Urban Heat Island effect and suistainable cities

The phenomeon of urban heat islands is not new. In Australia or the USA, this is a topic that is being discussed at all levels of government. Even the Czech architectural community is aware of the temperature differences between the countryside and the build-up urban areas and is trying to design new or reconstructed areas to mitigate them. New low-emission materials and green solutions play a major role in this fight.

Areas covered by the large-emission materials (concrete, asphalt), massive construction of buildings, lack of green areas (parks, trees), road surfaces and pavements with zero water permeability, inappropriately organized city, which produces reduced wind speed in the streets. All of these are aspects that significantly increase the temperature in cities and create urban heat islands (UHI). For the construction of roads and buildings, mainly low-reflectivity materials are used. Thanks to this, heat from the sun is accumulated in them. In cities, of course, is heat not produced only from sun, but as well from households is produced addition heat. Nonporous surfaces in cities do not allow rainwater to sink into the soil, contribute to the drainage of water from torrential rains to sewers, rivers and seas. So there is less groundwater, floods and consequently soil erosion. Water is contaminated with chemicals. Higher temperatures cause the formation of groundlevel ozone and change the climate to local and wider levels. Humidity and ecosystem disturbances occur. Urban canyons are formed - on streets lined by higher buildings, either air is standing or undesired drafts or dust-raising beliefs and a higher carbon dioxide concentration. Urban heat islands (UHI) are worsening the warmth of the population. In connection with this, the thermoregulation system of a person, which often causes cardiovascular and respiratory diseases, is demaged. UHI radiation also causes higher rainfall over urban areas, more frequent fog and in the winter intensifies the presence of smog.

Sustainability of cities can be promoted by limiting the remote import of energy, water and food, etc., as well as waste (heat, air and water pollution, etc.). Meanwhile, the sustainable city is just a paper concept, but it has to be said that cities like Paris and New York were supplied from nearby large gardens at the turn of the 21th century, and Havana returned to the model of city farms after the loss of cheap Soviet oil.

4. The potential of green roofs

There are several ways to cool the UHI. The first is planting greenery. Research suggests that greening (trees with dense crowns, hedges, shrubs, grass, vegetation roofs) reduces the temperature difference between the city and the surrounding countryside. Besides, the greenery has aesthetic character of the city. Greenery, or parks, helps to develop social relationships in the cities and influence the harmonious arrangement of the human psyche.

The effect of urban heat islands is also exacerbated by the inappropriate selection of material and color of roofing. The large topic is, for example, white roofs. Aesthetically pleasing, but above all very environmentally benefical, are roofs covered with greenery. The so-called green roofs regulate overheatings as well as standard greenery in cities, so they retain water at the point where it lands. So the water can gradually evaporate back into the atmosphere and take away the heat from the environment. This leads to cooling of the climate in which the building is located (although increased humidity may in part cause a sensational increase in temperature). Green roofs also isolate, shade, cool and therefore reduce the energy consumption of refrigeration equipment. The resulting quality of the green roof, of course, depends on the height and quality of the soil, but also on the particular climate in which the object is located. For example, strong and frequent rains exclude the use of some granular soils, etc. Green roofs regulate the drainage of surplus water away, but it does not already address its supply to deficient groundwater. Some studies also suggest that water flowing from green roofs into surface water can be contaminated with nitrogen and phosphorus from the substrate used.

5. Case study: EXPO 2000 Kronsberg

On the southeast edge of Hannover, the first part of the unique Kronsberg hausing estate was built on the occasion of the EXPO 2000 exhibition, which should have 15, 000 inhabitants after completion.

Basically, it was an exhibit that was to show the model sustainable living of the futore with the inclusion of not only urban and social planning, but also environmental considerations.

Kronsberg is characterized by a considerable amount of greenery. From the early stages, not only the gardens and private land tenants, but also park and green areas as community conters were also planned to be used for sports and games. The plan of the landscape counts with the planting of the woods at the top of the Kronsberg hill and with plenty of others extensive green areas for recreation, but also as a wildlife and plent environment.

The construction of Kronsberg has several objectives; serve as an accommodation capacity for visitors of exhibition and then to become an integral part of the city - and as such to be a model and a standard of sustainable urban construction with an emphasis on ecology, in the widest possible definition of the concept.

The ecological dimension of the housing estate begins with a thoughtful modeling of its landscape, where, among others things, natural water features (retention reservoirs, water absorption systems) have been applied and a natural element in the form of forests, wetlands and pastures, which form green corridors. Kroesberg, of course, has an ecological energy system tahat corresponds to the energy-saving nature of all the buildings in the housing estate.

If Kronsberg is written or spoken, usually facts and figures from the technical field of ecology usually appear. However, the psychological and social ecology plays role in the experimental Hannovers's hausing estate. The authors of the concept have not only tried to create a city space that will not only burden the environment but will also be pleasant for its inhabitants, it will be user friendly.

Hanover's attempt to create a city dor the 21st centrury is generally considered to be a success. The Kronsberg technical and environmental parameters have become benchmarks for similar sustainable projects, internationally. In Germany itself, Kronsberg is a model for similar housing estates, such as the Munich Nest Sola2. And the settlement itself has futher development before it.

6. Conclusion

Each municipality has other conditions; in modern urban planning we should stick to the saying „nothing must be exaggerated". The basis is ti work with all possible, affordable solutions that allow space and combine low-emission materials (high-reflex, light colors) with designs that work on the principle of consumption of evaporation heat (greenery, permable materials).

For the time being, the effect of urban heat islands in the Czech Republic, despite its provability, os often underestimated or ignored. Exceptions, however, confirm the rule, mentioning, for example, the international conference *Adaptace měst na změnu klimatu* (Adaptation of cities to climate change), which was held in Ostrava last September. Particularly green examples such as Ljubljana, Essen and Vienna were discussed.

7. References

[1] Široká H 2018 Městské tepelné ostrovy *Stavebnictví a interiéry* **10**
[2] Hájek P 2018 Udržitelná výstavba budov a její uplatňování ve střední Evropě *Stavebnictví* **11-12**
[3] Hájek P 2006 Metodika pro komplexní hodnocení budov v rámci životního cyklu *Sborník konference Vykurovanie* 2006 Tatranské Matliare Slovakia

IOP Conf. Series: Materials Science and Engineering **566** (2019) 012004 doi:10.1088/1757-899X/566/1/012004

Evaluation of the effect of concrete strength increase on buckling resistance of concrete filled tubes

P Bukovska[1]

[1] Brno University of Technology, Faculty of Civil Engineering, Department of Metal and Timber Structures, Czech Republic

Abstract. The paper deals with the behaviour of steel tubular columns filled by high-strength concrete at compression load. Attention is paid especially to the strength of the used concrete and its effect on the buckling resistance of column. Results of the experimental part of research have shown that increasing of concrete strength is efficient only to a certain extent. It was found in the experimental analysis that the use of concrete class C80/95 has only little effect on increase of buckling resistance as compared with the use of concrete class C55/67. In order to evaluate the efficiency of the concrete strength increasing, a theoretical study was carried out, the results of which are presented in this paper. The paper also discusses the possibilities of using other types of concrete (light-weight concrete or slightly reinforced concrete) with regard to their contribution to improving the bearing capacity and taking into account the production costs of the steel - concrete tubular column.

1. Introduction

In recent years, the development of civil engineering has a tendency to use of high-strength materials. Also, in the field of steel-concrete composite structures, the combination of materials is sought which would be the most effective from the point of view of strength, reliability and economy. The combination of circular tubes and high-strength concrete seems to be one of the most effective ways to use the high-strength concrete because of eliminating the lack of concrete tensile strength.

The problem in high strength concrete construction designing is that there are no valid design rules or standards. Within many experimental works dealing with tests on concrete filled circular tubes, behaviour of normal and high strength concrete was monitored and compared with behaviour predicted by the Eurocode 4 in order to evaluate the possibility of using the design standard for high strength concrete. The results of these researches show that the normative recommended calculation of buckling resistance corresponds to the actually observed values. American Specifications and Australian Standards were found conservative, while those of the European Code were for high strength concrete generally unconservative [5, 7, 8]. To be more precise, the Eurocode 4 shows to be safe for extending for high strength concrete if the load is applied eccentrically, i.e. when interaction diagrams are used. However, it is less accurate in predicting cases where eccentricity is null [8].

This paper relates to the research program which was carried out in order to investigate the effect of the concrete strength on buckling resistance of steel tubular columns filled by high-strength concrete. Experimental analysis of axially compressed circular tubes filled by concrete was conducted. Within the evaluation of results there was found that the increase of concrete strength leads to the increase of column buckling resistance only to a certain extend. Lower buckling resistance of high-performed concrete filled tubes was achieved than that would be expected, compared to normal concrete filled tubes [4]. This fact was confirmed also in [7].

The solution included development of numerical model in order to ensure consistency of the output of the numerical calculation and the data obtained at the test specimens. The aim of the numerical model is to replace some of the tests performed actually and provide data for developing of buckling curves for tubular columns filled by high-strength concrete [2]. For finding out which of input parameters has the greatest influence on the buckling resistance of composite column made by steel tube filled with concrete, the sensitivity analysis was performed. Four models were analysed differing in grade of steel and strength of concrete. Outcomes show that the strength of concrete has the greatest positive influence on buckling resistance. Influence of strength concrete is twice greater than influence of steel yield strength. As for the geometrical parameters the greatest is the influence of tube thickness which is greater by tubes of steel grade S355 than that of steel grade S235 according to the expectation. Negative influence on the buckling resistance has the length of tube which corresponds to the critical length of member when considering pinned ends of columns. The critical length has greater influence in case of models with using steel of grade S355. The results of sensitivity analysis confirm the influence of concrete strength on the buckling resistance of the column [4].

Evidently, concrete strength affects the increase in buckling capacity of slender columns. But it is unknown, if the use of higher and higher strength concrete makes a lot of difference. This paper presents the results of a theoretical study which was carried out in order to evaluate the efficiency of the concrete strength increasing. The paper also discusses the possibilities of using alternative types of concrete with regard to their contribution to improving the bearing capacity and taking into account the production costs.

2. Experimental analysis

Experimental analysis of compressed circular tubes filled by concrete was conducted within the research program. For the loading tests the tubes of the diameter of 159 mm and the thickness of 4.5 mm with the buckling length of 3 000 mm were used. One half of the test specimens were tubes of steel grade S235, other specimens were tubes of steel grade S355. Total of 18 test specimens were tested within the analysis. The specimens were divided into three groups of six pieces. The first group of specimens was tested without concrete, second group of specimens was filled by concrete of class C55/67 and third group of specimens was filled by concrete class C80/95. In combination with the two classes of steel strength six partial experiments were achieved.

The test specimens were gradually loaded by axial pressure in a vertical position. During loading process the loading force, vertical displacement in bearing and transversal deflection in middle of the length of specimens were measured. The maximum achieved load gives the buckling resistance of the specimen. More information about the loading tests see in [1]. The graph in figure 1 shows maximal measured load for each group of specimens (mean value), what is actually the buckling resistance. It can be seen that the buckling resistance increased by filling the tube with concrete C55/67 by approximately 100%, whereas the use of concrete C80/95 had only little effect.

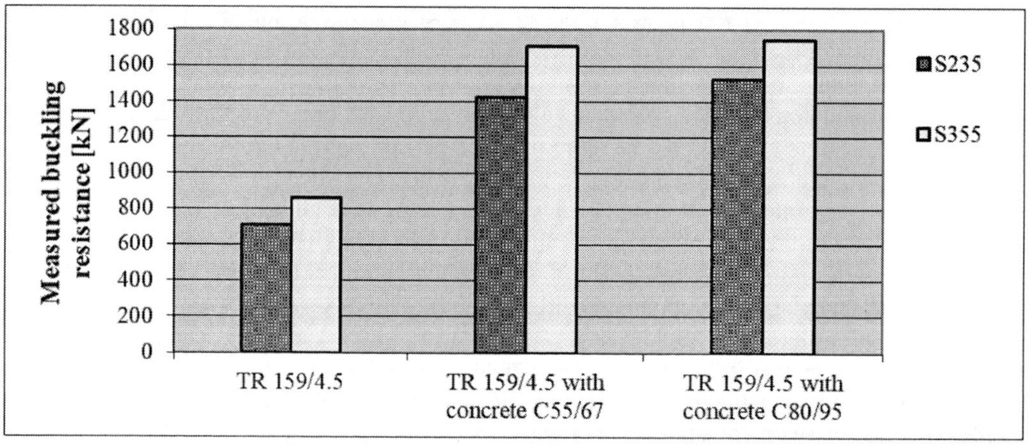

Figure 1. Experimentally proved buckling resistance (mean values).

IOP Conf. Series: Materials Science and Engineering **566** (2019) 012004 doi:10.1088/1757-899X/566/1/012004

Real material properties were measured on material specimens in order to compare values of buckling resistance achieved experimentally and numerically determined values. Within the comparison it was observed that the normative recommended calculation of buckling resistance corresponds to the actually observed values for all groups of specimens including those filled by high strength concrete [4]. Experimental analysis confirmed that the calculation methods given in Eurocode 4 can be used for the following theoretical analysis.

3. Theoretical analysis

The theoretical analysis was performed for finding the effect of increasing concrete strength on buckling resistance of concrete filled steel columns. Columns of length 3.0 m and 2.5 m were taken into account, hinged and axially compressed. As for the choice of diameter and thickness of the tube, cross-sections were selected from the assortment of the steel tubes, which allow the steel tube to be filled with concrete and which are utilized evenly in terms of both characteristic plastic resistance $N_{pl,Rk}$ and critical normal force N_{cr}. With regard to the parametric study in [1], tubes of diameter 133 mm, 152 mm, 159 mm and 168 mm are taken into evaluation. Grade of steel is S235 or S355. Concrete was considered of class from C20/25 to C90/105 with modulus of elasticity from 29 GPa to 44 GPa, respectively.

3.1. Evaluation of the effectiveness

Relations between buckling resistance (as the load capacity) and concrete strength are depicted on figures 2 – 4 for considered types of tubes with variable thickness. Within the solution the characteristic values of buckling resistance (N_{Rk}) are taking into calculation because the results of the experimental analysis have shown that the characteristic values better correspond to the actually achieved values. Concrete strength zero means that the tube is without concrete. It is clear from the graphs that the curves are concave - that is, the buckling load capacity growth gradually decreases. This fact can be seen in all investigated cases.

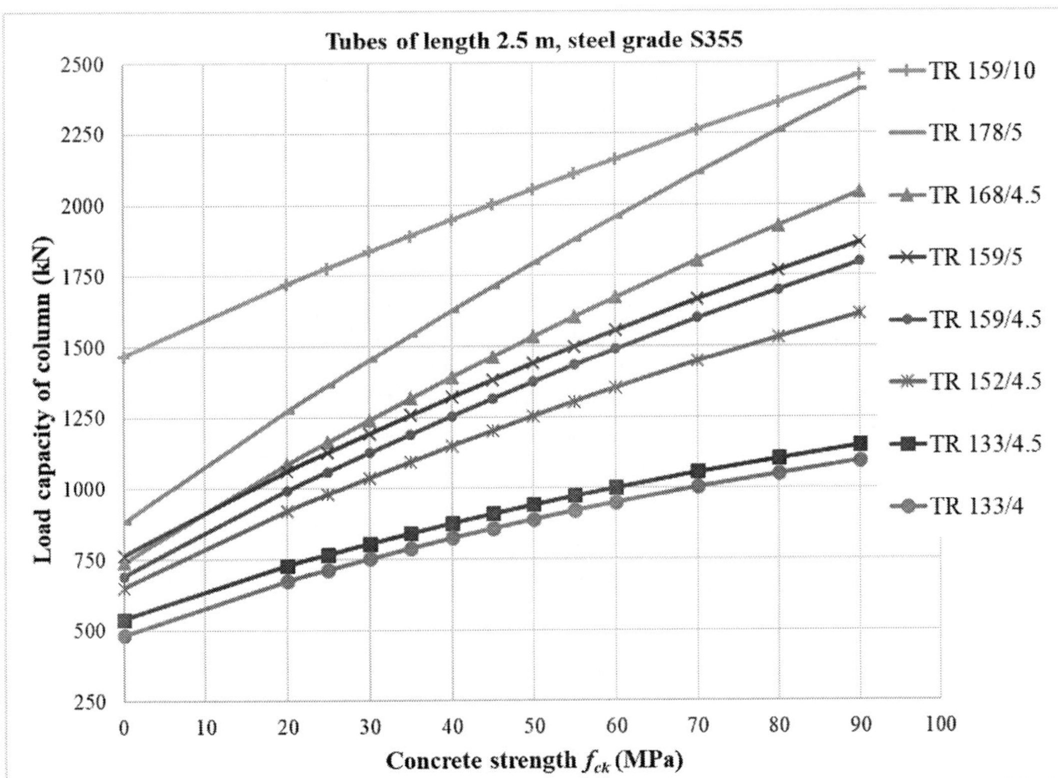

Figure 2. Effect of concrete strength on buckling load capacity of columns of length 2.5 m and steel grade S355.

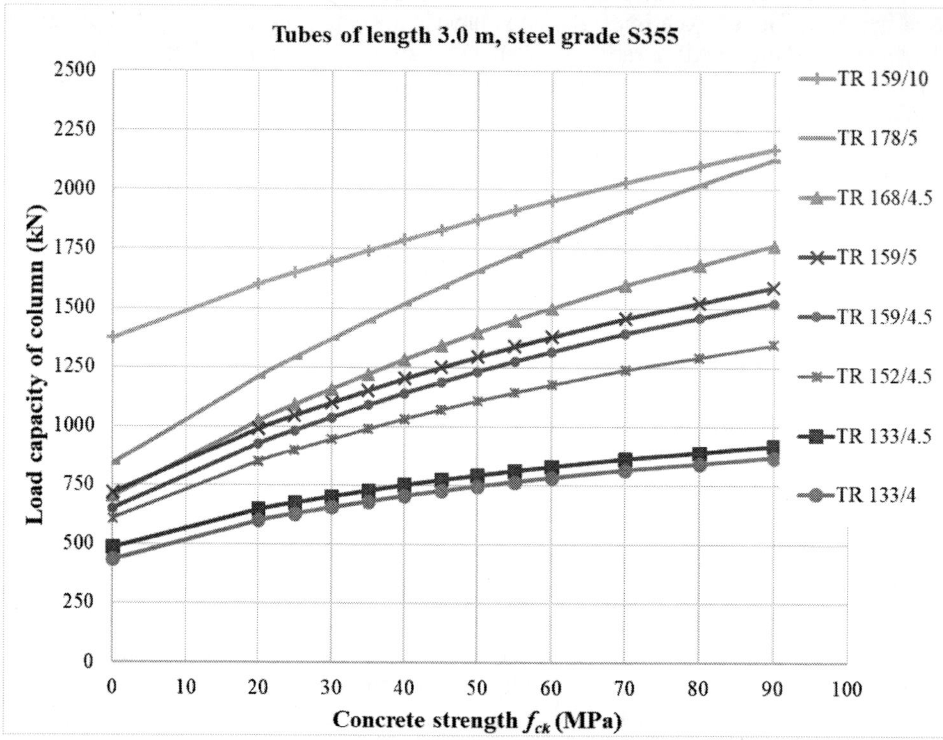

Figure 3. Effect of concrete strength on buckling load capacity of columns of length 3.0 m and steel grade S355.

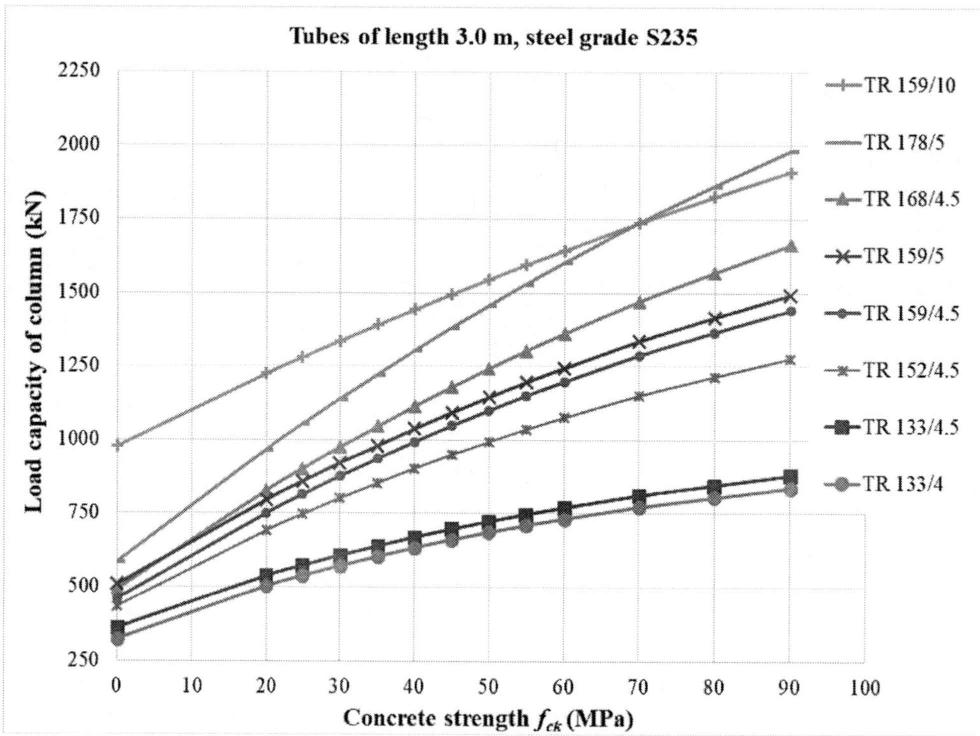

Figure 4. Effect of concrete strength on buckling load capacity of columns of length 3.0 m and steel grade S235.

Although the curves were plotted only for some cases of diameter and thickness, it is possible to imagine the form of dependence on other cases. When changing the thickness of the tube, the shape of the curve remains the same, it only moves in the y-axis direction - upwards with the increase of thicknesses and downwards when the tube thickness is reduced (see Figures 2, 3 and 4). Changing the column length has the opposite effect - the curve moves upwards when the column length is reduced and down when the column length is increased. For tubes with a larger radius (of 178 mm), the buckling capacity increases with increasing concrete strength more. In most cases, it can be clearly seen that the very filling the tube with concrete causes a relatively large increase in buckling strength, further increasing the strength of concrete does not have such a large effect, although it is not completely meaningless.

3.2. Economical study

Comparison of the price of monitored tubes filled with different types of concrete can be seen in table 1. Unit prices for individual variants of material were determined from current price lists on web sides of concrete plants in surrounding of Brno city. Prices are converted to Euro. The prices are exclusive of VAT. Concrete class is limited by current assortment, maximal class commonly available is C70/85. Prices of lightweight concrete are inaccessible, although some concrete plants can produce this concrete, the price depends on the total quantity and the required material properties. The use of lightweight concrete was therefore not included in the comparison. Every variants of column arrangement are calculated for tube of diameter 159 mm with thickness 4.5 mm, length 3.0 m and steel grade S235. In this arrangement is consumption of steel 51.4 kg and of concrete 0.053 m^3. The design values of buckling resistance are given (including coefficients of material reliability).

The price of concrete in the composite column can be reduced by replacing natural aggregate, whose price is around 13.6 €/ t for fraction 8/16. An interesting substitute could be concrete recycled aggregate, where aggregate of fraction 0/16 costs 4.3 € /t. However, the price of steel plays the most important role in total costs of this type of composite columns. From table 4 can be seen, that the lowest relative price is for tube filled with concrete C70/85 which is the highest strength class used in this study. This is probably because buckling resistance increases with the strength of concrete more than the price.

Table 1. Economical comparison of concrete strength.

Concrete			Steel			
	Class	Price for unit (€/m^3)	Price for unit (€/kg)	Total price (€)	Buckling resistance (kN)	Relative price (€/kN)
Concrete according the CSN EN 206 + A1:2018 a CSN P 73 2404	C20/25	76,0	1,55	83,76	647,6	0,129
	C25/30	76,9	1,55	83,81	688,2	0,122
	C30/37	91,1	1,55	84,56	727,4	0,116
	C35/45	101,7	1,55	85,12	765,1	0,111
	C40/50	104,3	1,55	85,26	801,4	0,106
	C45/55	107,2	1,55	85,41	835,5	0,102
	C55/67	121,9	1,55	86,19	899,1	0,096
	C70/85	127,1	1,55	86,47	984,1	0,088
STEELCRETE - fibre concrete	C20/25	73,8	1,55	83,64	647,6	0,129
	C25/30	82,0	1,55	84,07	688,2	0,122
Self-compacting concrete	C25/30 SCC	99,3	1,55	84,99	688,2	0,123
	C30/37 SCC	110,8	1,55	85,60	727,4	0,118
	C35/45 SCC	117,4	1,55	85,95	765,1	0,112

4. Conclusion
This paper discusses the influence of increasing strength of concrete on buckling resistance of circular steel tubes filled with concrete. The tubes are considered as columns subjected to axial compression. Results of experimental analysis shows that the increase of concrete strength from C55/67 to C80/95 has little effect. Therefore, theoretical analysis was performed to investigate this fact.
One part of analysis is the evaluation of effectiveness of concrete strength. Tubes of diameter 133 mm, 152 mm, 159 mm and 168 mm were taken in consideration. Grade of steel was S235 or S355. Concrete was considered of class from C20/25 to C90/105 with modulus of elasticity from 29 GPa to 44 GPa, respectively. The characteristic values of load capacity are used in the graphs. Graphs show that the buckling load capacity does not increase proportionally with the strength of the concrete, but its increment gradually decreases. Increasing the strength of concrete therefore gradually loses its significance, especially for tubes with a smaller diameter (133 mm).
Second part of theoretical analysis is economical study of the price of circular tubes filled with concrete where the prices of different arrangement of column were compared. It comes out from the study that use of higher strength concrete is still effective from the point of view of material price. The price for concreting is not included.
The results of the theoretical analysis can be used in practice. If the chart were extended to parameters of all tubes offered in the assortment of steel constructions and if the design values were used, a tool for design facilitating could be created.
With regard to efforts to mitigate the ecological impacts of building structures, it would be useful to monitor and evaluate the efficiency of concrete-filled tubes with respect to possibility of their recycling or reuse of building elements. This point of view will be included in next research.

5. References
[1] Bukovská P, Karmazínová M and Štrba M 2011 High-performance concrete-filled steel tube *Proc. European Conference on Steel and Composite Structures (Budapest)* pp 1749–54 ISBN 978-92-9147-103-4
[2] Bukovská P 2011 Numerical model of steel tube filled by high-performance concrete *In Proc. of the 3rd PhD. Student Conference of Civil Engineering and Architecture* 2011 Herlany ISBN 978-80-553-0639-1
[3] Bukovská P and Karmazínová M 2012 Effectiveness of high-strength concrete when used in circular steel tubes filled with concrete *Proc. Recent Advances in Engineering (Paris)* (WSEAS Press) ISBN 978-1-61804-137-1
[4] Bukovská P 2012 Influence of Concrete Strength on the Behavior of Steel Tubular Columns Filled by Concrete *Int. J. of Mechanics* 6 pp 149-57 ISSN 1998-4448
[5] Ellobody E, Young B and LAM D 2006 Behaviour of normal and high strength concrete-filled compact steel tube circular stub columns *J. of Constr. Steel Res.* **62** pp 706-15
[6] EN 1994-1-1 Eurocode 4: *Design of Steel and Concrete Composite Structures, Part 1-1: General Rules and Rules for Buildings* 2004 CEN: Brussels
[7] Ibañez C, Portolés JM, Romero ML, Espinos A and Albero V 2017 Ultra-high Strength Concrete on Eccentrically Loaded Slender Circular Concrete-filled Dual Steel Columns *Structures* **12** pp 64-74 ISSN 2352-0124.
[8] Portolés J M, SERRA E and Romero ML 2013 Influence of ultra-high strength infill in slender concrete-filled steel tubular columns *J Constr Steel Res.* **86** pp 107-14

Acknowledgments
The paper has been elaborated within the framework of the solution of MSMT project FAST-S-19-6009 and FAST-J-19-6056(Specific Research Projects).

IOP Conf. Series: Materials Science and Engineering **566** (2019) 012005 doi:10.1088/1757-899X/566/1/012005

Resistance of Steel-Concrete Composite Ceiling Beams of Multi-Storey Buildings

P Bukovská[1], V Kaňok[1], M Karmazínová[1] and M Štrba[1]

[1] Brno University of Technology, Faculty of Civil Engineering, Department of Metal and Timber Structures, Czech Republic

Abstract. The paper deals with the problems of steel-concrete composite structural members, which are applied as the ceiling beams of multi-storey buildings, and composed of steel beam of hot-rolled IPE cross-section, concrete slab and profiled sheets, here having a function of a lost formwork. In the paper the theoretical comparative study of the resistance of the particular selected type of steel-concrete composite ceiling beam, among others in the connection with its deflections is presented. In the study the accent is especially paid to the link between ultimate limit state and serviceability limit state, in parallel together taking into account the beam span, cross-section dimensions (steel beam cross-section, concrete slab cross-section and profiled sheet) and grades of steel and concrete used in the mentioned load-carrying structure. The attention has been paid mainly to the resistance of the beam, subjected to the flexure effects, in relation to the corresponding deflections and in this connection to the evaluation of the reserve in the cross-section capacity, with regard to the economy and efficiency of the structural design. The study has been, among others, inspired by the static and structural design of the real building with steel / steel-concrete load-carrying structure, and evoked by the need of more effective and more economical design because of inappropriate large cross-sections in origin design.

1. Introduction

On the workplace of Department of Metal and Timber Structures of the Faculty of Civil Engineering at Brno University of Technology, the long-term attention is paid to the investigation of steel-concrete composite structural members and parts. The subjects of this research, in particular experimental one, are steel-concrete composite beams, as well as steel-concrete columns, for that ones the problems of the actual behaviour and objective ultimate load-carrying capacity of those structural members are investigated. The results of that experimental verification are then usually the basis for the subsequent elaboration and evaluation within the theoretical analysis, like as, for example, numerical modelling, analytical solutions, and so on.

Within the frame of so far performed experimental and theoretical analysis of steel-concrete composite beams, the attention has been, among other, paid to the application of high-performance concrete, especially concrete with higher tensile strength than usually, hereto glass-fibre-reinforced concrete, and its behaviour in zones of negative bending moments of the continuous beams. One of the main solved topics has been oriented towards the investigation of the contribution of glass-fibre-reinforced concrete to the resistance and the rigidity of the composite beam subjected to the negative bending moment. Some partial outputs of that research mentioned are informatively and illustratively presented in [2], for example.

This paper is focused on the problems of load-carrying capacity (resistance) of steel-concrete composite beams subjected to the positive bending moment, not from the viewpoint of their maximal load-carrying capacity (resistance), that means isolatedly from the real static and structural design, but from the viewpoint of the real cross-section dimensions of ceiling beams of multi-storey buildings dimensioned for the given specific loading actions, and further regarding not only ultimate limit state, but also the serviceability limit state, and taking into account the real results of the static and structural design, satisfying all the design conditions.

2. Basic general principles

Generally, the load-carrying capacity of steel-concrete composite beam can be determined on the base of plastic or elastic behaviour and the subsequent calculating procedures, arising from that – see, for example [1], [2], [5].

In the case of steel-concrete composite ceiling beams subjected to the positive bending moment, the utilization of the plastic behaviour and corresponding plastic calculation is usual, because almost always the neutral axis of steel-concrete cross-section lies in concrete slab, so that all the steel beam is subjected to tensile stress. If steel-concrete composite cross-section has – in average – for ceiling structures standard dimensions, and in addition, if the slab is concreted to profiled trapezoidal sheets, then a substantial part of concrete slab is mostly subjected to the compression, and both parts of the cross-section – steel and concrete – may be approximately comparably utilized. The typically known stress distribution for the plastic behaviour and above cited case is shown in Fig. 1. However, the deflection of steel-concrete composite beam has always to be determined exclusively using elastic behaviour and corresponding analysis.

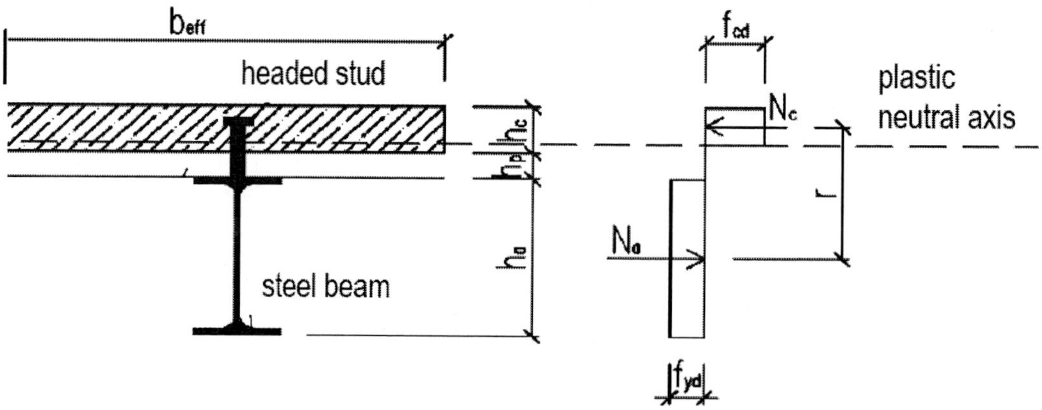

Figure 1. Typical stress distribution for plastic analysis of the beam resistance

3. Comparative study

For the finding, which cross-sections of steel beam – considering various steel and concrete grades, real loading actions, but without changing the concrete slab thickness and profiled sheet dimensions – will be sufficiently efficient for the given beam span, regarding the usual practically used parameters, the comparative study has been performed.

In this comparative study, the following parameters, taken from the real ceiling structures of multi-storey buildings, have been used:

- steel grades of S235 and S355 in combination with two concrete grades (as follows);
- concrete grades given by "concrete classes" of C20/25 and C30/37;
- profiled trapezoidal sheets with the height of 40 mm (Hacierco 40S/160);
- whole concrete slab thickness of 110 mm;
- beam span in the range from 5 to 8 metres (with the step of 1 metre) or eventually 9 metres (in one case only);
- effective width of concrete slab 1.5 m or 1.25 m (eventually), in dependence on the beam span and usual distance between ceiling beams, calculated according to the European Standard EN 1994-1-1 *Design of Steel and Concrete Structures* [5];

- steel beam cross-sections of IPE type, in the range from IPE140 up to IPE270 (see following tables – Table 1, Table 2, Table 3, Table 4), according to the need following the static design of particular beams;
- for all cases the same uniform loads, corresponding with the usual loads of ceiling structures, taken from the European Standard EN 1991 *Loading Actions*.

Within the frame of this study, the static assessment has been performed for all considered cases. The aim of that assessment was to obtain such steel-concrete composite cross sections, which satisfy both the ultimate limit state, and serviceability limit state, within the usage of commonly available cross-sections.

The partial results of the performed study are shown in Tables 1 and 2 for steel grade of S235 with concrete C20/25 or C30/37, and Tables 3 and 4 for steel grade S355 with the same concretes.
From Tables 1 and 2 for steel grade of S235 it is evidently seen, that for the same dimensions of profiled sheet and concrete slab, the higher concrete quality leads to the higher resistance in bending only in the case of larger spans of 9 metres. Respectively, only for the span of 9 metres it is possible to apply the smaller dimension of the steel beam. In the case of the span of 9 metres the usage of better concrete leads to the decrease of steel beam, but at a price of the very small reserve in the cross-section resistance. On the other hand, for the span of 9 metres and concrete C20/25 it may not be to use smaller steel profile than IPE270, although IPE240 exists.

Table 1. Partial results of comparative study – steel grade S235, concrete class C20/25

Steel grade ... S235, concrete class ... C20/25 sheet height ... 40 mm, concrete slab thickness ... 110 mm						
beam span L [m]	5.00	6.00		7.00	8.00	9.00
effective width b_{eff} [m]	1.25			1.50		
cross-section (profile)	IPE140	IPE160	IPE180	IPE200	IPE220	IPE270
uniform load [kNm^{-1}]	18.36	18.40	18.44	18.49	18.54	18.68
bending moment M_{Ed} [kNm]	57.38	82.81	82.99	113.26	148.34	189.09
moment resistance M_{Rd} [kNm]	64.23	83.16	103.24	127.40	154.45	230.26
ratio M_{Ed} / M_{Rd}	0.89	1.00	0.80	0.89	0.96	0.82
deflection w [mm]	8.0	11.8	9.0	12.8	17.1	16.3
limit deflection w_{lim} [mm]	20	24		28	32	36
ratio w / w_{lim}	0.40	0.49	0.38	0.46	0.53	0.45

Table 2. Partial results of comparative study – steel grade S235, concrete class C30/37

Steel grade ... S235, concrete class ... C30/37 sheet height ... 40 mm, concrete slab thickness ... 110 mm						
beam span L [m]	5.00	6.00		7.00	8.00	9.00
effective width b_{eff} [m]	1.25			1.50		
cross-section (profile)	IPE140	IPE160	IPE180	IPE200	IPE220	IPE220
uniform load [kNm^{-1}]	18.36	18.40	18.44	18.49	18.54	18.68
bending moment M_{Ed} [kNm]	57.38	82.81	82.99	113.26	148.34	189.09
moment resistance M_{Rd} [kNm]	65.98	85.34	106.35	131.79	160.48	194.87
ratio M_{Ed} / M_{Rd}	0.87	0.97	0.78	0.86	0.92	0.97
deflection w [mm]	8.0	11.8	9.0	12.8	17.1	21.7
limit deflection w_{lim} [mm]	20	24		28	32	36
ratio w / w_{lim}	0.40	0.49	0.38	0.46	0.53	0.60

Similarly, from Tables 3 and 4 for steel grade of S355 it is seen, that for the same dimensions of profiled sheet and concrete slab, the higher concrete quality does not significantly increase the bending moment resistance nor in the case of larger spans (of 8 or 9 metres). Respectively, for the spans of 8 and 9 metres the same dimensions of steel beam are needed, although the better concrete is used. As it turns out, from the viewpoint of practical usage of steel-concrete composite beams in ceiling systems of multi-storey buildings, steels and concretes of higher strengths in common are not so crucial in comparison with steels and concretes of normal strengths.

Table 3. Partial results of comparative study – steel grade S355, concrete class C20/25

Steel grade ... S355, concrete class ... C20/25 sheet height ... 40 mm, concrete slab thickness ... 110 mm						
beam span L [m]	5.00	6.00	7.00	8.00		9.00
effective width b_{eff} [m]	1.25	1.50				
cross-section (profile)	IPE140		IPE160	IPE180	IPE200	IPE220
uniform load [kNm^{-1}]	18.36		18.40	18.44	18.49	18.54
bending moment M_{Ed} [kNm]	57.38	82.63	112.71	147.54	147.93	187.74
moment resistance M_{Rd} [kNm]	92.96	94.97	120.56	148.78	182.29	219.36
ratio M_{Ed} / M_{Rd}	0.62	0.87	0.93	0.99	0.81	0.86
deflection w [mm]	8.0	15.9	21.8	17.1	21.8	27.4
limit deflection w_{lim} [mm]	20	24	28	32		36
ratio w / w_{lim}	0.40	0.66	0.78	0.53	0.68	0.76

Table 4. Partial results of comparative study – steel grade S355, concrete class C30/37

Steel grade ... S355, concrete class ... C30/37 sheet height ... 40 mm, concrete slab thickness ... 110 mm							
beam span L [m]	5.00	6.00	7.00	8.00		9.00	
effective width b_{eff} [m]	1.25	1.50					
cross-section (profile)	IPE140		IPE160	IPE180	IPE200		IPE220
uniform load [kNm^{-1}]	18.36		18.40	18.44	18.49		18.54
bending moment M_{Ed} [kNm]	57.38	82.63	112.71	147.54	147.93	187.22	187.74
moment resistance M_{Rd} [kNm]	96.97	98.30	125.55	155.86	192.31		233.12
ratio M_{Ed} / M_{Rd}	0.59	0.84	0.90	0.95	0.77	0.97	0.81
deflection w [mm]	8.0	15.9	21.8	17.1	21.8	34.9	27.4
limit deflection w_{lim} [mm]	20	24	28	32		36	
ratio w / w_{lim}	0.40	0.66	0.78	0.53	0.68	0.97	0.76

4. Conclusions

As generally known and here also seen from the results presented in tables above, in the case of the usage of usually produced hot-rolled profiles (hereto IPE), the needed bending moment resistance may be quite significantly influenced by the assortment of profiles. While in the case of small spans (about 5 metres, resp. less than 6 metres) the usage of steel-concrete composite beam is not very efficient, because the contribution of concrete has not been clearly demonstrated, in the case of larger spans (about 8 to 9 metres) the usage of steel-concrete composite cross-section may be very significant, because may lead to the decrease of the cross-section height, and thereby the reducing the overall ceiling structure height. How at least partially shown, the usage of better quality concrete does not probably have such meaning as often expected.

From the partial conclusions mentioned above, it may be stated, that investigated problematic still deserves attention, although it may be to seem, that many problems in this topic already have been solved, but with the development of better materials and new technologies (for example modern types and technological implementation of shear connectors) in civil engineering practice, there are still reserves in the static and structural design of steel-concrete composite beams, both in terms of more effective material utilization and more efficient cross-section configurations.

Because the comparative study mentioned above has been elaborated to the limited range of the parameters, it does not the predictive value, which could be more generalized, but it shows certain trends. For the more detailed analysis it will be needed to expand the range of parameters influencing the resistance of steel-concrete composite beams, of course with respect to the practical useability in the real structures.

5. References

[1] Johnson R P, Buckby R J 1984 *Composite Structures of Steel and Concrete* **1** (Collins: London)

[2] Melcher J J, Karmazínová M and Pozdíšek J 2009 Experimental verification of behaviour of composite steel and glass-fibre-concrete beam *In Proceedings of the 9th International Conference on Steel-Concrete Composite and Hybrid Structures „ASCCS 2009“* Leeds (Singapore: Research Publishing Services) pp 390-395 ISBN 978-981-08-3068-7

[3] Mindess S and Banthia N 1994 *Fiber Reinforced Cementitious Composites: Current Practice and Future Prospects, Concrete Technology, Past, Present, and Future* (Ed. P.K. Mehta), ACI SP-144, American Concrete Institute, Detroit pp 417-446

[4] Kaňok V 2019 *Hall for Metal Production with Administrative Building* (Brno University of Technology, Faculty of Civil Engineering) 315 pp (incl. attachments)

[5] EN 1994-1-1 *Design of Steel and Concrete Composite Structures*, CEN Brussels, 2008

Acknowledgment

The paper has been elaborated with the financial support of the Ministry of Education, Youth and Sports of the Czech Republic, under the solution of the following projects: (i) the project of the Specific Research FAST-S-18-5550, and (ii) the project of the National Sustainability Program I LO1408 AdMaS UP – Advanced Materials, Structures and Technologies.

IOP Conf. Series: Materials Science and Engineering **566** (2019) 012006 doi:10.1088/1757-899X/566/1/012006

Effect of water wall prototype on air humidity

K Čákyová[1], F Vranay[1]

[1] Institute of Architectural Engineering, Faculty of Civil Engineering, Technical University of Kosice, Vysokoskolska 4, 042 00 Kosice, Slovakia

Abstract. The presented document focuses on the effect of the water wall on air humidity (humidification and dehumidification of air). For the purpose of impact verification, a water wall prototype that was tested under laboratory conditions (climate chamber) was designed and built. Also a set of measuring sensors has been installed, these sensors are able to continually record air temperature and humidity and water temperature. In the case of air humidification, the water temperature and air temperature were equalized approximately after 2 hours and after 6.5 hours the air relative humidity reached 92%. This represents an increase in the specific humidity of 4.5 g/kg versus the initial state. The article also describes methodology and measurement of condensation ability of the prototype. In case of air dehumidification process, three different relative humidity was tested at the same air temperature. During the measurement periods, there was increasing the amount of condensate several times by increasing the relative humidity.

1. Introduction

The humidity of the indoor air is an important factor affecting the energy consumption of buildings, the durability of building elements and the perceived air quality. In summer, due to relatively high temperatures, their content in the outside air is significant, the higher the air temperature, the more it is able to absorb water. The air supplied to the interior is then, after cooling to an internal temperature, approximately saturated with water vapour, the relative humidity may be close to its saturation. In winter, due to low temperatures, their content in the outside air is low, because at low temperatures condenses or even freezes and fall to the ground. The air supplied to the interior after the heating to an internal temperature is dry, oftentimes the relative humidity falls below the 20%. The most common were indicated dry mouth and throat, also on the face, dry eye, lips, and also a cold [1]. Surveys show that the employees in office areas with and without air humidification suffer differently under the described symptoms caused by dry air. If air humidification is used, the discomfort is reduced by nearly half. In workplaces with air humidification, the discomfort caused by a dry respiratory tract is reduced by more than a third to 35 percent [2].

Indoor humidity depends on a several factors, such as sources of humidity, air flow, moisture exchange with a material or the state of water vapour in the outdoor environment. One of the sources of humidity can also be water elements, which are most commonly perceived as aesthetic elements. The design of water elements in the interior may vary, using different materials and shapes. The water wall when the water flows on solid material is a universal solution for existing buildings. The vertical position of the water wall does not require large spatial demands and brings the beauty of falling water into the interior. An environment containing water elements helps to reduce stress, increase the feeling of calmness and decrease heart rate and blood pressure, also improves concentration, memory and can lead to increased productivity [3].

To modification an indoor microclimate, the water wall can be used in two different ways. In the first case, the water wall is used for cooling and dehumidifying air in the space. The water film is cooled and its temperature is below the dew point, so a condensation process occurs. In the latter case, the water temperature is the same as the air temperature in the room, so evaporation occurs, which increases the humidity of the air [4]. Similarly oriented research was realized at FRAUNHOFER Institute for Building Physics. This research presents the dependence of water temperature to the condensation ability of the water wall [5].

2. Experimental prototype of water wall

For the purposes of verifying the effect of the water wall on the indoor environment, an experimental prototype was designed to define its essential parts. There are several possibilities to create a water film, in this case it is a perfect overflow, the water falls over the overflow edge, flows down the glass and forms a continuous water film.

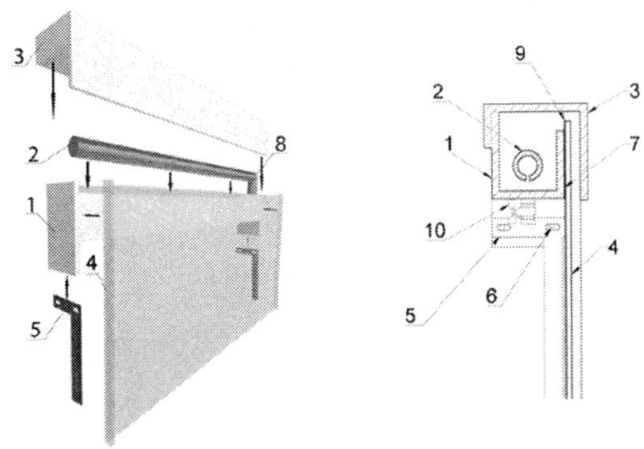

Figure 1. Arrangement scheme of main components of the upper part of prototype and section:

collection tank (1), perforated pipe (2), a cover (3), glass pane (4), supporting metal structure (5), screws (6), silicone adhesive (7), water supply (8), spillway edge (9), ball valve (10).

The formation and thickness of the water film affects, among other things, the geometry of the upper part of the water wall (Fig. 1). The upper part consists of a water collection tank of rectangular shape made from polypropylene material by melt-welded in order to achieve waterproof. A water supply pipe that is perforated is inserted into this tank. The downward perforation allows a steady flow of water into the tank to create a uniform water film across the whole width of the water wall. The effective water film area is $1m^2$. The overflow edge is formed by a glass plate which is connected to a collecting tank with a silicone adhesive. The overflow edge must be in the horizontal position. In order to achieve a united appearance of the water wall and at the same time, to prevent unwanted evaporation of water, the collecting tank is covered by a cover.

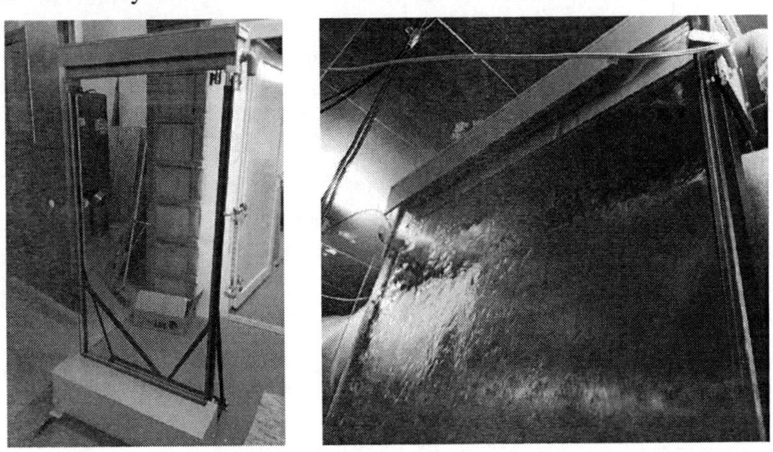

Figure 2. Final water wall prototype.

3. Methodology of measurement

Experimental verification of hypotheses take place under laboratory conditions in the climate chamber where steady conditions can be maintained, the temperature range is from - 20°C to +125 °C and the climatic range is from 20% to 95%. Its inner dimensions are 3.95x1.60 x 2.85m and the volume of the air in the chamber is 18.012 m3. The chamber walls are made of stainless steel, so during the experiment there isn't exchange of mass between air and chamber.

The set of sensors has been installed in the chamber. The temperature and relative humidity sensors are connected via the AHLBORN control unit. The AMR Win Control software was used to obtain and collect measured data. This provided continuous recording of the measured values in a 5 minute time step. Two sensors measured the water temperature in the lower tank and two sensors in the upper tank. From these sensors the average water temperature for the lower and upper part was determined. In front the water film, two sensors were placed to measure the temperature and the relative humidity of the air, and one behind the glass. The location of the sensors is shown on figure 3.

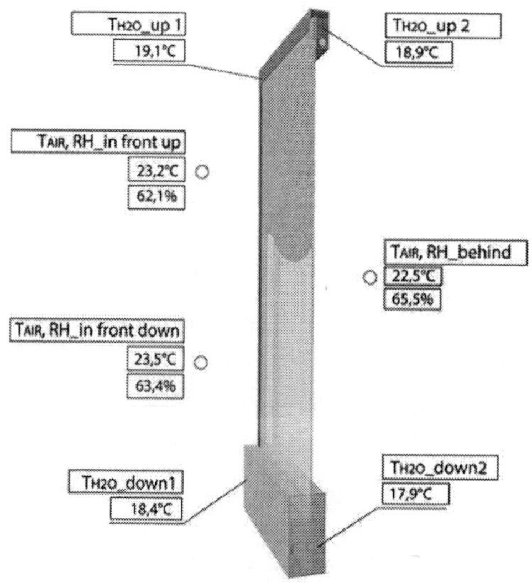

Figure 3. Set of measuring sensors.

For the measurements of influence of water wall to dehumidification and cooling of the indoor air, the system with using a measuring cylinder and weight increment measurement (fig. 4) was proposed using the RADWAG laboratory scale to an accuracy of 0.01g. This modification made it possible to accurately and continuously record the increase in the condensate formed on the water film. Also a plate exchanger was added to the water wall system which allowed a constant water temperature in the system around 14 °C. After the experiment is started, the parts of the water wall are flooded, and water needs to be added to the level of the overflow created in the measuring cylinder. Then, during the experiment the water from the lower collection tank flows into the measuring cylinder. From it the water is supply into the upper part of the water wall by the pump and cooled by the plate exchanger. The shut-off valve is used to control the flow rate velocity and the flow meter are used to monitor accurately flow rate. During the running of the water wall, it has been found that the optimal flow rate is from 300 l/h to 500 l/h, at lower velocity, water film was not formed in whole width and at higher velocity, water bounced off the water film.

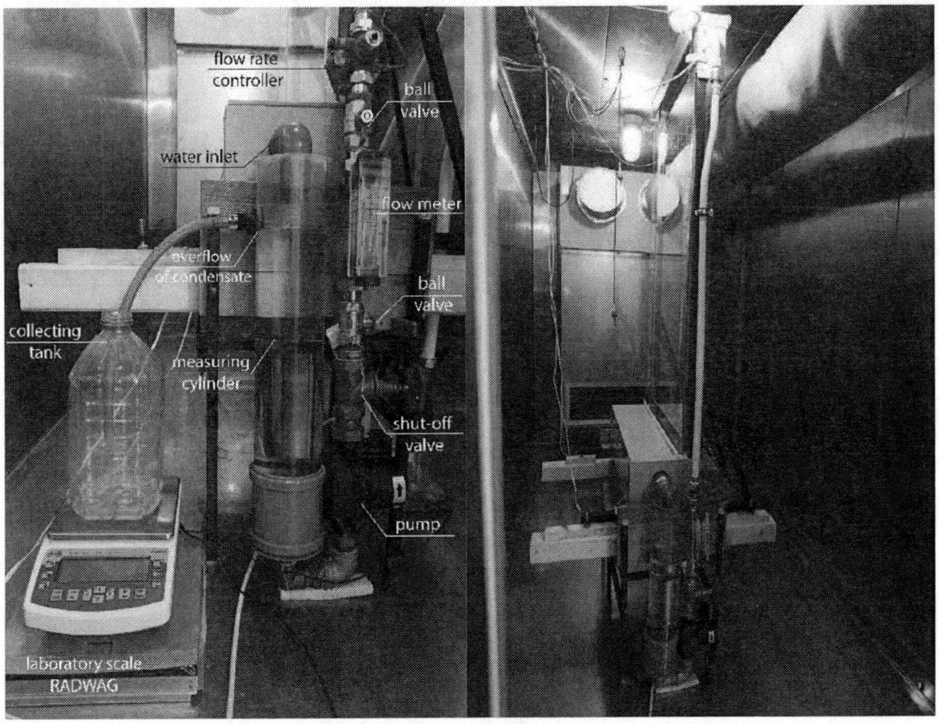

Figure 4. Measurement system of water wall.

4. Results and discussion

4.1. Evaporation process

Figure no. 5 shows the dependence of the monitored parameters during the evaporation process. The experiment took 6.5 hours, in two repetitions at a flow rate of 300 l / hr. The climate chamber was inactive, the experiment focused on the interaction of the water wall with a volume of air, with the possibility of excluding sorption of the surrounding components. The graph shows that the water temperature (TH2O) and air temperature (TAIR) equalled 2 hours after the experiment start.

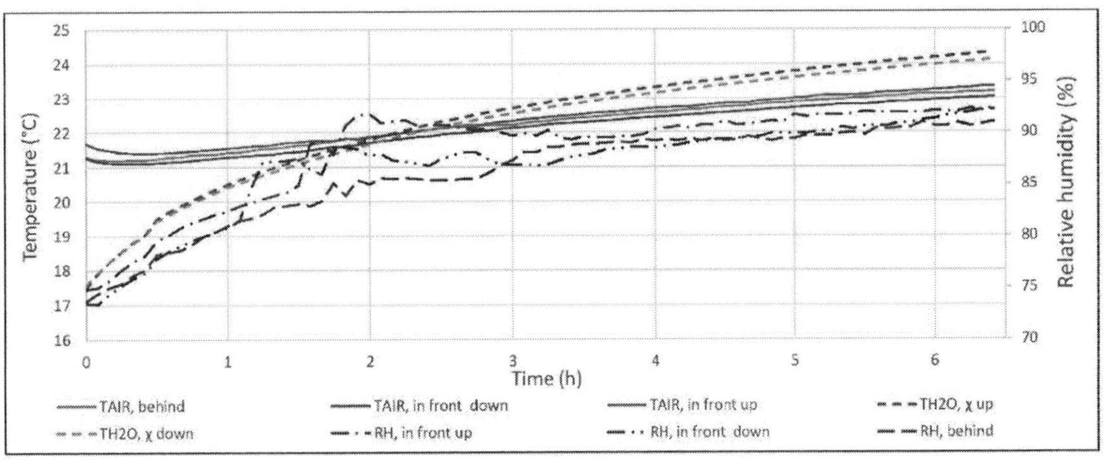

Figure 5. Evaporation process from water from water film.

At the beginning of the measurements, the water temperature (TH2O) had an average value 17.52 °C. The average air temperature (TAIR) was 21.42 ° C and a relative humidity (RH) was 74%, which represents a specific humidity 11.67 g/kg. At the end of the measurement cycles, the water temperature

(TH2O) reached 24.21 ° C, the air temperature (TAIR) was 23.19 ° C, the relative humidity (RH) was 91.83% and the specific humidity was 16.17 g/kg. Compared to the initial state, the specific humidity increased by 4.5 g/kg.

4.2. Condensation process

During the second series of the measurements aimed at determining the amount of condensate on the water film, the climate chamber was used actively, that provided steady boundary conditions (Figure 6) during the experiment. Two measurements were made, the mean water temperature (TH2O_down) in the lower tank was 14.58 °C, then it was cooled by the plate exchanger to 13.86 °C (TH2O_up), the mean air temperature (TAIR) was same in both cases 25.73 °C. In the first case, the mean relative humidity (RH case no.1) was 58.83%, in the second case the mean relative humidity (RH case no. 2) was 73.99%.

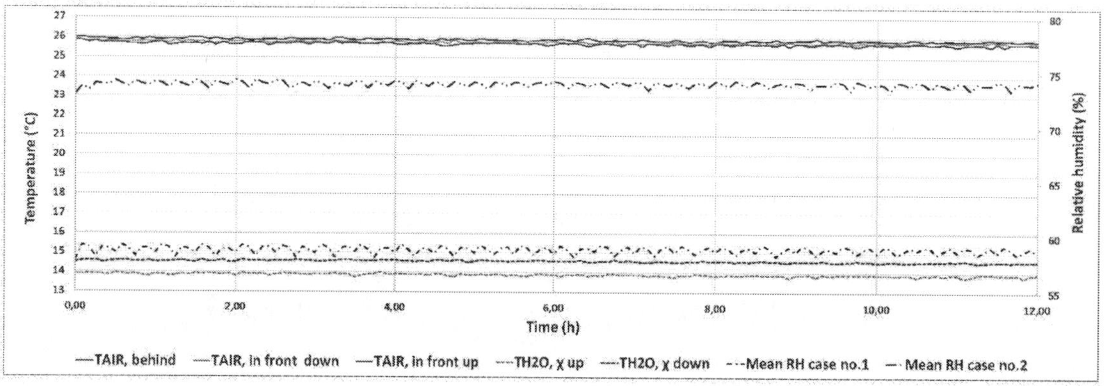

Figure 6. Boundary conditions in the climate chamber during the experiment.

The graph (Fig. 7) represents the courses of condensation on the water film in time dependence. In the first case, when the mean relative humidity was 58.83%, the hourly gain of condensate was 21.86 g/hr. In the second case, at the relative humidity of 73.99%, the hourly increase of condensate was 148.81 g/hr. The difference between mean relative humidity in the first and second cases is 15.16%, the hourly difference in the weight of the condensate is 126.95 g / hr. Compared to the second case, where the condensate gain curve is almost linear, in the first case the curve is stepped. This is caused by a low hourly gain of condensate.

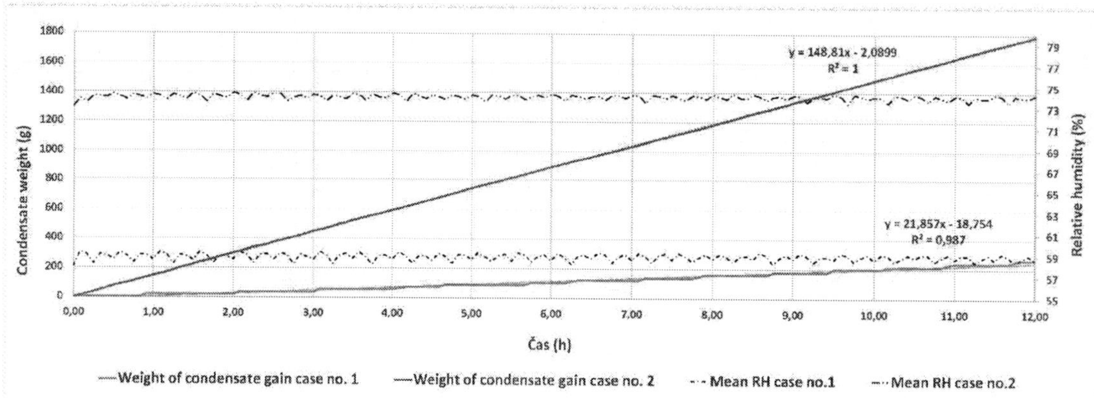

Figure 7. Courses of weight gain of condensate at constant air temperature.

5. Conclusion

Regulation of air temperature and humidity is a basic priority and a requirement to achieve an optimal environment. In addition to using conventional systems such as air conditioning systems, also water walls may be used for this purpose. The article showed the influence of the use of the water element in the interior of the buildings on the parameters of the indoor microclimate. Using the climate chamber to measure its impact has proven to be appropriate for laboratory analysis. The first part of the paper demonstrates the effect of the water wall to humidification the indoor air. During the experiment, the relative humidity increased from 74% to 91.83% and air temperature from 21.42 ° C to 23.19 ° C, it represents an increase in specific humidity of 4.5 g / kg. In the second part, the ability of the water wall to reduce the relative humidity of the air by the condensation process was presented. Supplementing of the original prototype by the plate exchanger and continuously weighing of the condensate exactly specified the results of laboratory experiments. During stationary conditions (water temperature, air temperature), the influence of relative humidity on the amount of condensate on the water film was investigated. With the mean relative humidity of 73.99%, the condensation ability of the water wall was 148.81 g / kg, that is more than 6.5 times higher the condensation ability of the air humidity on the water film compared to a relative humidity of 58.83%. The presented contribution is part of the doctoral study and it is necessary to define and verify other hypotheses for determination of the impact of the water wall under different boundary conditions.

6. References

[1] Jokl M 2011 *Teorie vnitřniho prostředi budov* <Available online: http://www.ib.cvut.cz/sites/default/files/Studijni_materialy/TVPB/Teorie_vnitrniho_prostredi.pdf

[2] Rief S and Jurecic M 2014 *Air humidity in the office workplace* Mediendienstleistungen Fraunhofer-Informationszentrum Raum und Bau IRB Stuttgart

[3] Browning W, Ryan C, Clancy J 2014 *14 patterns of biophilic design*, Terrapin Bright Green, LLC <Available online: http://www.terrapinbrightgreen.com/wp-content/uploads/2014/09/14-Patterns-of-Biophilic-Design-Terrapin-2014p.pdf >

[4] Bird R B, Steward W E, Lightfoot, W N 2006 *Tansport Phenomena*. John Wiley & Sons, Inc

[5] Mitterer Ch 2010 *Cooling and dehumidifying indoor environments using the chilled water wall.* IBP- FRAUNHOFER <Available online: https://www.ibp.fraunhofer.de/content/dam/ibp/en/documents/Information-material/Departments/Hygrothermics/Produktblaetter/IBP_255_PB_Klimabrunnen_neu_en_rz_web.pdf >

Acknowledgment

This work was supported by grant project 1/0217/19 Research of Hybrid Blue and Green Infrastructure as Active Elements of a 'Sponge City'.

IOP Conf. Series: Materials Science and Engineering **566** (2019) 012007 doi:10.1088/1757-899X/566/1/012007

The influence of aggregate gradation on the fresh and hardened concrete properties

R Figmig[1] and M Kovac[1]

[1] Technical university of Kosice, Faculty of Civil Engineering, Institute of Environmental Engineering, Department of Material Engineering, Slovakia

Abstract. This experimental study is focused on the aggregate gradation impact on the properties of fresh and hardened concrete. For the experiment, 9 mix design were prepared with the constant content of cement, effective water, plasticizer and aggregate. The variable factor is represented by different proportions of 4 fraction (0/0.7, 0/4, 4/8, 8/16). Consistency by slump and flow table in time, compressive strength, water absorption and density were measured. Some significant differences were observed. Consistency by the slump test varied from 0 to 170 mm after 5 minutes and from 0 to 50 mm after 30 minutes of concrete mixing. Consistency by flow table varied from 340 to 530 mm after 5 minutes and 320 to 460 mm after 30 minutes. Standard compressive strength varied from 50.8 to 61.6 MPa. Water absorption varied between 4.7 and 5.9 mass % and density from 2380 to 2440 kg.m^{-3}.

1. Introduction

The properties of the aggregate, as dimension, shape as well as the texture of the grains, have crucial impact on consistency as well as on the other rheologic properties and properties in hardened state of concrete [1]. This influence is based on mutual interactions between the aggregate grains (i. e. mutual wedging, frictional forces between grains etc.) and particle movement in liquid state of fresh concrete [2]. The knowledge about aggregate influence on the properties of fresh and hardened concrete connected with this study can be presented as follows [3-8]:

1. viscosity of the fresh concrete increases with increasing quantity of fine particles (higher viscosity protects fresh concrete against stability loss, but decreases the flow velocity),
2. the need of mixing water for reaching the required consistency decreases with increasing the volume of coarse particles,
3. fine aggregate as well as fine particles needs higher volume of mixing water to achieve the required consistency of fresh concrete,
4. optimal aggregate granularity provides denser arrangement of the aggregate grains and need less cement paste to achieve required consistency; it is based on the fact, that volume of gaps is lower in the concrete with denser arrangement of aggregate,
5. fractional forces between the aggregate grains have significant impact on the overall rheology of fresh concrete,
6. the grains with rounded shape and smooth surface are optimal from the achieving fluid consistency point of view,
7. as the result of above points can be predicted, that using round shape, well grading aggregate with minimum content of powder particles from the crushing, provides denser microstructure

with lower content of voids, lower need of mixing water to achieve higher consistency, what can result in higher compressive strength and lower water absorption.

The aim of this experimental study is the observation and measuring the selected properties of fresh and hardened concrete in connection to various proportions of aggregate fractions.

2. Material and methods

2.1. Input materials

Cement CEM I 42.5 R from Ladce production factory, tap water, superplasticizer on PCE base and natural pre-crushed aggregate, were used for the experiment.
The mineralogical composition of aggregate fraction 0/4, 4/8 and 8/16 was as follows:

- biotic granodiorite up to granite – 36%
- limestones – 12%
- dolomites – 12%
- limy sandstones – 8%
- arkose quartzite – 8%

Size fractions 0/4, 4/8 and 8/16 from locality Sučany were used. Size fraction 0/0.7 was natural quartzite with almost rounded grain from locality Šajdikové Humence (Figure 1a). This fraction was used to make the overall gradation finer due to relatively coarse and angular fraction 0/4. In the figure 1b is shown the shape of the grains 1,4 – 2,0 mm from the fraction 0/4 as comparation to fraction 0/0.7.

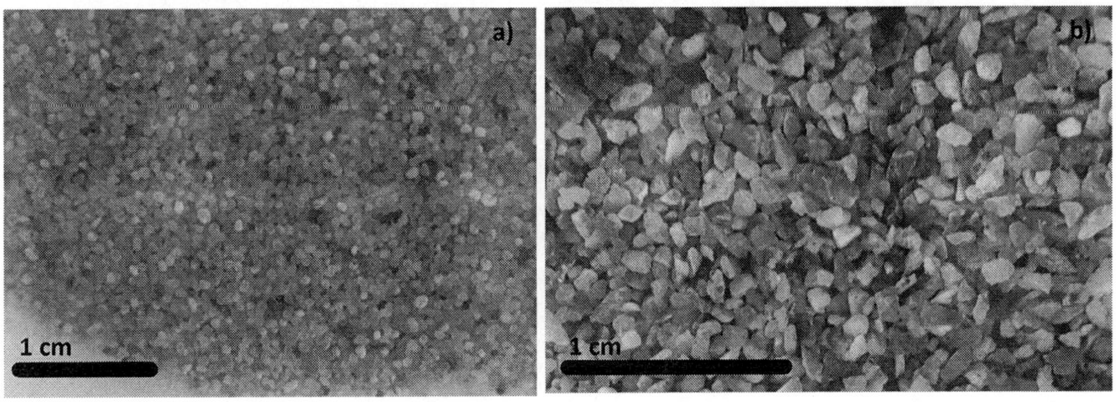

Figure 1. The shape features of used fine fractions: **a)** Siliceous sand fraction 0/0.7, **b)** Grains 1,4 - 2,0 mm from the fraction 0/4. [author]

2.2. Mix design

For the experimental study, 9 mix designs were prescribed (R1-9) with different proportion of aggregate fractions as noted in the Table 1. The amount of cement, effective water, plasticizer and overall aggregate amount were constant (Table 2) for each of the batches. According to practise, coarser granularity was chosen for the first batch. The reason for that was the experience, that for covering of finer aggregate of fraction 0/4 used in experiment is high need of cement paste (created either by cement or water or indirectly by plasticizer), what is subsequently reflected on the costs of mix design. R8 was designed as concrete with gap grading aggregate in according to prediction of better arrangement grains in microstructure to improve properties of fresh as well as hardened concrete. R9 had solely trial character. The amount of mixing water and plasticizer stayed the same for each of batches despite the fact, that the two first batches evinced the tough consistency.

Table 1. Proportion of aggregate fractions in the batches.

Fraction [%]	R1	R2	R3	R4	R5	R6	R7	R8	R9
0/0.7	0	0	10	10	10	15	15	30	25
0/4	50	55	40	40	50	55	40	0	25
4/8	0	0	0	10	0	0	0	30	25
8/16	50	45	50	40	40	30	45	40	25

Table 2. General mix design.

Material	Cement	$Water_{eff}$	Plasticizer	Aggregate	w/c [-]
Content [kg/m^3]	350	140	2.1	1915	0.4

2.3. Methods

At first, aggregate properties as water absorption (WA_{24}), apparent density (ρ_{ad}), proportion of crushed (C_c) and rounded grain (C_r), shape index (SI) and voidage V and grain-size distribution were found out. Subsequently, the amount of the fine particles under 0.25 mm was calculated. Grain-size distributions of each batch design were rendered in graph as percentage of passing materials through the sieve with specific size and compared in connection with the standard curves. The experimental study resided in the observation of changing the consistency (measured by slump (S_5 and S_{30}) and flow on the table with 15 impacts (F_5 and F_{30})) depending on the aggregate granularity of the batch in 2 time intervals (5 and 30 minutes after mixing). After measuring of consistency, specimens for testing the uniaxial compressive strength ($f_{c,150}$) (cubes, a = 150 mm^3) and for water absorption (WA) test (cubes, a = 100 mm^3) were prepared. The all samples were cured under water till the tests - 28 days. The apparent density in saturated state was calculated based on weight and dimensions. The samples for WA test were dried first at the temperature of 105±5°C till the time of fixed dried weight within the 24 hours. After cooling down they underwent the compressive strength ($f_{c,100,dried}$) test as well. This was performed as additional test for comparation with standard uniaxial compressive strength procedure.

3. Results and discussion

3.1. Aggregate properties

The results of water absorption, apparent density, proportion of crushed and rounded grain, shape index and voidage are given in the Table 3 and of sieve analysis in the Table 4.

Table 3. The properties of aggregate.

	0/0,7	0/4	4/8	8/16
WA_{24} [%]	0.5	2.0	1.0	1.0
ρ_{ad} [kg/m^3]	2650	2580	2660	2650
C_c [%]	0	83	51	75
C_r [%]	100	17	49	25
SI [%]	0	0	8	6
V [%]	-	28	38	37

Table 4. Gradation of used aggregate.

Sieve size [mm]	Overall passing material [%]			
	0/0.7	0/4	4/8	8/16
0	0	0	0	0

0.063	0.4	1.7	0.5	1.3
0.125	1.6	3.1	0.6	1.5
0.25	20.4	6.8	0.7	1.6
0.5	77.3	16.2	0.8	1.7
1	100	31.0	0.9	1.7
1,4	-	39.5	1.3	1.7
2	-	50.9	2.1	1.8
2,8	-	65.4	4.2	2
4	-	81.6	30.4	2.6
5.6	-	98.4	93.0	5
8	-	100	100	17.4
11.2	-	-	-	59.5
16	-	-	-	99.2
22.4	-	-	-	100
32	-	-	-	-

In the Figure 2, resulting grading curves of aggregate mixtures R1-R9, along with the standard curves as recommended by the technical standard for concrete [9], are given. In the figure 3, grains proportional content within each aggregate curve design is given.

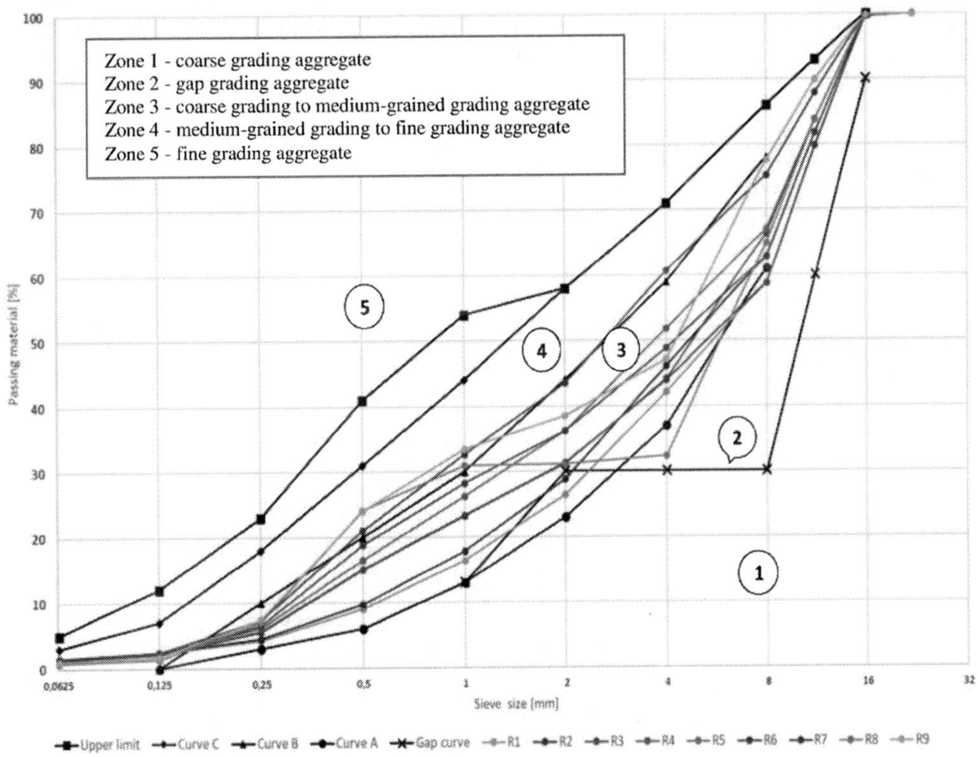

Figure 2. The aggregate grading curves of mix designs R1-9 along with the standard curves.

From the Figure 2 can be observed, that major aggregate mix design curves (R 1, 2, 3, 4, 5, 7) are located in Zone 3 what is in accordance with design stated in part 2.2. Curve R6 almost copy standardised curve B what can be considered to be optimal grain-size curve by technical standard. Curve R8 has gap-curve character. Curve R9 particularly extends to the Zone 4.

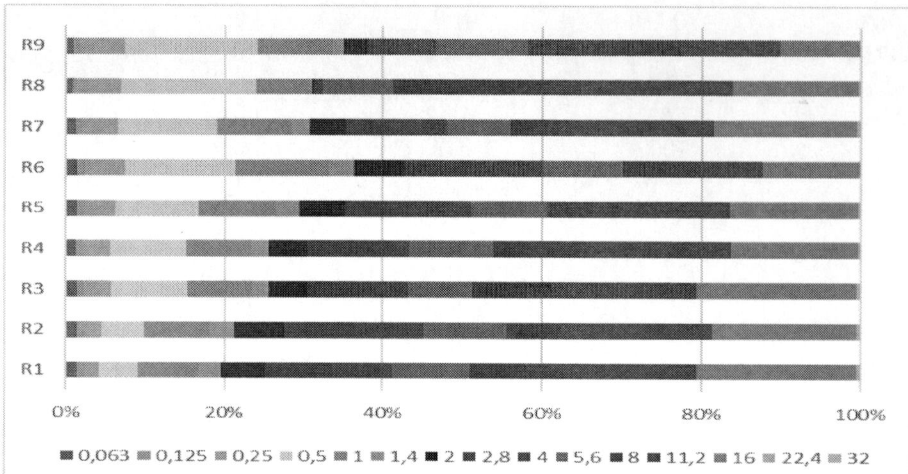

Figure 3. The particle [mm] (passing material) content distributions within the overall aggregate content.

The amount of the fine particles under 0.25 mm (including aggregate as well as cement particles) in individual aggregate designs R1 - R9 was found by calculation (Table 5) and it was confirmed that it met the recommended limit of the technical standard [9] for the concretes of standard composition (max. 530 kg.m^{-3} for D_{max}16).

Table 5. The amount of the fine particles for each aggregate mix design.

Aggregate Mix design	R1	R2	R3	R4	R5	R6	R7	R8	R9
Fine particles (<0.25 mm) [kg/m^3]	417	423	442	442	455	482	462	465	480

3.2. The fresh and hardened concrete properties

The results of measuring the consistency in time by slump (S_5 and S_{30}) and by flow table (F_5 and F_{30}) are given in the Figures 4 and 5.

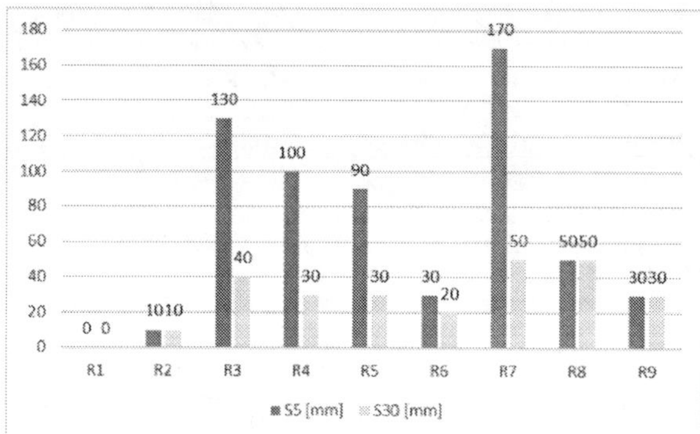

Figure 4. Results of slump test [mm] 5 and 30 minutes (S5, S30) after mixing.

The results for S_5 vary from 0mm (R1) to 170mm (R7) and for S_{30} from 0mm (R1) to 50mm (R7, R8). Generally, mix designs with replacement of the angular fraction 0/4 by the rounded fraction 0/0.7 achieved higher initial consistencies. The loss of consistency (evaluable) is the lowest (0%) at designs R8 and R9 and the highest (approx. 70%) at designs R3, R4 and R7

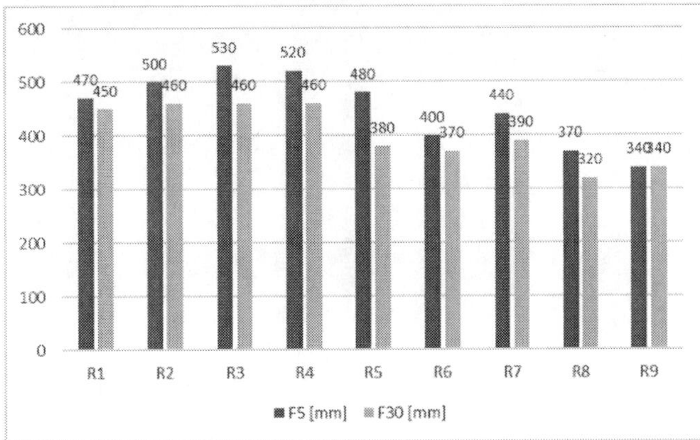

Figure 5. Results of table flow test [mm] 5 and 30 minutes (S_5, S_{30}) after mixing.

The results for F_5 vary from 340mm (R9) to 530mm (R3) and for F_{30} from 320mm (R8) to 460mm (R2, R3, R4). Lower flow can be observed at mix design with high content of 0/0.7 (25% and 30%), what could cause increasing viscosity. The highest values achieved mix designs with proportional content (50:50) of fine and coarse aggregate with 10% replacement of fraction 0/4 by fraction 0/0.7. Based on comparation of results from the slump test and from the flow on table test can be observed higher consistency loss at the slump test.

The results of compressive strength ($f_{c,150}$) and compressive strength after drying the samples for WA test ($f_{c,100,dried}$) are given in the Figures 6.

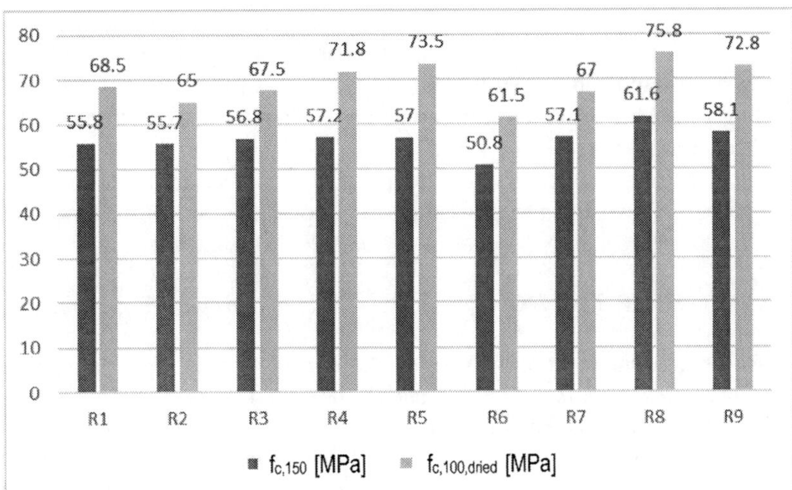

Figure 6. Results of compressive strength test.

The results of standard compressive strength ($f_{c,150}$) vary from 50.8MPa (R6) to 61.6MPa (R8). Compressive strengths of second set (dried samples – $f_{,100,dried}$) approximately (with taking the measurement and method errors into account) fit the standard test results. The lowest result is achieved with optimal grading curve (R6) by standard, the highest with gap grading curve (R8).

The results of water absorption test (WA) as well as density in saturated state (D_{sat}) are given in the Figure 7.

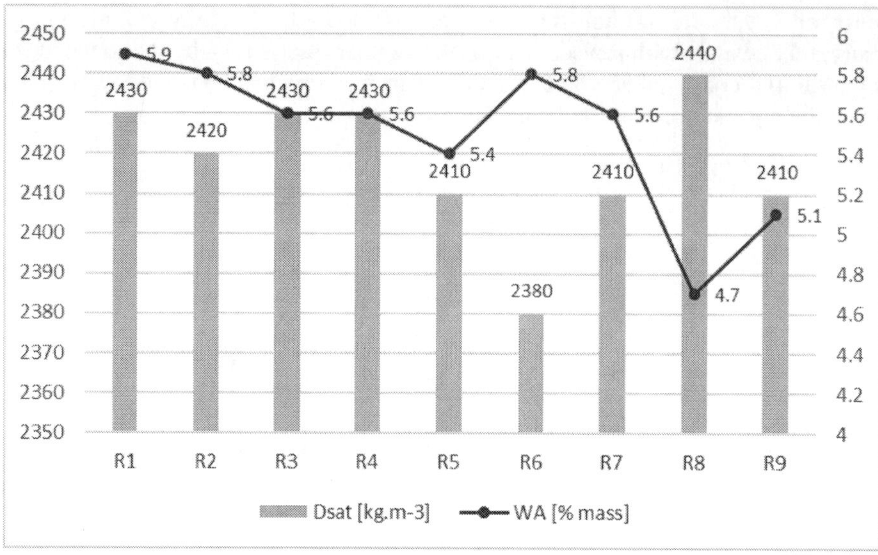

Figure 7. Results of water absorption and density.

Water absorption (WA) of tested samples vary from 4.7% (R8) to 5.9% (R1) and density in saturated state (D_{sat}) from 2380 kg.m-3 (R6) to 2440 kg.m-3 (R8). The mix design with optimal grading curve (R6) achieved the lowest density and high water absorption. The mix design with gap grading curve and with content 30% of fraction 0/0.7 (R8) achieved the highest value of density and the lowest value of water absorption what is in accordance with prediction in Introduction.

4. Conclusion

In this experimental study, various grading curves of aggregate in concrete were tested with the aim to observe the changes in fresh and hardened concrete. Based on results, following can be stated:

- properties of aggregate, especially the shape and surface of the grains as well as aggregate proportions, significantly impacted properties of fresh (consistency) and hardened concrete (compressive strength and water absorption),
- gap grading curve for aggregate into concrete can improve strength as well as durability of hardened concrete,
- optimal standardised grading curve did not provide proper values nor for fresh nor for hardened concrete properties, what can be consequence of not taking the grains properties into account.

5. References

[1] Mehta P K, Monteiro P J M 2006 *Concrete – structure, properties and materials* (New York: McGraw-Hill)

[2] Ferraris C F 1999 Measurement of the rheological properties of high-performance concrete: state of the art report *J. Res. Natl. Inst. Stand. Technol.* **104** 461–477

[3] Alexander M, Mindess S 2005 *Aggregates in Concrete* (London: CRC Press)

[4] Roussel N 2012 *Understanding the rheology of concrete* (Cambridge: Woodhead Publishing)

[5] Geiker M, Brandl M, Thrane L, Nielsen L 2002 On the Effect of Coarse Aggregate Fraction and Shape on the Rheological Properties of Self-Compacting Concrete *Cem. Concr. Aggreg.* **24** 3-6

[6] Jamkar S S, Rao C B K 2004 Index of aggregate particle shape and texture of coarse aggregate as a parameter for concrete mix proportioning *Cem. Concr. Res.* **34** 2021-2027

[7] Kurokawa Y, Tanigawa Y, Mori H, Nishinosono K 1996 Analytical study on effect of volume fraction of coarse aggregate on Bingham´s constants of fresh concrete *Trans. Japan Concr. Inst.* **18** 37-44

[8] Jiong H, Wang K 2011 Effect of coarse aggregate characteristics on concrete rheology *Constr. Build. Mater.* **25** 1196-1204

[9] STN EN 206/NA: 2015 *Concrete. Specification, performance, production and conformity. National Annex*

Acknowledgements
This paper has been prepared within the Grant No. 1/0648/17 of the Slovak Grant Agency for Science.

IOP Conf. Series: Materials Science and Engineering **566** (2019) 012008 doi:10.1088/1757-899X/566/1/012008

Determination of input parameters necessary for the design of the polder in the Horná Topľa basin, Slovakia

R Fijko[1]

[1] Technical university of Kosice, Faculty of Civil Engineering, Department of Environmental Engineering, Vysokoškolská 4, 042 00 Košice, Slovakia

Abstract. In the last decade, there have been many destructive floods in various parts of Slovakia. Despite the extensive investment in flood control works, there are only small changes and neither flood occurrences nor damages are decreasing. A possible consequence of climate change is an increased frequency of extreme meteorological events that may cause floods. The determination of the occurrence of these extreme hydrological phenomena in a particular catchment is directly subject to the input data in dependence on climatic and geomorphological processes. This article deals with the method of comparing the retention capacity of the polder by means of geographic information systems (GIS) compared to the results from map data and design of flood wave diagram for the river of Banský Stream. This proposal is part of the flood protection of the area of Čergov Mountain in eastern Slovakia where is designed the system of polders.

1. Introduction

The basis of the models forming is data acquisition using LIDAR and orthophoto maps techniques from which the relevant data for the project can be drawn. In the design of models, map data (data obtained by the classical method of data collection or GIS data), based on the principle of contour lines creation of specific structures, fractures, height uneven terrain, and thus created spherical splines are used [1]. The entire terrain model environment should be engineered to provide optimal terrain structure while maintaining full integrity between the architectural environment of the model (BIM) and its users [2]. Several 3D modeling software such as Autodesk, AutoCAD Map 3D, AutoCAD Civil 3D, Ecotect, Vasari, INOVA AreaCAD-GIS, Unity, Edificius, Google Maps, and ESRI with ArcGIS are available on the market. Ecotect and Vasari applications are now implemented in Autodesk Revit software solutions.

Conversion of data between CAD/BIM and GIS results in a relatively large data mismatch, which requires extensive manual work due to the limitations of data exchange and acquired knowledge [3]. The reason for the individual ambiguities, their deepening and delay was that the existing CAD and GIS platforms were developed independently for various purposes, leading to significant differences in the support of data formats, the terminology they used, the concept on which they were based, the different representation of degrees and the transformation of local coordinates into a reference system for horizontal and vertical coordinates (in Slovakia used the Trigonometric Network System in GIS S-JTSK Krovak East North). Additionally, CAD models usually do not store typological information (except geodetic measurements where x and y coordinates are defined and z need to be converted) that are characteristic of geospatial models [4]. As a result, various tools are being developed to transform models using 3D data model standards to enable visualization in 3D geospatial contexts using GIS software and BIM in conjunction with Industry Foundation Classes (IFC). In addition to the above-mentioned classic method of data collection using tachymetry or GPS [5], the most commonly used data is digital terrain model (DTM) or relief [6], photogrammetry in the form of ortho-photo maps or measurement using terrestrial or aviation laser scanning [7]. Today, one of the technologies is also LIDAR [1]. The use of laser scanning in combination with LIDAR in the CAD, GIS, and BIM implementation was used and compared in the case study of Banja Luka in the urbanized area, where the results demonstrated the complexity in individual steps and the focus of the investigators on a certain part of the problem. At each step, data rating requirements increased dramatically, and most users focused only on one of the above platforms [1]. When calculating the volume of the polder, the individual errors become more pronounced, as a lot of morphological parameters enter into the calculation. The process of processing the results is described in the paper [8], where four methods of data collection (ASTER, SRTM, LIDAR and DEM) were used to estimate the capacity of the Majdany Polder. The study showed different results and the impact of quantitative and qualitative source data in determining the polder retention capacity.

This paper reviews the design of the retention capacity of the polder in the area of Čergov Mountain in the Bardejov district in eastern Slovakia using two methods of data collection and final design of a diagrammatic course of the flood wave Q_{100}. Determination of retention capacity using the classic method of collecting data from map data using calculations in combination with CAD and determination of retention capacity using a digital elevation model (DEM). DEM is used to estimate the capacity of flooded valleys and the course of the flood wave [9], flood threats [10] and risk management to assess quantitative environmental impacts [11]. The exact estimate of the bulk of the polder is particularly important and the water in it is correlated with the reduction of the flood wave and the mitigation of the threat under its dike. After the flooded area, water from the polder is taken back to the river on the principle of gravity or mechanically by means of pumps [12]. The individual results were processed in the CAD environment implemented with ArcGIS and a final diagrammatic of the flood wave was created in excel. The results will be used by the author for the design of flood protection of the territory of interest.

2. Materials and methods

In the article are presented two available sources of information from the interest of area in a map format at 1:10 000 and the commercially available DEM 3,5-10. The map background was processed by CAD and a result was contour lines map of an interest area. From map was created the trigonometric 3D model (TIN) processed by GIS. DEM was processed in ArcGIS 10.3.

The area of interest (figure 1) is situated in the northeast of Slovakia in the upper section of the river Topľa, in the Čergov Mountains, in the Bardejov district, in the village of Livovská Huta. Polder of the Banský Stream is located on the rkm 0.570 and is the left tributary to the Topľa river. Topľa River belongs to the Bodrog catchment and is 136.8 km long (according to GIS), which is one of Slovakia's most significant watercourses, draining the area with a total area of 1 506 km^2. In last years, the area in question has been increasingly affected by lightning floods that threaten the inhabitants of adjacent municipalities, resulting in significant morphological changes and changes in bank lines [13].

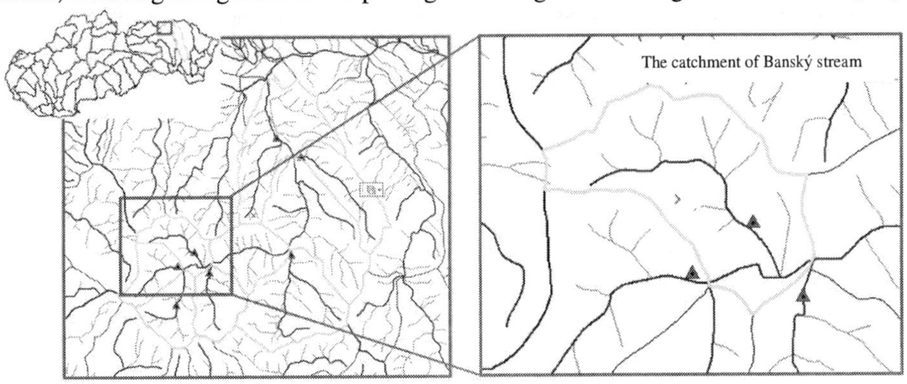

Figure 1. Area of interest Horná Topľa with the location of planned polders (left), the catchment of Banský stream with polder in rkm 0.570 (right).

IOP Conf. Series: Materials Science and Engineering **566** (2019) 012008 doi:10.1088/1757-899X/566/1/012008

2.1. Determination of the volume of the polder using a flood volume curve

The classic method of calculation of the total volume of the catchment is based on the determination of the volume of the dam (polder). In the case of reservoirs, the volume is divided into three main compartments, from the maximum level to the bottom, they are: retention, storage and permanently filled under the dead space outlet [14, 15]. The capacity of the dam in terms of the volume obtained and the flood area express the flood area curve $S=f_1$ (h) (m^2) and the flood volume line $V=f_2$ (h) (m^3). The floodplain area and the flood volume curve are basic characteristics of the reservoir and, according to [14], these curves can be determined by gradual planimetry of individual surfaces, bounded by contours in the map or graphically. The volume lines are gradually drawn from the flooded areas S_i and S_{i+1} and their height distances Δh_i by relation. The difference in the volume ΔV_i of the individual surfaces is then expressed by the equation (1).

$$\Delta \boldsymbol{V_i} = \frac{\mathbf{1}}{\mathbf{3}}\left(\boldsymbol{S_i} + \sqrt{\boldsymbol{S_i} \times \boldsymbol{S_{i+1}}} + \boldsymbol{S_{i+1}}\right) \times \Delta \boldsymbol{h_i} \qquad (1)$$

2.2. Determination of volume of the polder from DEM

ArcGIS 10.3 has been developed by ESRI to work with maps and geographic information and works as a comprehensive GIS platform that allows users to collect, organize, manage, analyze, communicate and distribute geographic information. ArcGIS is capable of reading CAD formats and integrating them into GIS as a layer. With the interoperability extension, convergence can be done for CAD and BIM formats [16].

The method for determining the basin volume (polder) in ArcGIS is typically done using the "3D Analyst" extension with the selected interval of peers (e.g. 5 m). Is necessary to create a separate layer contours.shp. After creating this layer, it is possible to specify the graphical properties and the marking of individual peers in the attribute table. Consequently, you need to create a TIN, and the quality of each model depends on the accuracy of the DEM input data. Enter the command "Open ArcToolbox> 3D Analyst Tools> TIN Management> Create TIN" or leave the GRID file. To determine the volume of the basin and the flood area, we derive from the maximum permissible height of the dike and the maximum water level read from it, assuming that the dam is divided into the above-mentioned main compartments. It is necessary to create a polygon in the form of the .shp flood area at the assumed elevation, when using the "Polygon Volume" tool to calculate the total volume of the polder. Then enter the necessary input parameters and run the calculation. As a result of the calculation, two new fields are generated in the attribute table: "Volume" and "Area_2D (3D)". If we work in an international SI system, the result is flood volumes in cubic meters and total flooded area in square meters.

3. Results

As mentioned above a map base for contouring the base in CAD, a map of the Banský Stream (tributary to Topľa) was used in M = 1: 10 000, with the hydrological number of catchment 4-30-09-003. The height difference of the individual planimetric surfaces spline a contour line was determined by $\Delta h = 5$ m from the map and supplemented by $\Delta h = 1$ m by interpolation. Since the data of maps are interpolated using spline the precision of the final results depends primarily on the quality of the map background. Figure 2 shows the processing of the map base and the final line of Banský Stream volumes.

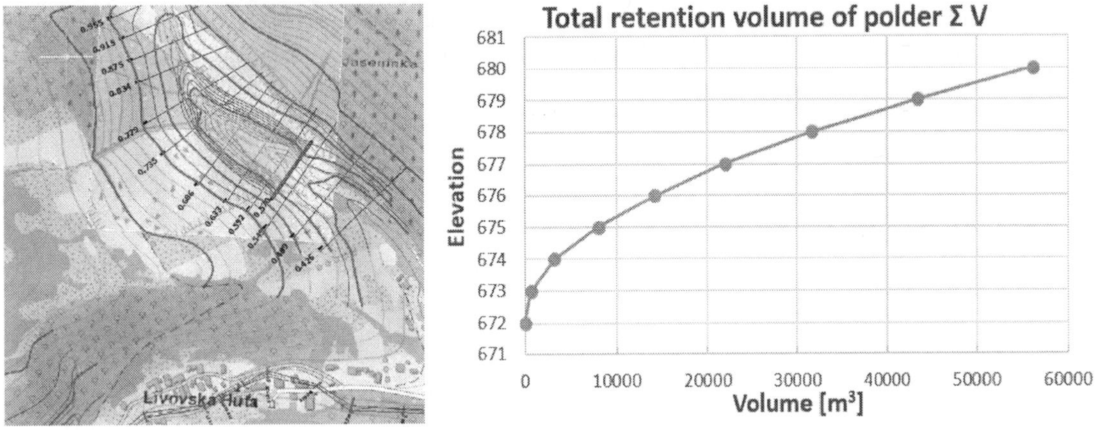

Figure 2. The resulting flooded area (left) and the total retention capacity of polder at a maximum dam height of 8 m (right).

The processing in ArcGIS software was less labor-intensive in terms of time and effort. The location of the polder was chosen at the same river of rkm 0.570 and as a verification tool served the CAD file. After processing the DEM and creating the TIN a polygon was defined and a .shp was created at the upper elevation of 680 m above sea level, which was used to define the flooded area and started the calculation. At this level, the maximum retention capacity of the polder is assumed. Figure 3 shows the flooded area of the polder in the TIN and the final depth map of the DEM after the calculation. At the bottom of the figure is a table of result of inundation and volume of polder.

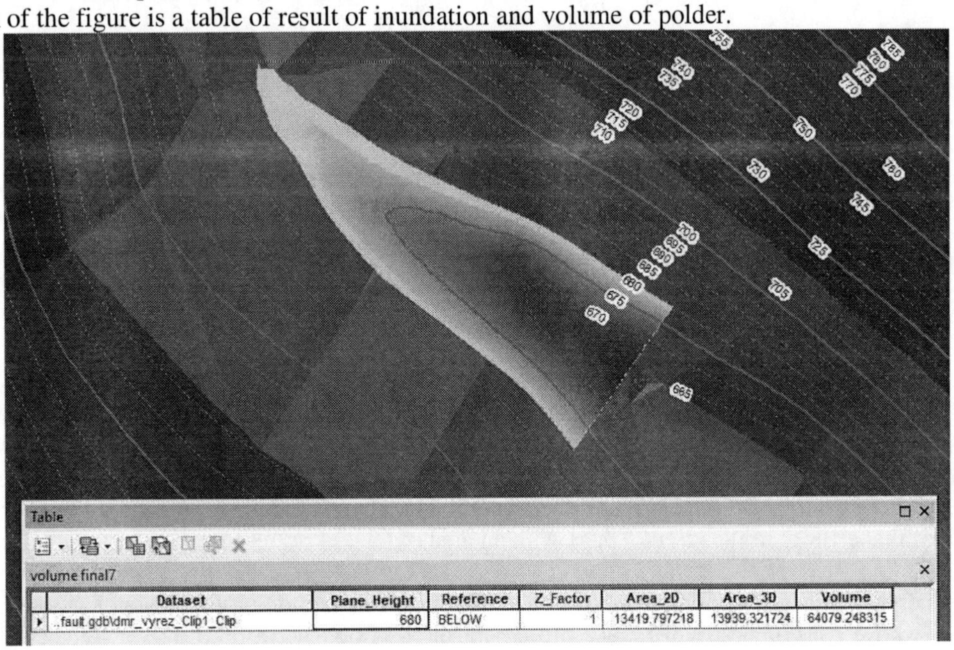

Dataset	Plane_Height	Reference	Z_Factor	Area_2D	Area_3D	Volume
..fault.gdb\dmr_vyrez_Clip1_Clip	680	BELOW	1	13419.797218	13939.321724	64079.248315

Figure 3. Total retention capacity of polder from GIS.

According to [8], it is assumed that there is a high correlation between the map profiles generated and the data provided with DEM or LIDAR-DEM. However, if the map and DEM are compared, the difference is more pronounced. The contour lines that were created and processed into CAD from freely available sources of maps are partly different from the models of DEM. This error was also confirmed in this case. This is mainly about defining CAD and GIS contour line and the initial location of the polder in the river in question. The elevation level of the bottom of the polder is assumed at the dimension of the CAD at the 672 asl. and in the case of the DEM at the elevation of 667 asl. This difference was also reflected in the final volume of the polder.

The resulting flood wave shown in figure 4 is generated from the map results. The retention capacity is derived from formula (1) and Δh = 1m. The total retention capacity is thus set at 56147.03 m^3 and total surface inundation is 13452.4 m^2.

Diagrammatic course of the flood wave Q_{100}

Banský stream

Time			*Q_{100}	Q_{median}	W -volume
[hod]	[min]	[sek]	[m^3/s]	[m^3/s]	[m^3]
0	0	0	0.0600	1.8622	216.00
1	60	3600	3.6644	3.6644	13191.77
2	120	7200	3.6644	6.7168	13191.77
3	180	10800	9.7693	14.6184	35169.52
4	240	14400	19.4674	23.8787	70082.67
5	300	18000	28.2900	26.8194	101844.00
6	360	21600	25.3489	23.0038	91255.89
7	420	25200	20.6587	18.7062	74371.34
8	480	28800	16.7536	14.9248	60313.13
9	540	32400	13.0960	11.2175	47145.77
10	600	36000	9.3389	7.9384	33620.19
11	660	39600	6.5379	5.1826	23536.44
12	720	43200	3.8272	3.8272	13778.07
13	780	46800	3.8272	2.3842	13778.07
14	840	50400	0.9411	0.8294	3387.98
15	900	54000	0.7176	0.7176	2583.39
16	960	57600	0.7176	0.7176	2583.39
17	1020	61200	0.7176	0.7176	2583.39
18	1080	64800	0.7176	0.7176	2583.39
19	1140	68400	0.7176	0.7176	2583.39
20	1200	72000	0.7176	0.7176	2583.39
21	1260	75600	0.7176	0.7176	2583.39
22	1320	79200	0.7176	0.7176	2583.39
23	1380	82800	0.7176	0.7176	2583.39
24	1440	86400	0.7176	0.7176	2583.39
25	1500	90000	0.7176	0.4084	2583.39
26	1560	93600	0.0992	0.0496	357.07

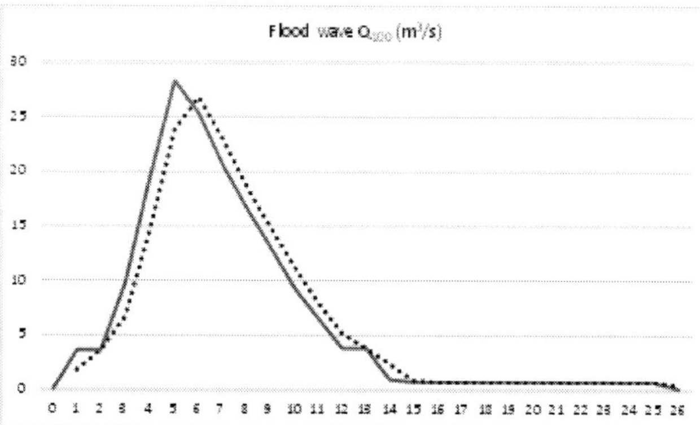

Retention capacity of polder

KH(asl.)	h	∑S	∑V
672	0	0	23.392
673	1	405.1	664.7389
674	2	1264	3247.463
675	3	5459.4	8014.411
676	4	5794.3	14128.19
677	5	7245.7	21884.6
678	6	9087.3	31596.09
679	7	11158.1	43489.18
680	8	13452.4	56147.03

Figure 4. The diagrammatic course of the flood wave from CAD.

From a geological point of view, the area covered by the Čergov Mountains is located in flysch zone. Even after mild rains is involved regular maintenance of the channel flow. This represents a considerable financial burden on the affected municipalities along the stream and forces local governments to find available solutions to this situation. The catchment area of the Banský Stream, with its area of 3.9 km^2, represents from the total catchment area of Topľa (1506 km^2), only 0.259 %, but from the regional point of view of the territory of the Horná Topľa with an area of 63.41 km^2 represents 6.15 % relatively large territory.

In general, water retention in the land is one the best for the community and one of the ways is also a polder whose construction will reduce the height of the flood wave on the river. One of the main responsibilities already in the preparatory phase of the project is the determination of the total volume of the basin, which was also the subject of this article. Two methods for determining the volume were used in the article. Determination of volume by planimetry from individual flood areas by CAD and determination by GIS calculations. From the volume curve (figure 2), which was determined by the calculation according to (1) in the CAD, it is evident that the volume of water in the created area of Banský Stream is 56147.03 m^3 and the flooded area is 13452.4 m^2. The total height of the dam is 8 meters in this case. Using GIS, the volume of the polder was set at 64079 m^3, which at the maximum capacity of the polder represents an area of 13419.79 m^2. In this case, it should be pointed out that in the same rkm profile a difference of 5 m was intended and the height of the dam, in this case, was set at 13 m. This difference can be verified by locating the GPS directly on site or repeating the process on a

IOP Conf. Series: Materials Science and Engineering **566** (2019) 012008 doi:10.1088/1757-899X/566/1/012008

more detailed DEM. The difference in volume between CAD and DEM 3.5-10 is 7932 m^3, corresponding to a difference of 12.37% and for an area of flooded is the difference of 0.25%. This has confirmed the assumption that map and DMR have a similar flooded area, but for the volume, we need measurement of the cross-section profile where the dam of the polder is located.

4. References

[1] Mijic N, Sestic M and Koljancic M 2017 CAD - GIS BIM integration - case study of Banja Luka city center *Advanced Technologies, Systems, and Applications* vol 3 (Cham: Springer) pp 267-281

[2] ACCA software S.p.A 2017 *Terrain modelling using contour lines and BIM software* citate: http://biblus.accasoftware.com/en/terrain-modelling-using-contour-lines-and-a-bim-software/

[3] El Meouche R, Rezoug M and Hijazi I 2013 Integrating and managing BIM in GIS, software review *International Archives of the Photogrammetry, Remote Sensing and Spatial Information Sciences* **2** W2

[4] Lapierre A and Cote P 2007 Using Open Web Services for urban data management: A testbed resulting from an OGC initiative for offering standard CAD/GIS/BIM services *Urban and Regional Data Management. Annual Symposium of the Urban Data Management Society (Electronic Materials)* pp 381-393

[5] Torge W and Müller J 2012 *Geodesy* (Walter de Gruyter)

[6] Daňová S 2010 Digitálne modely reliéfu na území Slovenskej republiky *Geodetická a kartografická podpora protipovodňovej ochrany* (Bratislava: STU v Bratislave SvF) pp 26-31

[7] Dolanský T 2004 *Lidary a letecké laserové skenování* (Ústí nad Labem: Univerzita JE Purkyně)

[8] Walczak Z, Sojka M, Wrózyński R and Laks I 2016 Estimation of Polder Retention Capacity Based on ASTER, SRTM and LIDAR DEMs: The Case of Majdany Polder (West Poland) *Water* **8** p 230

[9] Teng, J, Vaze J, Dutta D and Marvanek S 2015 Rapid inundation modelling in large floodplains using LIDAR DEM *Water Resources Management* **29** pp 2619-2636

[10] Fijko R, Zeleňáková M and Alkhalaf I 2016 Flood risk modeling of Stulianský stream in Lopúchov village, Slovakia *International Journal of Safety and Security Engineering* **6** pp 529–537

[11] Zeleňáková M and Zvijáková L 2017 Risk analysis within environmental impact assessment of proposed construction activity *Environ. Impact Assess. Rev.* **62** pp 76–89

[12] Gichamo TZ, Popescu I, Jonoski A and Solomatine D 2012 River cross-section extraction from the ASTER global DEM for flood modeling *Environ. Model. Softw.* **31** pp 37–46

[13] Frandofer M and Lehotský M 2013 Stupňovitá štruktúra dna koryta vodného toku s výrazným pozdĺžnym sklonom (na príklade horného toku rieky Topľa) *Geographical Journal* **65** pp 141-159

[14] Švecová A and Zeleňáková M 2005 *Vodné stavby* (Košice: Stavebná fakulta TU v Košiciach) p 190

[15] Ondrejka Harbulakova V and Lechamn M 2018 Flood control measures - dry basins - case study, *Public recreation and landscape protection - with nature hand in hand!*, 2.- 4.5.2018, Křtiny-Brno: Mendel University, pp 168-173

[16] El Meouche R, Rezoug M and Hijazi I 2013 *Integrating and managing BIM in GIS, software review* (Istanbul: Turkey) pp 31-33

Acknowledgments

This work has been supported by the Slovak Research and Development Agency by supporting the project SL-PL-18-0033.

IOP Conf. Series: Materials Science and Engineering **566** (2019) 012009 doi:10.1088/1757-899X/566/1/012009

History of vegetation roofs in the Czech Republic

M Hejl[1],

[1] Brno University of Technology, Faculty of Civil Engineering, Institute of Technology, Mechanization and Construction Management, **Czech Republic**

Abstract. This article is mainly focused on the development and history of vegetation roofs in the Czech Republic. In the article is summarized history of green roofs in Czech republic from the oldest vegetation roof. Subsequently, it contains the significance of the function of vegetation roofs and the general classification according to the types of buildings including their description and price. Last but not least, there has been described the reconstruction of the oldest vegetation roof in the Czech Republic on the building. Also there has been described a modern design and reconstruction of the roof garden. At work I have been able to summarize all the advantages and disadvantages of vegetation roofs compared to classical roofs. The results of this work provide a detailed view of the development of vegetation roofs and on the basis of the work it is possible to suggest possibilities for accelerating work on vegetation roofs.

1. Introduction

The population of the planet is still increasing, and so the areas in our neighbourhood are increasingly being used for human settlements and the construction of industrial, commercial or administrative buildings. At the outskirts of the city, new quarters are emerging that degrade fertile soil and destroys the ability of the landscape to naturally retain water. The water that we come **across** is therefore missing throughout the chain of natural cycles, and people are forced to pump away more and more water from underground supplies that are not naturally replenished to they original volume. However, we have other variant of how to rehabilitate these areas despite the green areas, especially thanks to green architecture. Whether using parks, green facades or above all, using vegetation roofs.

There are many questions about vegetation roofs, such as: Are they useful? Do they have a tradition in our country? Can we use their benefits or are they perceived as unnecessary luxury? How is the construction of green roofs going in Czech Republic? Is it fast enough and effective or slow and unnecessarily expensive? These are just a few questions that probably can not be answered satisfactorily and adequately within one article. Nevertheless, the task of this article is to clarify at least a few of these questions and to show how the construction of green roofs within the Czech Republic has developed in the last decades or hundreds of years. What success has been achieved in this field and what progress has been made in building materials and time necessary for build this type of roof.

2. Theory

2.1. History of green roofs

In the very beginning, it is necessary to look into the history of vegetation roofs in the Czech Republic. There is no vegetation roofs from past times, as is the case in the northern part of Europe where, at the turn of the antiquity and the Middle Ages, the first human dwellings with green roofs were created and their tradition is still alive in these countries. Where they were built mainly due to good thermal insulation properties, when it was necessary to keep the dwellings with the most stable temperature as possible during hot summers and cold winters. As far as the territory of the Czech Republic is concerned, the oldest known green roof is dated to a time that is much closer to us. It was designed in year 1863 by the architects Josef Žák and Ferdinand Wenzel. This roof was designed as a terrace above the building of horse stables at the renaissance chateau in Lipník nad Bečvou. The roof was further upgraded to its present form, which corresponds to the 1910 modifications, when the roof terrace was rebuilt into a reinforced concrete slab. Because the roof was originally terrace of the castle, the roof about which we are talking here was constructed as intensive green roof. On the surface of which are found many paths created with the help of a mixture of sand, clay and crushed limestone. At present, the terrace serves as part of the chateau park in Lipník nad Bečvou. The castle itself is used as the seat of the city office. On this roof can be well presented the development of green roofs in the Czech Republic because roof has been recently reconstructed by using modern materials and procedures to continue to serve its original purpose. Since 1863, other green roofs have been created, especially as roof gardens at castles and for buildings meant to nobility. Perhaps for this reason, green roof are seen as unnecessary luxury. For example, it is worth mentioning the terrace at Konopiště Castle, where was a terrace with vegetation - lawn and trees. However, this roof garden had to be reconstructed several times due to frequent leaks. Further green roofs then began to emerge in the first half of the 20th century. Many of them have been preserved until today, such as the Roof Garden at the Union Bank in Brno or at the Baťa factory in Zlín. In the second half of the 20th century other green roofs were built mainly in public and representative buildings such as the Prague Castle Riding Hall, the Thermal Hotel Terrace in Karlovy Vary, Prior House in Brno ...). In the years since the end of the 2nd world war by the 1990s, about 15,000 ha of green roofs had been created in the Czech Republic. Since the beginning of the nineties there has been an increase in the number of green roofs in our country. However, their growth can be calculated in percentages, and we are still lagging behind neighboring countries like Germany and Austria. [2]

2.2. Benefits and disadvantages of green roofs

Before process of building the vegetation roof, it is necessary to focus on its importance and its function. Vegetation roofs have many categories. First of all, it is worth mentioning the function of urban and landscaping. This means that green roofs are places created for relaxation, which are increasingly lacking in city centers. This type of roofing can also be used to extend the living space for other animals and therefore not only for humans. Last but not least, they are also involved in increasing the number of greenery in the countryside. Functions that should be mentioned includes the environmental functions of vegetation roofs as roof vegetation significantly contributes to slowing air movement and thus reduces dustiness of the environment. It is also involved in the retention of rainwater and thus its return to the cycle of water in the nature. Plants are able to absorb pollutants from rainwater - especially copper and lead, which are thus prevented from further penetrating into the lower waters and rivers. They also contribute to the increase of air humidity due to retention and subsequent evaporation of water, which further contributes to the reduction of temperature fluctuations. [4]

It is also necessary to address with the building advantages and disadvantages of green roofs. As an obvious advantage, is their aesthetics function within the city environment. However, this is not the only benefit of the green roofs. It is also necessary to mention the other benefits of green roof constructions. Vegetation layer is an additional thermal insulation of the roof, where the decrease in thermal losses of the roof in the case of planting is counted by 10 to 30%. The vegetation layers have a significant role in the protection of waterproofing layer against degradation due to ultraviolet radiation, infrared radiation and temperature fluctuations (with temperatures ranging from -3.6 ° C to 65 ° C on conventional roofs). In the case of vegetation roofs this range ranges from - 0.8 ° C at 30 °C. Effect of this is a significantly lower stress of the waterproofing layer due to temperature variation has been demonstrated. The vegetation itself can also be used to improve the acoustic properties of the roof where the sound absorption of the roof structure is better by 3dB. The substrate itself has a function as an additional impact insulation. The retention capabilities of green roofs are used to reduce the load on the sewerage network, where a reduction in load on extensive vegetation roofs is measured by up to 30% and in intensive vegetation roof even up to 50%. Last but not least, it is necessary to mention the protection, which the structure provides for the load bearing and the waterproofing layers in front of mechanical damage or for the better efficiency of the photovoltaic panels However, thanks to these benefits the green roofs can be seen as ideal solution of many problems but they also have some problems and disadvantages. workload in their construction. Work as such can then be associated with a further disadvantage, and this is a need for a high-quality execution where in the case of a poor performance of the vegetation stack there is a risk of water leakage into construction structures. Separate risk is then vegetation where some types of plants could cause allergies, and the effect of leaves, flowers and branches is subject to biological contamination. In particular, in the Czech Republic is no financial support for construction of green roofs, as opposed to, for example Germany, where are many programs for support of creation of new green roofs. To defend the Czech Republic there is one grant program which is called Zelená úsporám (Green Savings) which the last stage was supports construction of vegetation roofs thanks to their water retention. Between the other disadvantages of green roofs which we can call purely as constructional or aesthetic ones is their higher cost compared to the classical roof and higher workload in their construction. Workload as itself can then be associated with a further disadvantage, and this is a need for high-quality execution where in the case of a poor execution of vegetation layers is there a risk of water leakage into the construction of building. Separate risk is a vegetation as itself, where some plant species can cause allergies, and the fall of the leaves, flowers and branches are biological contamination of surroundings. [11]

2.3. Types of green roofs

Another very important parameter for construction of green roof is need of choose right type. Which is very important due to different prices of every type of green roof. There are four basic types of green roof as biotope green roof, extensive green roof, semi-intensive green roof and intensive green roof.

For the biotope vegetation roof, we are not counting with foundation of vegetation and area of roofs is free for natural raid of plants seeds. There are savings of the purchase of seeds and care of vegetation layer. The optimum substrate height is from 6 to 12 cm. Maintenance consists of removing unsuitable plants once or twice per year. The approximate price of biotope green roof is by 55,50 to 91,56 EUR/m^2. [2]

In the case of extensive green roof is being considered with minimum maintenance of vegetation (approximately two times per year). Considered substrate height is from 2 to 20 cm. This type of roof is mostly used for non-walkable roofs. Plants usually used for this type of roofs are mosses, sedums, nonskarks, some types of herbs and grass which likes dry climate. Price of these roofs is approximately 56,00-128,60 EUR/m^2. [2]

For the construction of semi-intensive green roof are also used with unpretentious greenery but roofs are mostly designed as walkable. Vegetation on these roofs are primarily different types of grasses, herbs and shrubs. Substrate height is between 15 and 30 cm. Price of semi-intensive roof is around 67,40-179,20 EUR/m^2. [2]

Intensive vegetation roofs are characterized by the need for constant care and maintenance. Height of substrate is more than 30 cm, where the only limitation is possible depth for roots of vegetation and height of the trees which shouldn't be higher than 10 m. Intensive green roofs are possibly used even in agriculture. Orienting costs of intensive vegetation roof is about 89,60-307,80 EUR/m^2. [2]

A lot of other, diversification of green roof are mostly influenced by designer of roof. Whether it is roof accessibility, its function or the slope of supporting structure where is necessary to say that the vegetation roof can be applied to almost any building and any pitched roof but with higher slopes it is more difficult to make a good design which also increasing cost and length of building. There, are also different possible diversification of vegetation roofs but these are almost impossible to change at the construction stage. For example, it can be roofs in line with the terrain. [4]

2.4. Case study

If we will talk about construction of green roofs as itself, we need to go back to the beginning of this article and of the green roofs in Czech Republic, where is mentioned chateau in Lipník nad Bečvou where in the year 2005 was reconstructed the first green roof. For this reconstruction was used modern procedures in order to preserve roof garden after the construction work. At the beginning of the works had to be pumped out the existing soil. That revealed waterproofing layer which was made by asphalt bands (height of layer was around 30 cm) which were found still in condition. As a result, the original waterproofing was repaired with a vapor permeable primer and PE foil. Thermal insulation was choosed as a foam insulation EPS 100 and the last layer was from waterproofing membrane which was led through balustrade to the drainage where it was finished by copper plating. Then was finalized the composition of intensive green roof by separation geotextiles, nano-foil hydroaccumulation layer, other geotextile separation, topsoil and vegetation substrate. In order to prevent the soil from being channelled into the drainage was created a vertical dilatation layer around the perimeter of drainage from a shaped hot dip galvanized expanded metal. An important detail which was constructed was dilatation to prevent lateral pressure to balustrade which was made by applying of vertical stripes by compressible material in the composition of individual layers. Because roof is an intensive green roof, it is possible to find sidewalks on its surface, which were formed by mixture of clay, sands and crushed limestone to prevent growth of the grass. Sidewalks were separated from soil by vertical hot dip galvanized sheet metal (8 mm). With dilatations by 3 m. In the middle of the area there is a fountain that has been dismantled and replaced by a replica. Garden was also designed for easy maintenance. Reconstruction of roof garden which has area 680 m^2 was finished in the September of 2006. [8]

3. Conclusion

As we can see from the data collected in Czech Republic there is still growing number of realizations of green roofs. Especially extensive green roofs for family houses. If we will talk about case of intensive vegetation roofs than this type of green roof is mostly constructed on the representative or commercial buildings. If we will talk about private construction sector, however it is still only the realization within the framework of several tens of houses per year in the whole country, which is still a few compared to the countries like Germany or Austria. If we would like to compare costs for vegetative roofs to classic types of flat roof construction than we can´t make a green roof under 54,55 EUR/m^2 but cheapest type of flat roof with waterproofing layer made by mPVC foil and thermal insulation by EPS 100S with thickness 180 mm has cost 30,20 EUR/m^2. In the subject of pitched roof is costs of typical construction even smaller. We can see that the price of vegetation roofs is still an important factor for their wider expansion in Czech Republic. However, at present it is possible to apply for a subsidy of up to 19,50 EURč/m^2 form the state grant program Zelená úsporám, which reduces the price of the vegetation roof to 35,06 EUR/m^2. The green roof is still the most expensive variant, but it is not almost double of costs against other types of roofing. And benefits which green roof providing to house and his surrounding are their biggest benefit.

Other problem which green roofs has is their hard construction and construction time. In the case of construction classic roofs, they are completed at the last day of realization but in the case of green roof we have to count also with time which we need to growth of grass or other greenery to have a fully functional green roof. Of course, we can use the grass carpets, but this solution is not financially effective and not everyone can afford them. Also, the transport of the material itself is very expensive and laborious, especially on roofs that are not constructed in contact with the terrain, where the construction can be helped by heavy machinery transporting the vegetation substrate and soil directly to roof. But if we doesn´t have this possibility, then is needed to transport all of soil and substrate on the roof manually. In almost every case we also have to transport manually all plants on the roof. In the Czech Republic is not currently any possibility of speeding up and making these constructions more efficient. If we will look at abroad, we can see the effort to simplify whole process, which is only at the beginning and leads especially to the prefabrication of the individual parts. These are prefabricated panels that are planted with sedum and herbs in pre-treatment process. These parts are transported to a building where they can be simply moved on the roof and make a finalized look of vegetation layer. That means a great saving of time within the time of construction.

Unfortunately, these technologies are not available within Czech Republic, so the time labor and financial demands for construction of vegetation roofs are still relatively higher than for classical roofs. However, if we will talk about benefits the green roofs overcomes the classical roofs thanks to the need of water retention and cooling of the city centers in the summer period, it is possible to count with their constant development. If we can simplify their construction, then we can be convinced about that their popularity will rise.

4. References

[1] Igra, International Green Roof Association 2009 Green roofs - bringing nature back to town proceedings ed. R Appl a W Ansel (Berlin: International Green Roof Association) pp 1-181

[2] Čermáková B and Mužíková R 2009 *Ozeleněné střechy* (Praha: Grada) pp 1-248

[3] Bohuslávek P, Horský V and Jakoubková Š 2009 Vegetační střechy a střešní zahrady (Praha: DEKTRADE) pp 1-71

[4] Svaz zakládání a údržby zeleně Standardy pro navrhování, provádění a údržbu: Vegetační souvrství zelených střech 2016 (Brno: Odborná sekce Zelené střechy při Svazu zakládání a údržby zeleně) pp 1-36

[5] Svaz zakládání a údržby zeleně Svaz zakládání a údržby zeleně 2014 (Brno: Svaz zakládání a údržby zeleně)

[6] Greenville GreenVille service 2017 (Brno: GreenVille)

[7] Earch.cz Nejstarší zelená střecha v České republice 2007 (Praha: Earch)

[8] Asb-portal.cz Nejstarší zelená střecha v České republice 2014 (Praha: Jaga Media) 2014

[9] Green roof technology History of green roof technology *Green roof technology* 2017 (Batimore: Jörg Breuning & Green Roof Service LLC.)

[10] Stamaco engineering Orientační ceník střech 2017 (Ostrava: STAMACO - stavební materiály a constulting)

[11] Krajská pracoviště Státní fond životního prostředí Ministerstvo životního prostředí Nová zelená úsporám: nově na výstavbu nízkoenergetických bytových domů, zelené střechy a využívání tepla z odpadní vody 2016 (Praha: Krajská pracoviště Státní fond životního prostředí Ministerstvo životního prostředí)

Acknowledgement

This paper has been worked out under the project No. LO1408 "AdMaS UP - Advanced Materials, Structures and Technologies", supported by Ministry of Education, Youth and Sports under the „National Sustainability Programme I" .

IOP Conf. Series: Materials Science and Engineering **566** (2019) 012010 doi:10.1088/1757-899X/566/1/012010

Hydraulic assessment of flood protection measures proposal

A Janík[1]

[1] Slovak University of Technology in Bratislava, Faculty of Civil Engineering, Department of Hydraulic Engineering, Slovakia

Abstract. Ensuring flood protection is a never-ending task, therefore there is always something to improve or add. Nowadays there is a big boom in detention reservoir projecting for smaller river basins in Slovakia. This paper focuses on flood protection proposal for village Veľká Lúka located in the central part of Slovakia. The contribution deals with the hydraulic modelling of present state of river Lukavica channel and its capacity and also proposed flood protection measures. According to geodetic survey and flow rate measurements a 5 km long partially calibrated mathematical model of the Lukavica River has been created. For this purpose, the HEC-RAS software has been used. After evaluation of the flow rate capacity through the village, a flood protection measure – detention reservoir – has been proposed using designed flood wave provided by Slovak Hydro-meteorological Institute (SHMI). According to transformation of flood wave computation, specific parameters of necessary flood protection measures were determined.

1. Introduction

A typical example of problems faced by the water managers in Slovakia is the individual residential construction in the flood plains of the watercourse. People buy land, they get a building permit from the village authorities (sometimes despite negative statement from the water management enterprise), they build their dream house, but the river or stream soon shows that it was a mistake. These people then appear in various television sessions and demand protection from floods. The water manager then faces a very difficult task.

The case of the village Veľká Lúka is not other. In this village, especially in the recent past, there have been several cases of flooding of the surrounding areas of the Lukavica River (figure 1). This problem has only emerged in the last few years after the larger development of individual houses along the Lukavica River occurred. These houses are in direct contact with the inundation area and there are other builders demanding solution for the flood protection situation because they want to build houses on their own land close to the river. This problem is particularly hampered by the fact that there is no space for increasing the capacity of the riverbed in the village, which capacity is somewhere at Q_2 to Q_5 level, not to mention Q_{10} at 28 $m^3.s^{-1}$. For this reason, it was necessary to approach the solution by reduction of the flow rate in front of the municipality by proposing a detention reservoir that would reduce the peak flood wave to the capacity flow in the village [1].

Another problem in this village is also the crossing of the Lukavica River by the railway led between towns Zvolen and Banská Bystrica. The capacity of the railway bridge is very limited; at the time of the construction of the bridge, it was located hundreds of meters below the village, what is the reason for its insufficient capacity nowadays.

Due to the fact, that no morphological measurements of the riverbed were available, the research team had to realize a survey that resulted into 38 measured cross-section profiles including the structures on it - from the mouth of the Lukavica stream into the Hron River to the profile of the proposed detention reservoir. Afterwards it was possible to determine the capacity of the most critical places of the Lukavica stream in the Veľká Lúka area.

Figure 1. Flooding of the MDD street (right bank of the Lukavica River) in February 2016.

2. Current state of the Lukavica River in the village of Veľká Lúka

The river basin of Lukavica itself is not very extensive. Its area drawn to the profile in front of the village of Veľká Lúka is approximately 34 km^2. The Lukavica River itself is the left-hand tributary of the Hron River northerly of town Sliač with total length of 14 km. It stems in the Zvolen valley in the territory of the village Horná Mičiná at altitude of 490 m a. s. l. It flows through the districts of Banská Bystrica and Zvolen via the villages of Dolná Mičiná, Lukavica and Veľká Lúka. Its most significant tributaries are the Jasenica brook (right-hand tributary in rkm 7.420) and the Samporský stream (left-hand tributary in rkm 4.900) [2].

The last major floods that triggered the need for flood protection occurred in 2009 (25.12), 2013 (13.3.) and 2016 (10.2) when the floods reached the second and third degrees of floods activity [3]. During the first field survey problematic areas were identified in terms of the capacity of the Lukavica River, which were subsequently confirmed after the survey and calculation by the mathematical model. It was the mentioned railway bridge (rkm 1.185) and a section from this bridge upstream along the new district of housing of the village Veľká Lúka.

The distance of the railway bridge supports is 4.5 m and the bottom edge of the bridge deck is approximately 1.3 m above the level of the river bottom. The bridge structure appears to be inadequate, according to the previous study, it is able to transfer the flow rate of only 14 m^3.s^{-1} and in the pressure regime approximately 21 m^3.s^{-1}.

In the section upstream of this bridge, the channel has an untreated appearance with a variable width of 1.5 to 4.5 m in the bottom. The channel itself is very narrow in this reach with a dense vegetation on the banks what significantly reduces the flow area at higher water levels (figure 2). In this part, on the right bank of the river, there are new houses located in the floodplain zone of the river and the inhabitants of adjacent building lots report frequent flooding of their lands. Due to the planned construction of family houses on the left bank of the river flow, this section of the river is the most problematic and later proved to be the most critical in terms of the capacity of the Lukavica River in the village residential area.

Figure 2. Railway bridge (a) and channel of Lukavica river (b) in village Veľká Lúka along the MDD street.

2.1. Field measurements of the morphology of the stream of Lukavica

Since there has not been any field terrain survey in this location before, the research team had to create a geodetic data of the channel (cross-section profiles in the range from 0.000 to 3.455 km), and then a detailed measurements of the area that would be affected by the construction of the detention reservoir and its retention basin.

The measurements were proceeded from the confluence of Lukavica River with Hron River and gradually 38 cross-section profiles were measured. Subsequently, in April 2018 the geodetic survey of the river basin between the municipalities of Veľká Lúka and Lukavica was carried out. The survey was realised for detention reservoir proposal in this area. Therefore, it was made from the specified cross section in the upstream direction and was bounded by a contour line at 235 m a. s. l. but in a matter of fact, the terrain measurements were even one meter higher. The whole measurement in form of spatial points above the proposed detention reservoir profile are shown in figure 3.

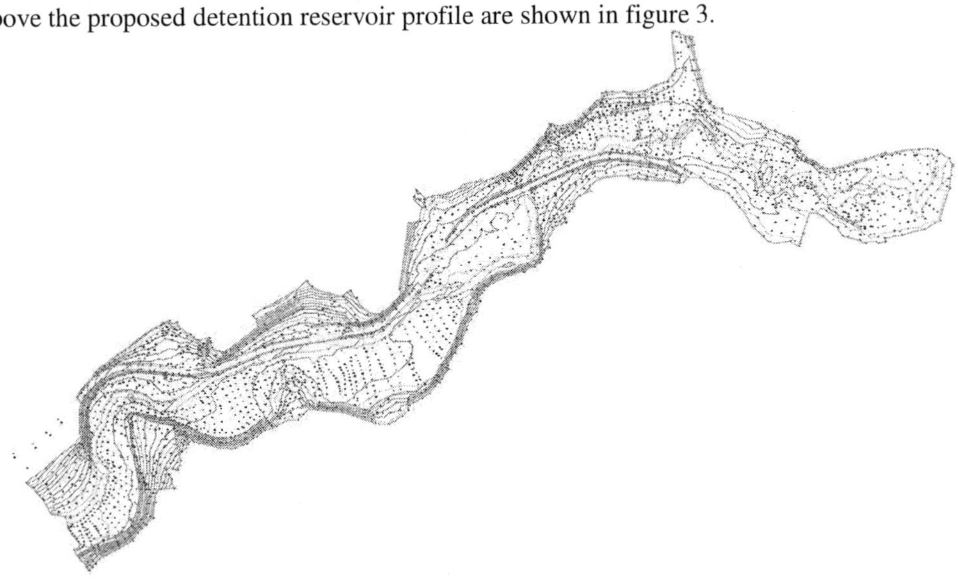

Figure 3. Measured points along with digital terrain model (blue – watercourse, yellow – road between Veľká Lúka and Lukavica, green – boundary of the measured area).

Later on, the intercepted volume lines for the two selected profiles were constructed from the digital terrain model. Profile 1 was taken from a flood protection study of this site done by local water management enterprise (WME), profile 2 was selected based on the field survey because it appeared to be more advantageous in terms of intercepted volume in detention reservoir [4,5,6]. Both profiles are located above the village Veľká Lúka and were located in such a way that the eventual dam of the detention reservoir at the planned height of about 10 meters did not reach the road that leads across the valley between the municipalities Lukavica and Veľká Lúka (figure 4).

Figure 4. Chosen profiles for calculation of intercepted volume lines.

In the selected profiles, flooded area lines were at first constructed from a digital terrain model with a 0.25-meter contour lines spacing from which the intercepted volume lines were calculated. The assumption from the terrain proved to be correct because in the profile 1 at the considered height of the 10-meter dam the retention volume would be 618 thousand m^3 and in profile 2 the volume would be 633 thousand m^3 (figure 5). However, in the case of the construction of a dyke, the actual retention volume would be smaller by the volume of the dyke body, which reaches into the flood area since the profile was considered as a vertical wall during the calculation.

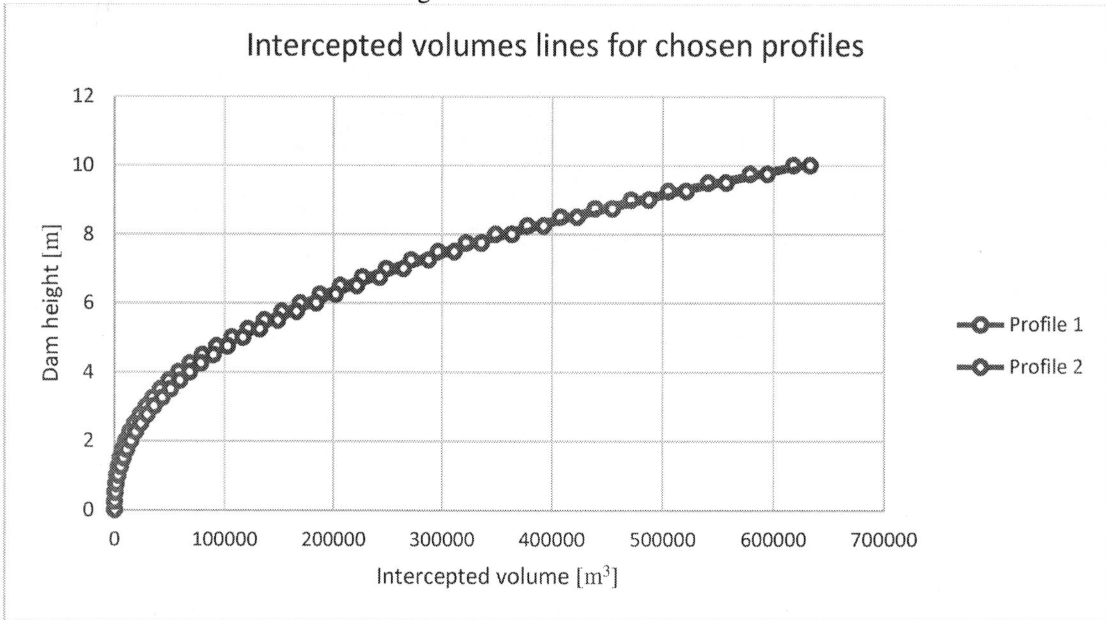

Figure 5. Graphical comparison of intercepted volume lines for profiles 1 and 2.

IOP Conf. Series: Materials Science and Engineering **566** (2019) 012010 doi:10.1088/1757-899X/566/1/012010

3. Hydrological data

Based on the WME order, the SHMI team worked out the design flood wave for a 100-year flow rate of $Q_{100} = 49\ m^3.s^{-1}$. The volume of the flood wave W_{PV100} was, according to SHMI, estimated at 1.8 million m^3 with the total duration of the flood t_c = 20 h 25 min, increasing for t_{st} = 8 h 10 min and decreasing for t_{kl} = 12 h 15 min (figure 6).

In order to solve the task, another profile, which was considered suitable for the implementation of the detention reservoir, has been selected. It was located above the village of Lukavica in profile rkm 7.100. Due to the fact, that there should be variant solutions developed in the study the client was asked to obtain the hydrological data for this river profile, as well. The design flood wave was provided for this profile by SHMI. Its culmination occurs after approximately 6 hours at a 100-year flow rate of Q_{100} = 39.5 $m^3.s^{-1}$ and the volume of the W_{PV100} flood wave was set at 1.02 mil. m^3.

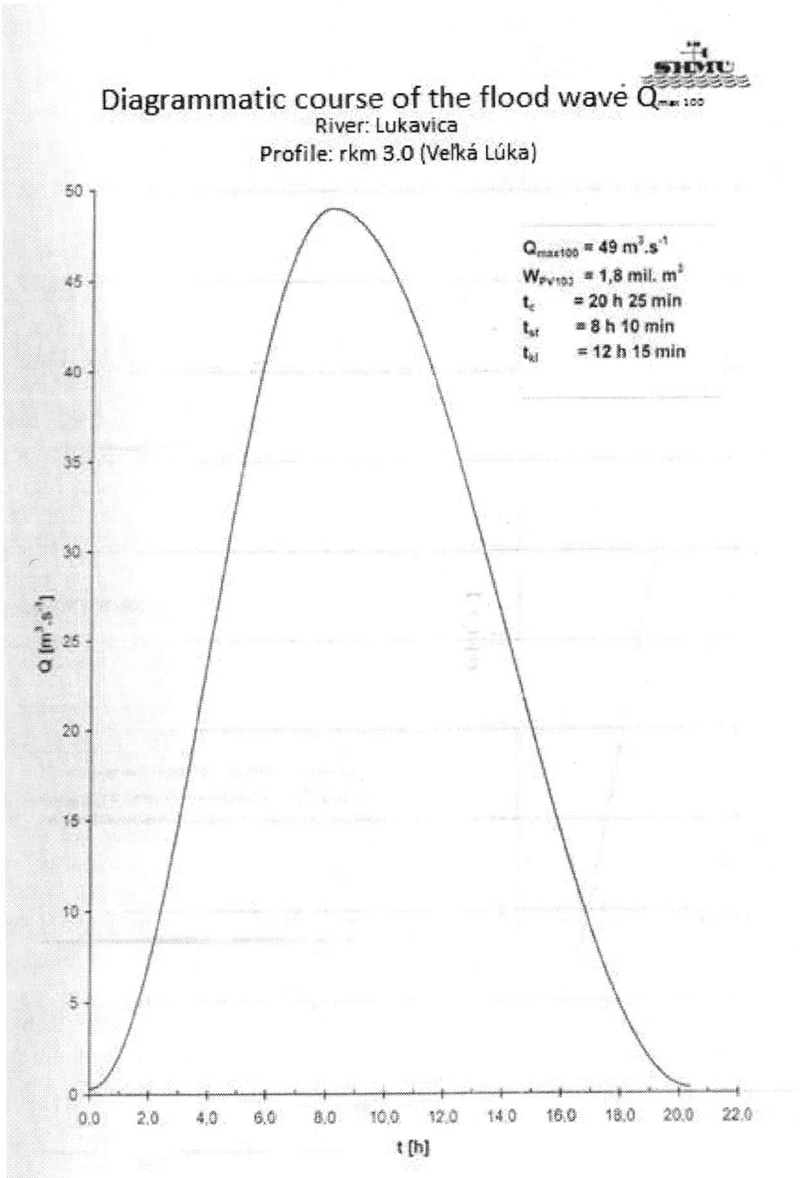

Figure 6. The design flood wave diagram on Lukavica river in rkm 3.0 [7].

4. Modelling the flow regime on the Lukavica River

In order to determine the flow regime of the Lukavica River section with planned flood protection measures, the one-dimensional mathematical model HEC-RAS (US Army Corps of Engineers Hydrologic Engineering Centre's River Analysis System) was used in which the steady flow regime is calculated from profile to profile by solving the equation of energy iteratively (method by segments). The geometry of the Lukavica River section was schematized using cross-section profiles (figure 7) which were generated based on geodetic survey with stationing according to the previous survey. At the same time, the water level was measured in three profiles using FlowMate device. The measurement served then for calibration of the current state model.

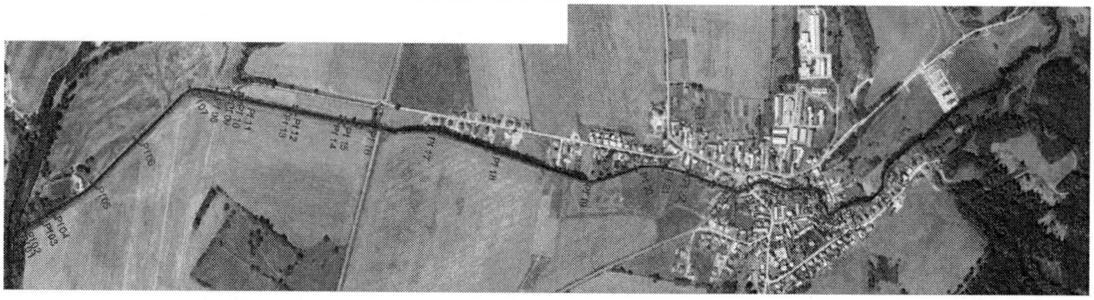

Figure 7. Scheme of measured cross-section profiles on Lukavica river.

From modelling solution point of view, several variants of the mathematical model were considered. For the channel defined in the individual models, calculations of the steady non-uniform flow were performed with a mixed flow type, due to the height changes of the bottom where the flow regime was assumed to change.

The result of the simulations were the elevations of the water levels in the interest area what showed the problematical localities during the flood passing. Following the results of performed simulations, proposals were carried out to increase the river capacity, resp. they also served as a basis for determining the value of maximal flow rate from the proposed detention reservoir above the village of Veľká Lúka. Coefficients of roughness were calibrated to reach the measured water levels at the profiles with real positions of water level. Unfortunately, the flow rate was very low ($Q = 0.19$ $m^3.s^{-1}$), what allowed only the bottom part of the cross sections to be calibrated. Even during repeated visits to the interest site there were not any higher water levels in the river and therefore the roughness coefficients of the banks were estimated only on the base of field reconnaissance [8,9].

Calibrated model of the present state was used to determine the capacity of the channel in a state with huge amount of sediments in the Lukavica River which were, in the meantime, removed from the central part of the village. The capacity of the channel with sediments has reached the value of $Q_{cap} = 7$ $m^3.s^{-1}$, with the problematic railway bridge transferring $Q = 12$ $m^3.s^{-1}$ in open channel flow regime (water level did not reach the deck) [7].

Subsequently, the model with all the sediments cleared was created in the municipality's residential area. It was formed by modification of previous profiles (profiles with sediments) at the original station. The cleaning of the banks was taken into account in the Manning degree of roughness, which was considered $n = 0.030$ for the entire interest section, that represents grass-covered, almost weed-free, clean, straight banks [8].

Capacity of this modified channel without deposits of sediments reached $Q_{cap} = 15.9$ $m^3.s^{-1}$, which was also considered in the study as the limit value of the flow rate for the planned detention reservoir upstream.

5. Proposal of flood protection measures and calculation of flood wave mitigation

For the needs of hydraulic calculation of the flood wave mitigation a detention reservoir was proposed in the valley of the Lukavica River with a dike height of 10.5 m [10]. Its location is shown in figure 8. The dimension of the bottom outlet has been created according to the dam height (about 10 m) and the flood wave Q_{100}. The diameter was modified so that the water level in detention reservoir would not reach the level of the emergency spillway. In this case, a circular outlet with a diameter of 1.8 m was proposed. The ground elevation in the profile of detention reservoir is 314.5 m a. s. l. and the elevation of the top of the dike is 325 m a. s. l., the total height of the detention reservoir dam is 10.5 m. This proposal resulted in a flood wave mitigation with a peak flow of 49 $m^3.s^{-1}$ reduced to a maximum outflow of 21.7 $m^3.s^{-1}$. This proposal has been evaluated as insufficient due to the capacity of the river channel in the municipality.

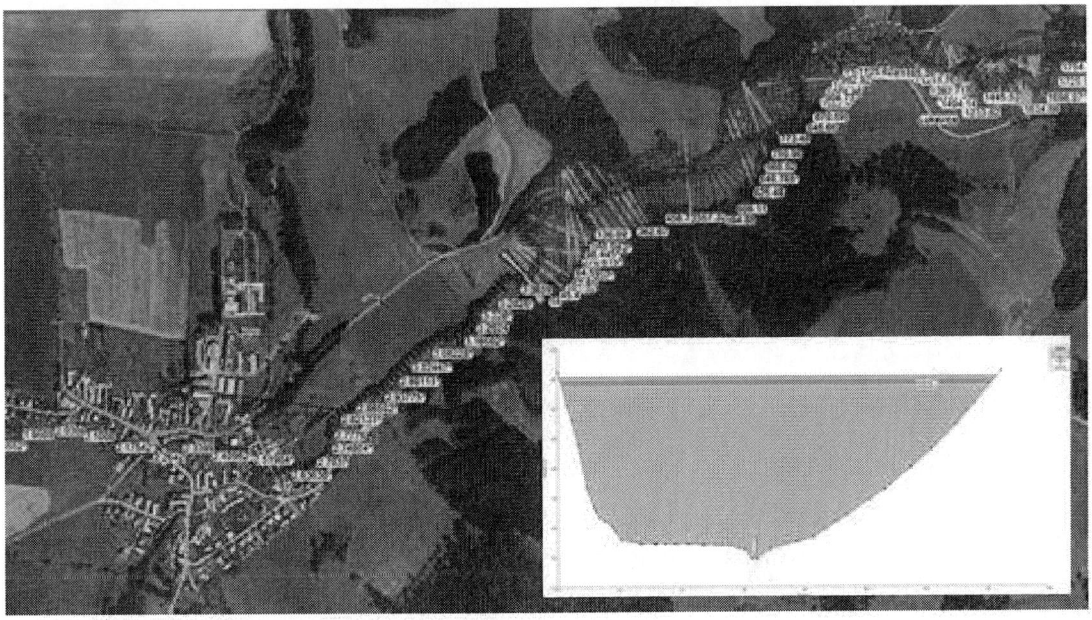

Figure 8. Modelling of detention reservoir on Lukavica in HEC-RAS programme.

Another draft was to increase the detention reservoir dike to the limit of the geodetic survey, i. e. to the elevation of 326 m a. s. l. It showed that increasing the dike by only one meter has a great impact on the retention volume of the reservoir, which allows to reduce the diameter of the outlet and to achieve better results in terms of flood wave mitigation. Thus, a dike with a height of 11.5 m and a bottom outlet diameter of 1.5 m was modelled. As a result of this design, a 15.9 $m^3.s^{-1}$ peak outflow (figure 9) has already been achieved, which is, due to the adjusted capacity of the channel in the area from the planned building to the confluence of the Lukavica and the Hron rivers appropriate.

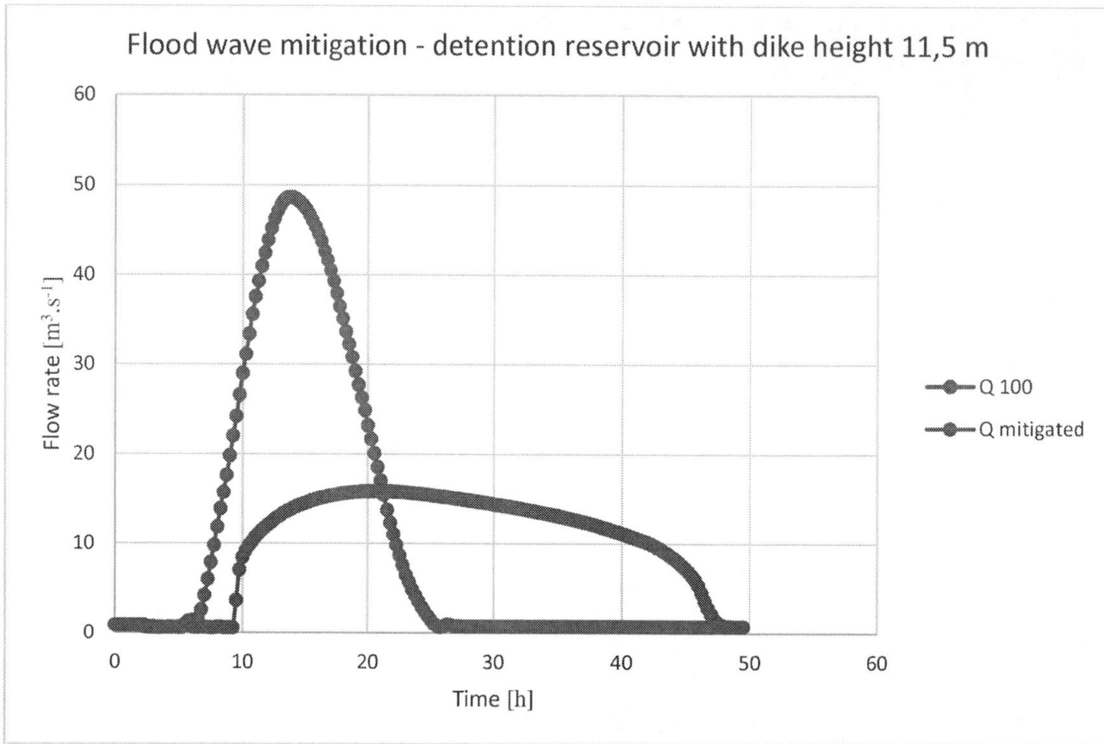

Figure 9. Flood wave mitigation of design flood wave in case of proposed dike height 11.5 m and bottom outlet diameter 1.5 m.

Another considered alternative was to shift the detention reservoir stream upwards to handle with a smaller design flood wave. This proposal would be dependent on implementation of measures in the riverbed area between these two profiles, especially on the mainstream tributaries. Therefore, the second flood wave according to SHMI was used with a peak flow rate of $Q_{100} = 39.5\ m^3.s^{-1}$. The parameters of the flood wave allowed the proposal of the detention reservoir with the height of the dam of 8.5 m and the diameter of the bottom outlet 1.7 m. The peek outflow was $17.3\ m^3.s^{-1}$ in this case. Several variants of calculation were performed with different heights of the reservoir dike and the different diameters of the bottom outlet, but this variant appeared to be the most suitable in terms of flood wave transformation. The resulting detention reservoir proposed for the Lukavica River above the village of Veľká Lúka in rkm 3.366 is schematically illustrated in figure 10.

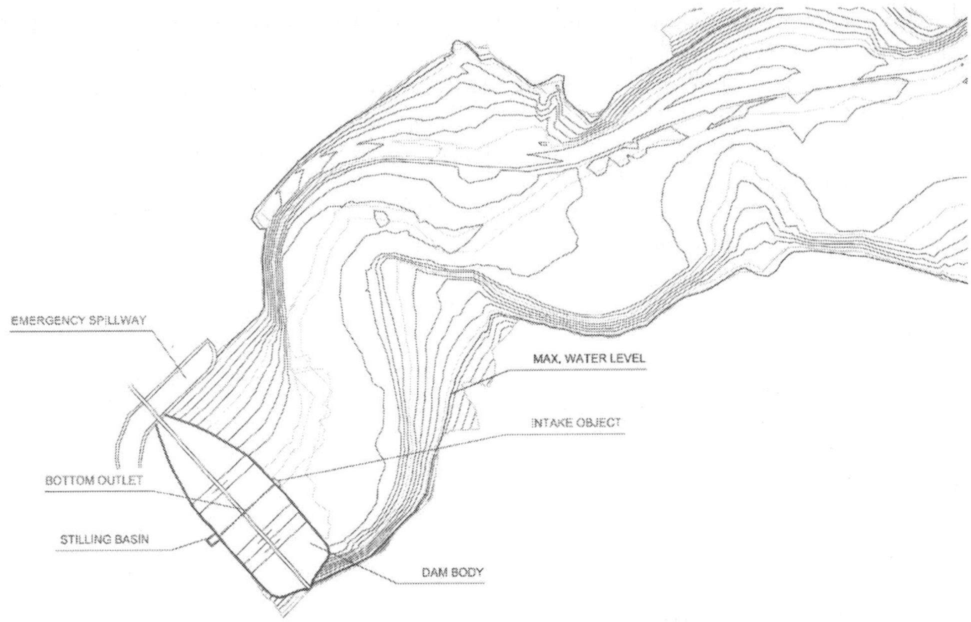

Figure 10. Schematic plan of the detention reservoir dike on the Lukavica river in rkm 3.366.

The reservoir dike itself is designed as an earth fill dam in the profile no. 2 (figure 4) with the crown of the dyke at an elevation of 326 m a. s. l. Its volume was set to be 25 650 m^3. The bottom outlet was proposed as a circle with diameter of 1.5 m. It was calculated hydraulically as a culvert. The emergency spillway, which is also proposed according to $Q_{100} = 49$ $m^3.s^{-1}$, is designed in the left part of the reservoir dam body at a sufficient depth below the roadway. The length of the spillway rim reaches a value of 50 m at overflow beam height of 0.5 m.

6. Conclusion

The task within the framework of this study was to design the possible flood protection measures for the Lukavica River focusing on the village of Veľká Lúka, to assess them hydraulically and to choose a single variant of flood protection that would prevent flooding in the village during higher flow rates. This term has been chosen because for all documented floods of 2009, 2013 and 2016, the flow rate in the Lukavica River was not evaluated or measured.

Due to the fact, that the cross sections of the riverbed that could be used for the hydraulic calculation were not defined in the previous materials, it was necessary to proceed to the geodetic survey of the Lukavica River at the beginning of the solution. For this purpose, 38 cross sections in the location from the junction of the Lukavica River (rkm 0.000) to the Hron River to the profile of the proposed detention reservoir (rkm 3.455) had been measured. Subsequently, another measurements of the geodetic group focused in detail on the flooded area that would be created by detention reservoir construction to obtain a digital terrain model and the volume and area of the floodplain area during the flood wave mitigation. Based on these data and the control flow measurements in the Lukavica River, a mathematical model of open channel flow was created in both steady and unsteady conditions. The model was only partially calibrated, despite of multiple attempts to measure flow rates at higher water levels during the project duration. However, these did not occur. The mathematical model was generated in the HEC-RAS program as a one-dimensional mathematical model. It was primarily used to determine the capacity flow of the Lukavica River inside the village of Veľká Lúka and at the same time, it was used under unsteady conditions to calculate the mitigation of the design flood wave provided by SHMI.

As the clearing of the channel from sediments in the central part of the village was recorded during the project solution, a model of the water flow in the Lukavica River with a modified geometry with removed deposits has been created, as well. Capacity of the treated watercourse with no deposits reached

Q_{cap} = 15.9 $m^3.s^{-1}$, what was considered in the study as the most critical value of the flow rate for the above-lying planned detention reservoir.

The most critical section in terms of flood threat is the part of the modelling area between the road and the railway bridge along the MDD street. The individual construction of the family houses is adjacent to the Lukavica River on the right hand side, and the construction of the family houses is planned on the land on the left side of the channel.

For the reduction of Q_{100} = 49 $m^3.s^{-1}$ it is necessary to build a flood protection measure that intercepts larger part of the volume of the flood wave above the municipality to such an extent that the flow rate flowing to the village will meet the condition of the capacity flow in the village Veľká Lúka. For this purpose, a detention reservoir was proposed for the river in rkm 3.366 (figure 10) with a height of 11.5 m and with a circular bottom outlet with a diameter of 1.5 m. The result of this proposal was the mitigation of the design flood wave to a 15.9 $m^3.s^{-1}$ peak outflow (figure 9), which is in line with the adjusted channel capacity in the area from the proposed construction to the junction of Lukavica to river Hron.

7. References

[1] Šoltész A, Čubanová L, Janík A and Baroková D 2018 Flood protection measures to protect urban areas (*Acta hydrologica Slovaca* vol 2) pp 296-303

[2] Beczányiová E 2014 *Návrh protipovodňových opatrení na toku Lukavica v k. ú. Veľká Lúka a Sliač* (*Diploma thesis, Faculty of ecology and environmental sciences, Technical University in Zvolen*) p 84

[3] Živčicová K 2018 *Návrh protipovodňových opatrení v povodí toku Lukavica* (*Diploma thesis, Faculty of civil engineering, Slovak University of Technology in Bratislava*) p 63

[4] Department of development and planning, Slovak Water Management Enterprise 2009 *Štúdia – Veľká Lúka, návrh protipovodňových opatrení* (Banská Bystrica) p 15 s. and appendixes

[5] Department of development and planning, Slovak Water Management Enterprise 2010 *Štúdia – Veľká Lúka, návrh protipovodňových opatrení* (Banská Bystrica) p 14 s. and appendixes

[6] Department of development and planning, Slovak Water Management Enterprise 2012 *Veľká Lúka - návrh protipovodňových opatrení na toku Lukavica, Doplnok štúdie* (Banská Bystrica) p 6 s. and appendixes

[7] Šoltész A, Čubanová L, and Janík A 2018 *Veľká Lúka - protipovodňové opatrenia v povodí toku Lukavica*, Research report, SvF STU Bratislava, p 39

[8] Raplík M, Výbora P and Mareš K 1989 Úprava tokov (Bratislava: Alfa) p 639

[9] US Army Corps of Engineers 2016 HEC-RAS River Analysis System User´s manual (Version 5.0.) Institute for Water Resources, Hydrologic Engineering Center p 960

[10] Janík A and Šoltész A 2017 Flash flood mitigation modeling — case study small Carpathians. Pollack Periodica, 12(2), 103–116. doi:10.1556/606.2017.12.2.9

Acknowledgments

The contribution was developed within the frame of and based on the financial support of the VEGA 1/0800/17 project "Optimization of the flood protection of municipalities in river basin of mountain streams".

IOP Conf. Series: Materials Science and Engineering **566** (2019) 012011 doi:10.1088/1757-899X/566/1/012011

Water transport in materials of historical buildings

P Jaroš[1], M Vertaľ[1]

[1] Technical University of Kosice, Faculty of Civil Engineering, Institute of Architectural Engineering

Abstract. The historic buildings renovation is one of the unwritten obligations of every nation. In the territory of the Slovak Republic there is quantity of the historical buildings which are from different architectural periods, which we can find not only in the squares of the bigger cities, but also in a wide areas. The technical conditions of these buildings describe our relationship with which we look at them. At the present, many historic buildings are restored, but more are left untouched and without maintenance. Incorrect construction details, but also inappropriate additional building adaptation, cause degradation of the masonry and surface layers. Absence of the water-insulating waterproofing against capillarity allows the water to be transported into vertical structures and with it transport water-soluble salts from the subsoil. Increasing mass of the humidity by capillary transport and crystallization of water-soluble salts in porous materials causes visible defects on the surface of the structures. Laboratory measurement of the humidity and capacity of the water-soluble salts in a historic building point to the impact of the segmentation and using of the building. Different values of the taken samples define critical places that necessitate additional remediation. Suggestion of the remediation influences diversity of the materials which are used and their material characteristics. For the water transport analysis, it is necessary to know the transmission parameters of the water of the historical structures. Determination of the humidity coefficient is possible by the water absorption coefficient and the value of the capillary humidity. The measurement is carried out by an experiment of the unidirectional capillarity by the laboratory instrument.

1. Introduction

The most used building material in the history of constructing buildings was natural stone. When constructing buildings and cities, only the stones from surrounding areas which were influenced by geological composition were used. Sandstone from flysch belt and from sub-tatra group is used in eastern Slovakia. Absence of the water-insulation waterproofing against capillary water are enabling transport of water influenced by capillary forces in porous materials. Measuring of moisture and the content of water-soluble salts is done by laboratory measurement. Established valuesrepresent extensive damage in certain sections of the building. To be able to do proper constructional changes it is essential to know specific material parameters.

2. Stone as building material

2.1. Stone as building material in Slovakia

Natural stone was used in locations where suitable geological conditions were created. Hundreds of stone-pits surrounding cities show us, its massive usage in constructions. In the Middle Ages mining natural stone wasn't limited by any means. Geological diversity of minerals on Slovak soil allowed mining of andesite (240 stone-pits), limestone (312), sandstone (336), travertine (87) and many others 0. The most used material was hewn and quarried stone. Foundations and cores of vertical constructions were made from quarried stone and hewn blocks were generating only masonry cheeks. In the gothic ages, stone was used in great manner to decorate constructions too. Railings, window frames, cornicles and arches were made in stone workshops.

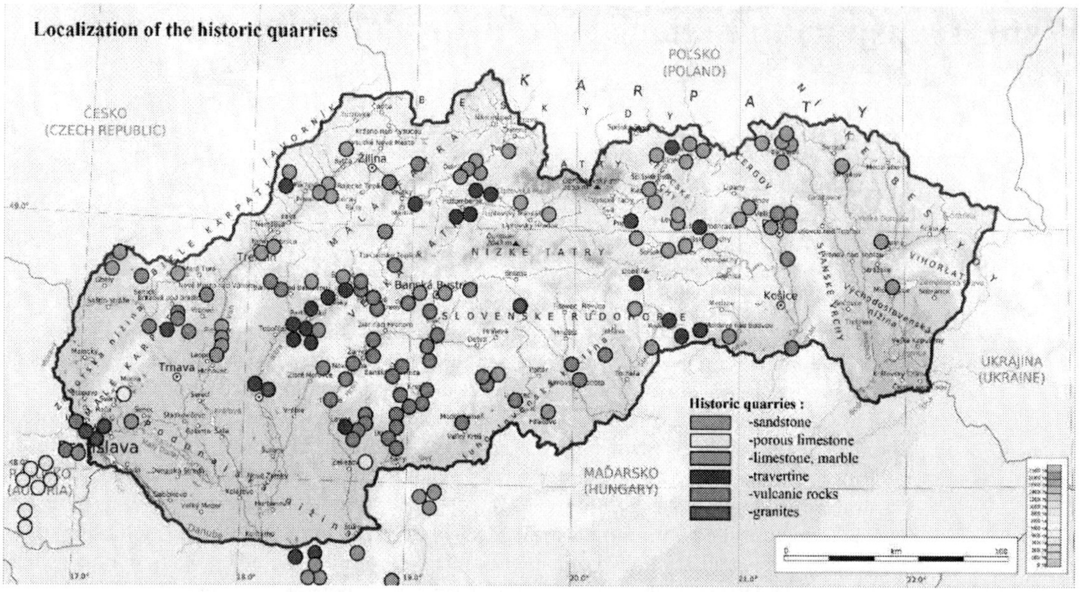

Figure 1. Localization of historic stone quarries0

2.2. Sandstone as the main building material in eastern Slovakia

Layers of sandstone are located on Slovak republic's territory mostly in areas of flysch belt, including Orava, Kysuce and north-eastern parts of Slovakia. This sedimentary mineralwas mostly used for constructing buildings not for their properties, but for the lack of others, more suitable minerals. Sandstone was the most used mineral with and 336 registered stone-pits are proving it. The mining of the mineral raw materials was not limited by any legislation therefore stone-pits were created. Many of them were abandoned after mining all of the suitable material. The great amount of stone-pits created one big advantage, that is the ability to select minerals according to their quality. The most suitable sandstone was used mostly for decorative purposes 0.

2.3. Sandstone as the main building in Spiš region

The most exhibitory and known city in this area was Levoča, which contained many of building, carving and stone-cutting workshops. Levoča is the most sandstone city in Slovakia. Town houses, portals, churches, city hall, city walls, and also decorative features are built of sandstone.When entering the historical parts of the city you can see authentic fortifications from 13th century made of sandstone. Walls, towers and gates were built from quarry stone. Renewed city walls are creating the Middle age nature of Levoča, which contributed to inclusion of it to UNESCO World Heritage. The centre of historical part of the town is formed by rectangle square. The core of the square is surrounded by classic town homes. The second largest cathedral after St. Elizabeth's Cathedral in Košice is located in the centre of the city. It is St. Jacob's Cathedral. Its valuable decorative feature, the west entrance portal has been damaging for centuries.Today, the portal is renewed. From stonecutter's and historical point of view was the most valuable and decorative feature the epitaph. This unique feature from sandstone was carved by Martin Urbanowitz. It is placed in front of the vestibule 0.

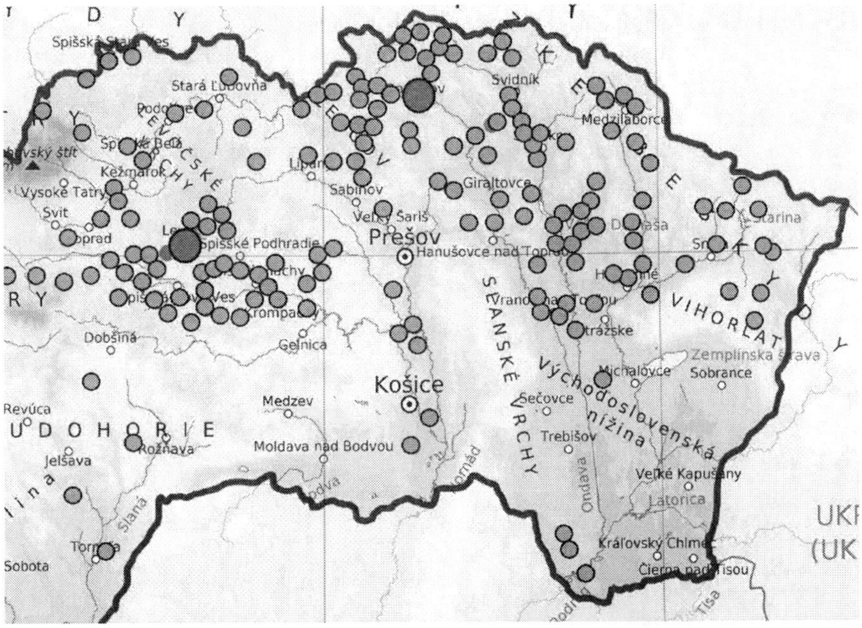

Figure 2. Location of sandstone quarries in eastern Slovakia0

Figure 3. a, The sandstone epitaph on the church of St. Jacob. b, Bell Tower and City Hall

3. Moisture measurement and water soluble salt content

Increased moisture is one of the most common cause of damage of historical buildings, which is caused by absent waterproofing and incorrect constructing details. Stone as a building material is capillary active and allows transport of water and water-soluble salts in porous materials. Crystallization of salts causes degradation of plaster's surface and carrier framework. For the correct design of the masonry, it is necessary to know the moisture content and content of water-soluble salts. The subject of the assessment moisture content and content of water-soluble salts is the town house on the St. Paul's Square in Levoča. The building went through multiple constructional changes. In 1963 it was included to national cultural monuments. It has two floors and a basement. The dominating building material was stone. Defects caused by moisture and water-soluble salts are visible by eye. For laboratory measurement, samples from defected areas were detracted. The valuating concerns front façade that is facing the square. Because of creating remediation measures, there is regulation STN P 730610, which states adequate and also increased content of water in constructions. The moisture of building materials is generally expressed as moisture mass.

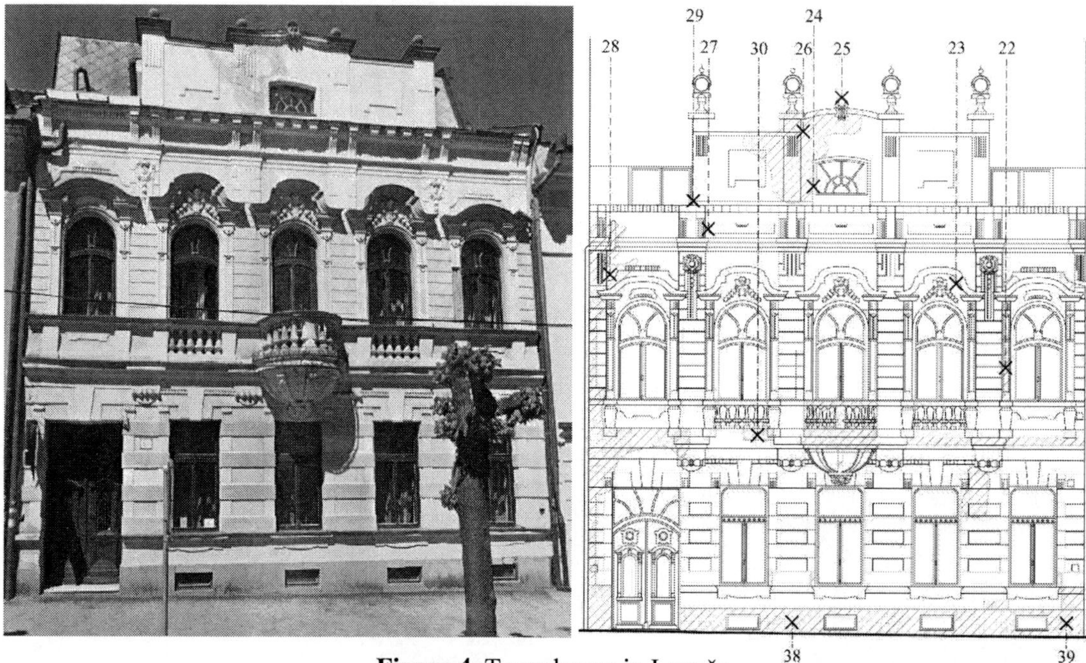

Figure 4. Town house in Levoča

The addressed facade of the town house is oriented towards the east side of St. Paul's square. Near to the house there is a road, a parking lot and a pathway which is connected to the building. There are visible defects of the front facade. Incorrect constructing details of sheet-metal lining of cornice, balcony and rain spout pipe are causing degradation of its surface design. The samples were detracted from the most affected parts. The content of water and the contentof water-soluble salts were measured in laboratory.

Table 1. Moisture classification STN P 7306100

		Measured value		(%)	MOISTURE CONTENT
	≤	Mv	≤	3,0	VERY LOW
3,0	≤	Mv	≤	5,0	LOW
5,0	≤	Mv	≤	7,5	INCREASED
7,5	≤	Mv	≤	10,0	HIGH
10,0	≤	Mv	≤		VERY HIGH

Detracted samples were reviewed in terms of its pH values and content of water-soluble salts. The displayed classification is established according to WTA 2-9-04.

Table 2. Classification of water-soluble salts WTA 2-9-04

Content of water-soluble salts	Cl – chloride (%)	NO3 - nitrate (%)	SO4 – sulphate (%)
LOW	< 0,2	< 0,1	< 0,5
MEDIUM	0,2-0,5	0,1-0,3	0,5-1,5
HIGH	> 0,5	> 0,3	> 1,5

4. Results

Moisture content of detracted samples is demonstrated in Figure 5. The measured values according to STN P 73 0610 are in the first category, which means they have very low moisture content.

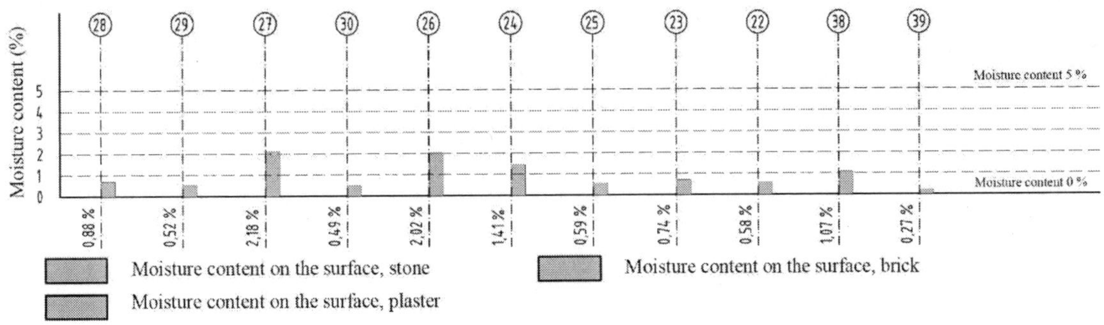

Figure 5. Content of water in samples

Measured data of water-soluble salts are demonstrated in Figure 6. Detracted samples are indicating that they have average content of water-soluble salts. Two samples contain high content of chlorides. Four of the samples show too high values of nitrates, which originated from decay of organic materials. For setting the level of salinization it is important to set intensity of concentration, of any kind of it from our lab-discovered concentration. It can be set by measuring the concentration of sulphate, chloride or nitrate in reviewed samples.

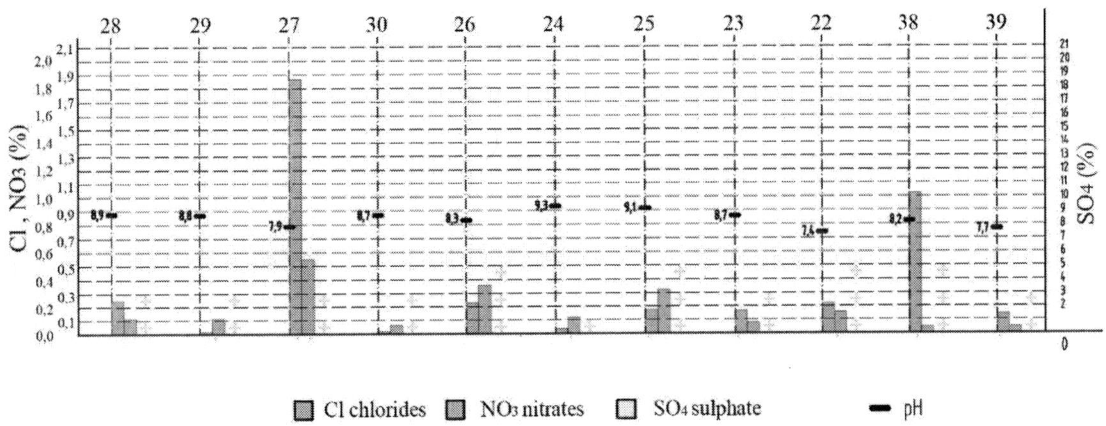

Figure 6. Content of water-soluble salts in samples

5. Conclusion

The most used building material in our history was stone. It was used as building material and also as decorative component. Sandstone was used for building purposes in eastern Slovakia. The most known sandstone rich cities are Levoča and Bardejov. Cathedrals, city halls and town houses are built exactly from this sedimentary mineral. The most occurring problem of historical buildings is transport of water from subsoil and water-soluble salts. Laboratory measurements of detracted samples from the front facade show very high volume of concentration of water-soluble salts in certain areas. Based on discussed measurement a sanitation of existing surface modification is needed.

6. References

[1] Pivko D *Stavebný a dekoračný kameň a jeho opracovanie na stredovekom Slovensku* Univerzita Komenského v Bratislave

[2] Čabalová D 2013 *Krása kameňa v životečloveka* Veda vydavateľstvo Slovenskej akadémie vied Bratislava pp 1-374 ISBN 978-80-224-1291-9

IOP Conf. Series: Materials Science and Engineering **566** (2019) 012011 doi:10.1088/1757-899X/566/1/012011

[3] Pivko D *Extraction methods in historical quarries in Slovakia nearby areas for dressed stone products* Actageologica Slovaca
[4] STN P 730610-12: 2001 *Hydroizolácia stavieb – Sanácia vlhkého muriva – základné ustanovenie*
[5] WTA 2-9-04-2006 *Smernica WTA: Sanačné omietkové systémy*

Acknowledgements
This work was supported by the VEGA: 1/0674/18Theoretical and experimental analysis of architectural-structural shapes and fragments of building envelope structures designed for harsh conditions.

The search for an optimal architectural shape using wind performance analysis

L Kabošová[1]

[1] Technical University of Kosice, Faculty of Civil Engineering, Department of Architectural Engineering, Slovakia

Abstract. The rapid enhancement of computers' power and the boom of open-source software development have offered a broad range of possible applications in architectural design. Grasshopper is a visual programming language developed for the 3D CAD modeling program, Rhinoceros. Utilizing Grasshopper's free software extensions, it is possible, among other things to evaluate the environmental performance of architecture in the early conceptual stage. This paper introduces a novel approach to architectural designing presented in a case study in the extreme environment of Reykjavik's airport. The shape of the airport building formed within this case study can be parametrically changed, and the performance of the various shape modifications in the wind can be tested. Using Ladybug and Swift for Grasshopper, the wind situation at the airport is analyzed, and the aerodynamic shape of the proposed airport building is evaluated in the prevailing easterly winds. The proposed bio-climatic architectural design approach incorporates the surrounding micro-climatic conditions, wind in particular, as a driving factor already in the phase of first design ideas. The parametrically developed airport is formed with regard to the pedestrian wind comfort, as well as airport traffic safety.

1. Introduction

Due to the rapid development of digital design tools, it becomes possible for architects and designers to relatively seamlessly incorporate simulations such as environmental [1, 2] or structural analysis [3, 4] to their projects in the early conceptual phase of architectural design. Moreover, some of the recently developed analysis software is accessible free of cost which unfolds new perspectives in design. Computer simulations utilized in the early conceptual stage, predicting the behavior of environmental phenomena such as solar or wind fluxes can reveal a lot about the performance of the future building, a building cluster, or a building's structural system. The results of such analyses can be employed to adapt the design, the building envelope or its structure. Such a design loop can be applied until an optimal solution is achieved creating a basis for nature-adapted designs. When applying such a design approach in the early stages of design, the goal is to do it in the most time and cost-effective way and simultaneously not to lose on precision.

1.1. Wind-driven design

The air movement is usually perceived as a negative phenomenon in the building design, although sometimes the effects of wind can be beneficial. On the one hand, the air movement in interiors is necessary for natural ventilation, on the other hand, the speed of moving air has to stay within limits for a comfortable indoor climate. The wind speed and the flow pattern are shaped by the relative position and form of the buildings [5] in the outer environment which can result in an acceleration of the flow or unpleasant turbulence. By the lower ambient temperature, the pedestrians perceive the wind flow more negatively. From the structural point of view, the wind acceleration and turbulent flow intensify the loads on structures. It might be fair to assume that the wind-driven design could not only contribute to regulating the wind flow, as well as wind pressure on buildings and around them, but also positively affect the comfort of their inhabitants.

Design strategies incorporating wind flow suggest that an improved, optimal natural ventilation of interiors can be achieved [6], or buildings can be adequately placed and rotated to benefit from the cooling effects of the wind [7], or the negative effects of the wind can be attenuated using small architectural interventions [8]. The following case study demonstrates an architectural design strategy that incorporates the specific wind conditions of Reykjavik's airport and regulates the wind flow pattern using the parametrically created airport terminal building. Three design variants of the streamline-shaped airport terminal are evaluated based on their aerodynamic properties.

2. The case study in Reykjavik

Designing with the wind in the very early design phase urges to utilizing a digital approach, including Computational Fluid Dynamics (CFD) simulation tools. Such digital approach is necessary for a relatively fast evaluation of the various design options. In the early conceptual stage, the time of the wind simulation, the cost of the software and the precision of the results are the key factors. The most suitable combination for the analysis in the conceptual stage is a fast, but precise enough analysis (Figure 1). In the later design phases, a professional CFD analysis or wind tunnel experiments can be applied.

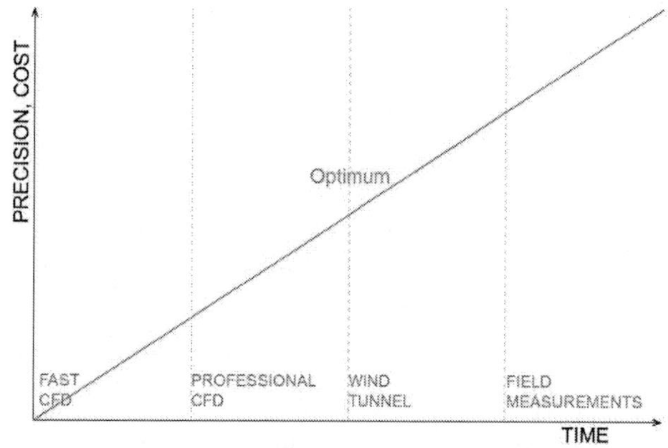

Figure 1. The relation of the time, and the precision and cost requirements for various wind analysis techniques.

The design approach introduced in this paper consists of the following steps: (1) the wind situation of the case study site is analyzed based on the data obtained from the EnergyPlus web database, (2) the 3D shape of the airport terminal is developed parametrically using Grasshopper for Rhinoceros (Rhino), (3) various shape modifications can be created thanks to the parametric approach, (4) a new open-source wind analysis tool for Grasshopper called Swift is used for evaluating three shape alternatives in the prevailing easterly wind flow, (5) Paraview is employed for post-processing the obtained results outside the working environment of Grasshopper and Rhino, (6) in the last step, the performance of the designed shapes in the wind is evaluated. Looping the design steps directs the process into finding an optimal shape option (Figure 2).

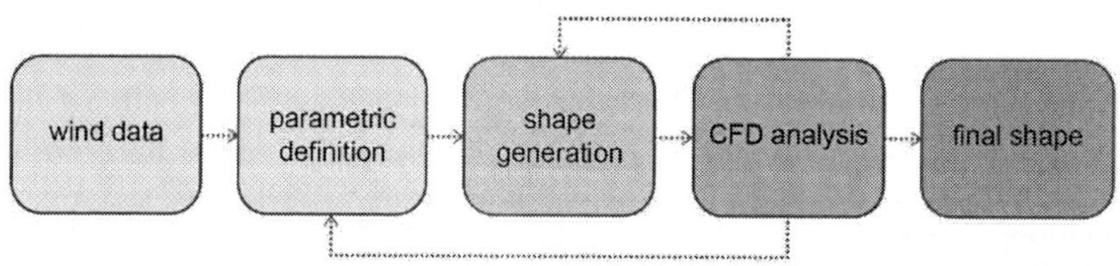

Figure 2. The steps in the proposed design approach. The arrows indicate the loop for achieving the optimal final shape.

2.1. Prevailing winds in Reykjavik

EnergyPlus digital data source [9] is used to obtain representative wind data for Reykjavik's airport in the *epw format. A plug-in for Grasshopper called Ladybug requires such *epw file to generate a graphical wind rose (Figure 3). The wind rose suggests that the easterly winds are dominant in the given locality for, on average, 15.64 % of hours per year. The easterly wind's average wind speed is 6.2 m/s. This value is used in the CFD simulations. The average wind gust speed for all wind directions reaches 25 m/s. Moreover, the wind temperature is usually not exceeding 10°C. This combination of ambient conditions is very unpleasant for pedestrians and workers at the airport.

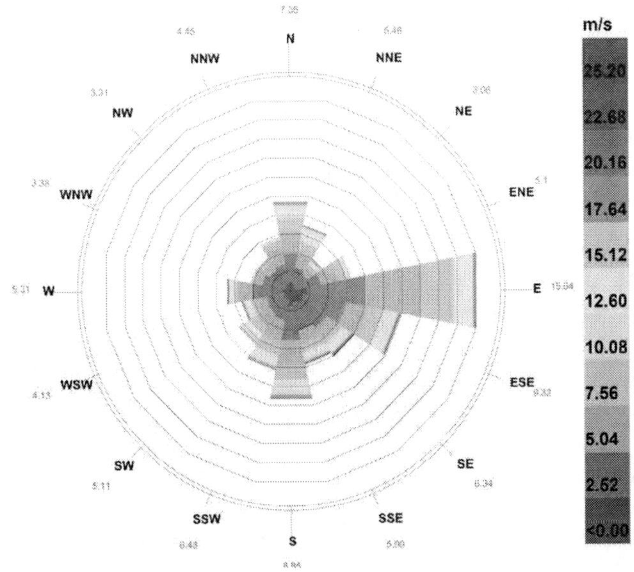

Figure 3. The wind rose of Reykjavik's airport.

2.2. Parametrically-created airport terminal building

The benefits of parametric designing lie in the possibility to experiment with the selected variables and obtain a new final shape for every change in parameters. In this case study, two parametric variables influence the final shape: a planar closed curve and an arbitrary number of attractor points. The representation of the final shape is a result of the form-generation process decided by the designer. In this paper, the curve is drawn by the designer and represents the footprint of the future airport terminal building. Subsequently, three attractor points are determined in the same plane. The proximity of attractor points to the planar curve influences the final form and forces it to close towards the wind (Figure 4).

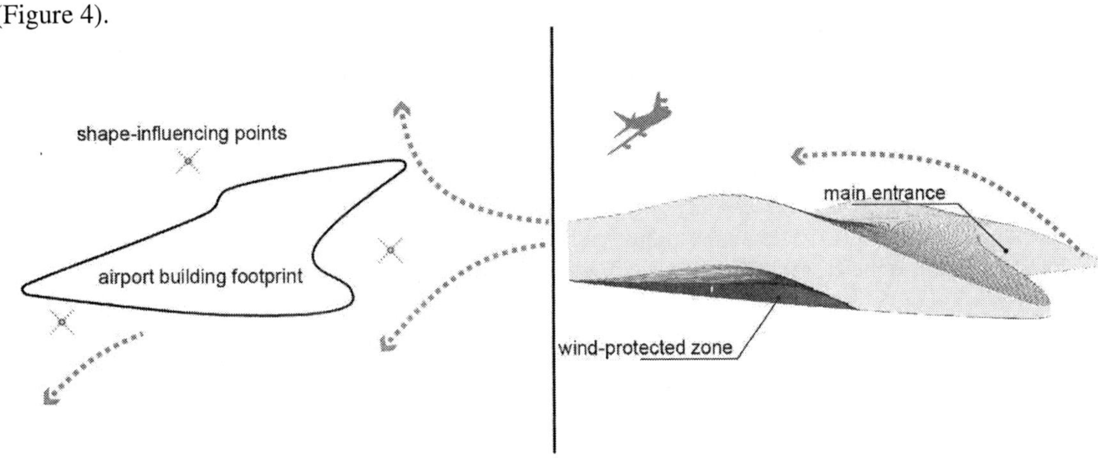

Figure 4. The variables that influence the final form in the parametric definition (left), the final shape (right).

2.3. CFD simulations in Swift for Grasshopper

The goal of the research is to simplify the wind analysis process in the early design stages. Architects tend to use Grasshopper for a wide range of environmentally-related design tasks where multiple design aspects are considered. In the field of parametric CFD analysis, Swift for Grasshopper is a recent extension that works on the platform of the powerful CFD tool OpenFOAM which guarantees the reliability of the results. 'SimpleFoam' (semi-implicit method for pressure-linked equations) algorithm executes the calculations of the Navier-Stokes equations. RAS (Reynolds-averaged stress) turbulence model is used for steady-state simulations [10]. Three designed shape variants are analyzed using Swift. The following input conditions are defined: (1) the wind speed is set to 6.2 m/s, (2) the simulations run in 250 iterations to meet the convergence criteria, i.e., the convergence tolerance of 10^{-5} for the velocity components Ux, Uy, and Uz. The results of the simulations are subsequently post-processed in Paraview external visualization software. For a better visual comparison, the data range of velocity and pressure is customized for all three design options. The range of values for the wind velocity is set from the minimum value 0 m/s, represented in blue color, to a maximum value 10 m/s, displayed in red color. The range of pressure values is set from -40 to 20 m^2/s^2, again on the color scale from blue to red (Figure 5).

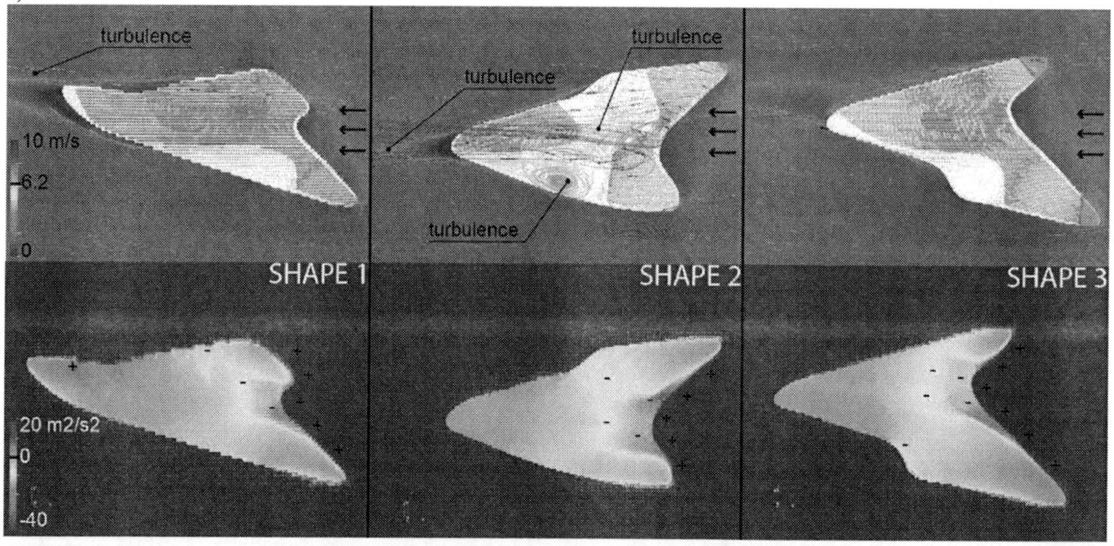

Figure 5. Three examined shapes: bottom row – surface pressure [m^2/s^2], upper row – velocity [m/s].

3. Conclusions

The proposed bio-climatic architectural design approach combines 3D modeling software Rhino with open-source extensions for a smooth integration of wind analysis into the conceptual design stage. It offers a fast and precise enough way for architects and designers to evaluate the performance of their designs in the specific wind conditions. By incorporating parametric designing, such technique leads to finding an optimal building shape based on its desired interaction with the wind flow. The design approach is presented in a case study in Reykjavik, Iceland. The proposed parametric definition in Grasshopper enables to create shape modifications of the designed airport terminal building. Three shapes are tested in the easterly winds, prevailing in the area. The wind flow around the proposed shapes, as well as wind pressure on the buildings' envelope, is observed. All variants are aerodynamic; the shape 1 causes mild turbulence on the leeward side. Shape 2 alters the wind flow causing turbulence on the leeward side, above the building, and in the intended wind-protected zones. Shape 3 performs as the most streamlined and aerodynamic in the observed wind situation. The surface pressure as a result of the wind loads is examined too. The results prove the aerodynamic properties of the three shapes. The wind flow altered by shape 2 causes the highest negative surface pressure, i.e., suction when compared to the other two options, whereas the positive surface pressure is the highest in the case of shape 1. It might be debated, how precise is the interpretation of the results based on pure observation. However, the proposed design loop is developed for the very early stages of design with an intention to direct the wind-influenced design to the next design stages. The search for an optimal shape guarantees that the wind fluxes will not be interfered by the designed building shape and the pedestrian wind comfort around the building in the wind-protected zones, as well as airport traffic safety, will not be affected.

4. References

[1] Pellitteri G, Lattuca R, Concialdi S, Conti G and Amicis R De 2009 Architectural shape generating through environmental forces *Joining Languages, Cultures and Visions: Proc. 13th Int. CAAD Futures Conf.* (Montréal, Canada: Les Presses de Université de Montréal) p 875–86

[2] Wang L, Tang Z and Ji G 2016 Toward the wind-related building performative design *Living Systems and Micro-Utopias: Towards Continuous Designing: Proc. 21st Int. Conf. of the Association for Computer-Aided Architectural Design Research in Asia CAADRIA* ed Chien S et al. (Hong Kong) pp 209–18

[3] Hu Y and Li Q 2014 Study on an architect-oriented approach of structural performance-based design *SIGgraDi 2014: Design in freedom* (Montevideo, Uruguay: Blucher Design Proceedings) p 111–4

[4] Białkowski S 2016 Structural optimisation methods as a new toolset for architects *Complexity & simplicity: Proc. 34th eCAADe Int. Conf.* ed Herneoja A et al. (Oulu, Finland: Oulu School of Architecture) p 255–64

[5] Tsou J Y 1998 Applying computational fluid dynamics to architectural design development Proc. 3rd Conf. on Computer Aided Architectural Design Research in Asia, CAADRIA (Osaka, Japan) p 133–42

[6] Stavridou A D 2015 Breathing architecture: Conceptual architectural design based on the investigation into the natural ventilation of buildings *Front Archit Res.* **4** p 127–45

[7] Taleb H and Musleh M A 2015 Applying urban parametric design optimisation processes to a hot climate: Case study of the UAE *Sustain Cities Soc.* **14** p 236–53

[8] Moya Castro R A 2015 Wind analysis in the early design stage: An empirical study of wind visualisation techniques for architects RMIT University

[9] EnergyPlus Weather data by region Available online: https://energyplus.net/weather-region/europe_wmo_region_6 (accessed on 18.12.2018)

[10] OpenFOAM: The open source CFD toolbox Available online: https://www.openfoam.com/documentation/user-guide/ (accessed on 03.02.2018)

Acknowledgments
The author wishes to acknowledge the financial support from the Grant Agency of the Slovak Republic. The project numbers are VEGA 1/0302/16 and VEGA 1/0674/18. For the encouragement, guidance and support the author wishes to thank the supervisors, prof. Katunský and prof. Kmeť.

Cyclic load triaxial test procedures for soils in transportation engineering

J Kaluder[1]

[1] Faculty of Civil Engineering and Architecture Osijek, Republic of Croatia

Abstract. Unbound granular base course is built in pavement structures between subgrade and bound base course. Its main function is bearing capacity - load transfer from upper layers to subgrade. For its successful design, one of the required parameters is resilient modulus. Resilient modulus describes behaviour of unbounded materials at various load conditions, and is defined as the ratio of the applied cyclic stress to the elastic strain. It can be determined with cyclic triaxial test that gives good simulation of stresses and strains which are result of traffic loads. In this paper resilient modulus testing procedure will be described with the emphasis on European Standard (EN 13286-7). Furthermore, it will be given comparison of European standard with other test procedures.

1. Introduction

Large deformations (rutting) and cracks on the pavement surface can lead to traffic safety risk due to possibility of water collection, extra costs from faster tyre wear and higher fuel consumption due to higher friction between tyres and pavement, higher road maintenance costs and the reduction of the driving comfort. AASHO road test [1] have shown that rutting is affected by base and subbase course deformations by 59 %. Hence, good design of those layers is necessary. Until the Second World War, road thickness was designed empirically considering traffic load as static [2]. Increase in traffic quantity and traffic loads resulted with numerous tests and traffic loads were soon observed as dynamic. According to [3], first repeated load tests were conducted with plate load tests with smaller amount of load repetitions. However, limitations were noticed shortly, therefore the laboratory testing soon began. Since triaxial test offered the possibility of application of different stress levels in time, respectively reproduction of critical soil conditions, first triaxial tests with repeated load have started [3]. The first comparisons of results gained in repeated load tests and static tests [4, 5] have shown differences in obtained moduli, and it was concluded that repeated load tests should be carried out for soils under traffic loading [3].

2. Resilient modulus and permanent deformations

The results of cyclic triaxial test are resilient modulus, elastic and plastic deformations. Resilient modulus is crucial for mechanistic-empiristic approach used for pavement design procedures. It was introduced by [4] as ratio of dynamic deviator stress and axial resilient deformation. Plastic deformations are permanent deformations that important parameter for flexible pavement design. They are accumulated with every stress cycle, and after certain number of cycles their accumulation becomes negligible. This was shown in AASHO road test [1], where larger ruts were recorded during the first year of the test compared to the second year. Haynes and Yoder [6] have tested base layers and recorded accumulation of permanent deformations even after 100,000 load cycles. Morgan [7] has carried out tests on sand specimens and recorded negligible accumulation of permanent deformations after 2 million loading cycles, while [8] found the stabilization of permanent deformations after 100,000 cycles (according to [9]).

Resilient modulus and permanent deformation values are under impact of several properties, some of which are density, grading, moisture content, grain shape and surface roughness, stress level. Seed and Chan [10] were among first that started repeated triaxial load tests to determine impact of load duration on deformation development of silts. They concluded that the resilient modulus increased as the load duration decreased (according to [3]). The impact of water content on the values of resilient modulus and permanent deformation was studied to [6, 11-13]. The tests showed that resilient modulus decreases and permanent deformation increases as the water content increases. The resilient modulus increases as the specimen density increases according to [12-15]. Hicks and Monisimith [15] have shown increase in resilient modulus with increasing particle angularity or surface roughness. The same have shown [12] comparing rough, angular materials with rounded gravel. The results of the research of fine content impact on resilient behaviour and/or development of permanent deformations, conducted by [11, 15], have shown that resilient modulus decreases and permanent deformations increases with the increase of the fine content in the material.

3. Test procedures

Although numerous triaxial repeated load tests were carried out, only in 1982. first standard for resilient modulus test was adopted – AASHTO T274. Nowadays, after several revisions, AASHTO T307-99 is currently valid. Following the adoption of first standard, new standards and regulations were created. NCHRP 1-28A [16] gives procedure for laboratory determination of resilient modulus for unbound granular base and subbase materials, which is harmonized with AASHTO and FHWA test procedures. Australia gives AG:PT-T053-07 [17] for resilient modulus and permanent deformations testing for unbound granular materials with repeat load triaxial test, while Europe in EN 13286-7 [18] standard gives procedure for cyclic load triaxial test for unbound mixtures. Several researchers have proposed recomendations or test procedures for determination of resilient modulus and/or permanent deformations [19-21].

3.1. EN 13286-7

European standard EN 13286-7 [18] defines procedure for resilient modulus and permanent deformations testing at lower or higher stress levels and with constant or variable confining pressure. For test conduction, the triaxial cell is needed along with loading frame for axial load, displacement transducers (Figure 1), as well as pressure control box, test control and data capture system.

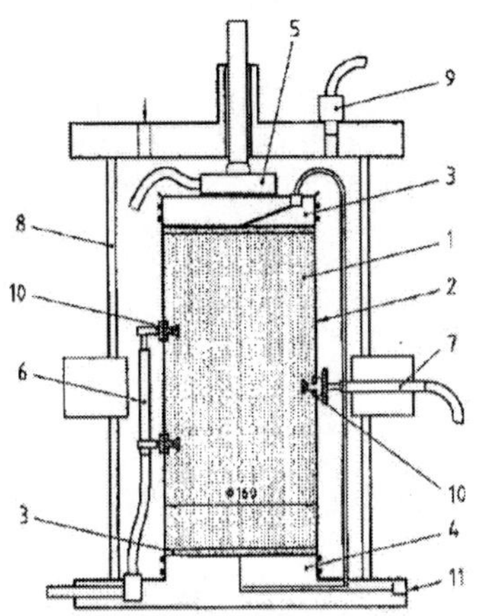

Figure 1. Triaxial cell with displacement transducers [18].

Cyclic load triaxial test is carried out in phases – the conditioning of the specimen phase and the resilient testing phase, and, if permanent deformations are to be determined, a third phase for permanent behaviour testing can be added. The test begins with conditioning phase, which aim is stabilization of

IOP Conf. Series: Materials Science and Engineering **566** (2019) 012013 doi:10.1088/1757-899X/566/1/012013

deformation and achievement of stable resilient behaviour of the specimen. In that phase large number of cycles are applied at defined confining pressure and deviator stress. Some researchers have investigated the required duration of the conditioning phase. Barksdale and Hicks [22] found that 1,000 load cycles are sufficient for cohesionless specimen conditioning (according to [3]). The European standard [18] gives a higher number of load cycles (20,000) with the possibility of lesser number of cycles if stable resilient behaviour of the material is achieved. The resilient modulus test phase is carried out at greater number of stress paths with lower number of load cycles. Barksdale and Hicks [22] concluded that, after conditioning, 50 to 100 load cycles is sufficient for resilient modulus testing of cohesionless materials and that with such small number of load cycles, specimen can be tested at different stress states (according to [3]). Depending on the test method, European standard [18] gives various stress paths in duration of 100 cycles each. Permanent deformations are tested at larger number of cycles and according to [18] can be tested for one stress path at 80,000 cycles, or at several different stress paths (multi-stage loading) in duration of 10,000 cycles each.

3.2. Comparison of test procedures

The test procedures have its differences in regard of sample dimensions, maximum particle size limitations, sample formation, duration of the conditioning phase, number of stress paths applied in resilient modulus test phase along with its levels (confined and deviator stress), response measurement, number of cycles used for result analysis (resilient modulus computation). Comparison of test requirements of EN 13286-7 with several other test procedures is given in Table 1.

Table 1. Comparison of test procedures.

Procedure	Specimen	Specimen preparation	Load wave form and duration	Stresses and number of cycles	Response measurement and results
EN 13286-7 [18]	$Ø_s$ > 5xD_{max} H_s= 2x$Ø_s$	Vibro compression (1 layer, without membrane) or vibrating hammer (≈6-7 layers, with membrane)	No waveform and no duration, frequency of loading between 0.2 Hz and 10 Hz	Conditioning 20,000 cycles (variable or constant σ_3 with high or low stress levels); M_R testing 100 cycles each stress path (17 stress paths for variable or 29 for constant σ_3 with high or low stress levels); ε_p testing single-stage loading 80,000 cycles (variable or constant σ_3) or multi-stage loading 10,000 cycles each stress path (30 stress paths for low stress levels, 28 for high)	1) 3 internal* axial LVTDs and 3 internal* radial LVDTs 2) 2 internal* axial LVDTs and 2 flexible epoxy resin rings M_R computed as avg. of last 10 cycles
AASTHO T307 [23]	$Ø_s$ > 5xD_{max} H_s = 2x$Ø_s$	Type 1 vibratory compaction (6 layers, with membrane) Type 2 vibratory compaction or static loading or kneading (5 layers, if needed with membrane)	Haversine waveform, cycle duration 1-3.1 s with load duration 0.1 s	Conditioning 500-1000 cycles (σ_3= 41.4 kPa, σ_d= 27.6 kPa subgrade soil, σ_3= 103.4 kPa, σ_d= 103.4 kPa base/subbase soil); M_R testing 100 cycles each stress path (15 stress paths; σ_3= 13.8-41.4 kPa, σ_d= 13.8-68.9 kPa subgrade soil, σ_3= 20.7-137.9 kPa, σ_d= 20.7-275.8 kPa base/subbase soil)	2 external* axial LVDTs M_R computed as avg. of last 5 cycles
NCHRP 1-28A [16]	D_{max}= 25.4 mm H_s = 2x$Ø_s$	Type 1 impact (6 or 8 layers) or vibratory compaction (6 layers, with membrane), Type 2 vibratory compaction (6 layers, with membrane) Type 3 impact (6 or 8 layers) or kneading (5 layers, if needed with membrane)	Haversine waveform, cycle duration 1 s with pulse 0.1 s (base/subbase) or 0.2 s (subgrade)	Conditioning min. 1,000 cycles (σ_3= 27.6 kPa, σ_d= 53.8-227.7 kPa - depends on course base, subbase, subgrade); M_R testing 100 cycles each stress path (20 stress paths for granular subgrades, 16 fine grained subgrades, 30 for base/subbase; σ_3= 13.8-138 kPa, σ_d= 9.7-993.6 kPa - depends on course base, subbase, subgrade)	2 internal* axial LVDTs – middle half of the specimen M_R computed as avg. of last 5 cycles
AG:PT/T053 [17]	D_{max} = 19 mm (up to 5% > 19 mm) $Ø_s$ = 100 mm H_s = 200 mm	Hammer with standard (5 layers) or modified effort (8 layers), Compaction without membrane	Cycle duration of 3 s, rise and fall up to 0.3 s and load pulse 1 s	Conditioning 1,000 cycles (σ_3= 50 kPa, σ_d= 100 kPa); MR testing 200 cycles each stress level (66 levels; σ_3= 20-150 kPa, σ_d= 100-600 kPa)+ 50 cycles preconditioning; εp testing 30,000 cycles (σ_3= 50 kPa, σ_d= 150-550 kPa - depends on course base, subbase)	2 LVDTs internal or external* for δr, 1 LVDT external* for δp M_R computed as avg. of last 6 cycles

$Ø_s$ – specimen diameter, H_s – specimen height, D_{max} – maximum aggregate diameter, M_R – resilient modulus, ε_p – permanent strain, σ_3 – confining pressure, σ_d – deviator stress, δ_r – vertical resilient displacement, δ_p – vertical permanent displacement, * internal/external – load cell internal/external, LVDT – linear variable displacement transducers

4. Conclusion

The resilient modulus and permanent deformations are one of the most important properties for pavement design. They can be determined in cyclic load triaxial test for which exist several test procedures, one of which is European standard EN 13286-7. The test provides excellent traffic simulation on tested materials where different conditions can be observed (for example gradation, density, moisture content). In this paper important aspects of testing procedure following EN norm are discussed and compared to other procedures suggested by other norms.

5. References

[1] Highway Research Board 1962 The AASHO Road Test: Report 7, Summary Report *Special Report 61G* National Academy of Sciences (Washington, D.C.)
[2] Groeger J L, Rada G R and Lopez A 2003 AASHTO T307 – Background and Discussion *STP 1437 Resilient Modulus Testing for Pavement Components* ASTM International
[3] Young M A and Baladi G Y 1977 Repeated Load Triaxial Testing State of the Art *RR-477* (Michigan Department of Transportation)
[4] Seed H C, Chan, C K and Lee C E 1962 Resilience characteristics of subgrade soils and their relation to fatigue failures in asphalt pavements *Proc. Int. Conf. on the Structural Design of Asphalt Pavements* (Univeristy of Michigan)
[5] Ahmed S B and Larew H G 1962 A study of the repeated load strength moduli of soils *Proc. Int. Conf. on the Structural Design of Asphalt Pavements* (University of Michigan)
[6] Haynes J H and Yoder E J 1963 Effects of repeated loading on gravel and crushed stone base course materials used in the AASHO road test *Highway Research Record* No 39 pp 82-97
[7] Morgan J R 1966 The response of granular materials to repeated loading *Proc. 3rd Conf. Autralian Road Research Board* Vol 3 Part 2 No 255 pp 1178-1190
[8] Lashine A K F, Brown S F and Pell P S 1971 Dynamic properties of soils *Progress Report No. 2* (Koninklijke-Shell Laboratorium, Amsterdam)
[9] Hyde A F L 1974 Repeated load triaxial testing of soils PhD thesis (University of Nottingham) p 368
[10] Seed H B and Chan C K 1961 Effect of duration of stress application on soil deformation under repeated loading *Proc. 5th Int. Congress on soil mechanics and foundation* Vol 1 (Paris) pp 341-345
[11] Thom N H and Brown S F 1987 Effect of Moisture on the Structural Performance of a Crushed-Limestone Road Base *Transportation Research Record 1121* Transportation Research Board, (Washington D C) pp 50-56
[12] Barksdale R D and Itani S Y 1989 Influence of Aggregate Shape on Base Behaviour *Transportation Research Record* 1227 pp 173–182
[13] Kolisoja P 1997 Factors affecting deformation properties of coarse grained granular materials *Proc. 14th Int. Conf. on Soil Mechanics and Foundation Engineering* (Hamburg) pp 337-342
[14] Trollope D H, Lee I K and Morris J 1962 Stresses and deformation in two-layer pavement structures under slow repeated loading *Proc. Australian Road Research Board* Vol 1 pp 693-721
[15] Hicks R G and Monismith C L 1971 Factors influencing the resilient response of granular materials *Highway Research Record* No 345 pp 15-31
[16] NCHRP 1-28A 2004 Laboratory Determination of Resilient Modulus for Flexible Pavement Design Issue 285 Transportation Research Board
[17] AG:PT/T053 2007 Determination of permanent deformation and resilient modulus characteristics of unbound granular materials under drained conditions
[18] EN 13286-7: 2004 Unbound and hydraulically bound mixtures - Part 7: Cyclic load triaxial test for unbound mixtures Comite Europeen de Normalisation
[19] Claros G, Hudson W R and Stokoe II K H 1990 Modifications to Resilient Modulus Testing Procedure and Use of Synthetic Samples for Equipment Calibration *Transportation Research Record* 1278 pp 51-62
[20] Andrei D, Witczak M W, Schwartz C W and Uzan J 2004 Harmonized Resilient Modulus Test Method for Unbound Pavement Materials *Transportation Research Record: Journal of the Transportation Research Board* No 1874 (Washington D C) pp 29–37
[21] Kim D and Siddiki N Z 2006 Simplification of Resilient Modulus Testing for Subgrades Publication FHWA/IN/JTRP-2005/23Joint Transportation Research Program (Indiana Department of Transportation and Purdue University) p 176
[22] Barksdale R D and Hicks R G 1972 Evaluation of materials for granular base courses *Proc.3rd Int. Conf. on Materials Technology* (Rio de Janeiro)

[23] AASHTO T 307-99 2012 Standard method of test for determining the resilient modulus of soils and aggregate materials

Analysis of an experimental tension structure

L Kapolka[1],

[1] Department of Structural Engineering, Faculty of Civil Engineering, Technical University of Kosice, 042 00 Košice, Slovakia

Abstract. This article builds on the article written last year. The paper is the comparison between the results from the first numerical pre-analysis of the construction of an adaptive hyperbolic paraboloid, the results of which were used as input data for the design of the experimental device in the laboratory at the Faculty of Civil Engineering of the Technical University in Košice with the newest results gained from the numerical model created according to documentation from the manufacturer of this device. The structural numerical model was modeled in regard to geometry, materials and cross-sections characteristics, weight distribution, existing eccentricity and it was created in Dlubal software based on finite element method.

1. Introduction

Following the results of my previous post [1], the results of which were used as input data for the design of experimental frame of adaptive hyperbolic paraboloid structure in a laboratory, it was necessary to verify and compare the results of pre-analysis with the results of exact prototype, which was calculated in finite element method software according to specification after delivery by the external contractor, Inova Praha. Creating more precise calculation model consisted of studying the documentation from the manufacturer of device, followed by creating of an axis-spatial model in AutoCAD software according to the corresponded documentation and afterwards the geometry was imported into the DLUBAL RFEM 5.17 calculation software [2] and subsequently to individual members were assigned their cross-sections with their associated eccentricities. The joints were modeled at the joint sites, the dimensions of the anchor bars and the action members were adjusted to fit the specified geometric-physical model, and the shape and dimensions of the membrane surface were preserved in accordance with the initial requirements of technical textiles fabrication and also technical specifications of the fabric were included.

2. Specification of construction

The dimensions of the frame in the axes are 3.34 x 3.34 x 2.50 m, cross sections of beams were used according to the documentation as rectangular hollow sections (80/100/6; 140/80/6), square hollow sections (80/80/6 ; 100/100/8; 140/140/8; 160/160/8), sheet metal (PL6; PL30; PL45), anchored to the frame through the "U" profile (135/150/45/50) linked with rods full circular section (D46). The action members were modeled as full circular section (D145). There were used modulus of elasticity $E_B = 210$ GPa. Edge ropes were modeled as rope cross-section members (D8) with modulus of elasticity $E_R = 130$ GPa and initial curvature s = 0.20 m. The membrane surface was modeled with parameters of the Serge Ferrari Precontraint 502 technical fabric with a thickness of 0.56 mm, modulus of elasticity in the warp direction is $E_x = 1057$MPa and in weft direction is $E_y = 612$ MPa. The modulus of shear was established as $G_{xy} = 240$ MPa and Poisson's ratio is $\upsilon_{xy} = 1.01$. The floor plan dimensions of the technical fabric are

2.00 x 2.00 m, with the height of 0.50 m in the saddle of hyperbolic paraboloid and a total height of 1.00 m.

3. Finite element analysis

There were used 2D triangular membrane surface finite elements with a mesh size of 50 mm and 1D beam finite elements with 10 divisions per member for beams and also 1D cable finite elements with 10 divisions per member for cables. The form-finding of membrane surface was calculated with the value of axial force of surface $n_x = n_y = 4.0$ kN / m width. Due to the stiffness of the frame, had to be increased the robustness of the computational solution in order to find the initial state of equilibrium for n_x and n_y = 4.0 kN / m width. Subsequently, this model was loaded in a similar way as pre-analysis model and afterwards the combinations of these loads were created.

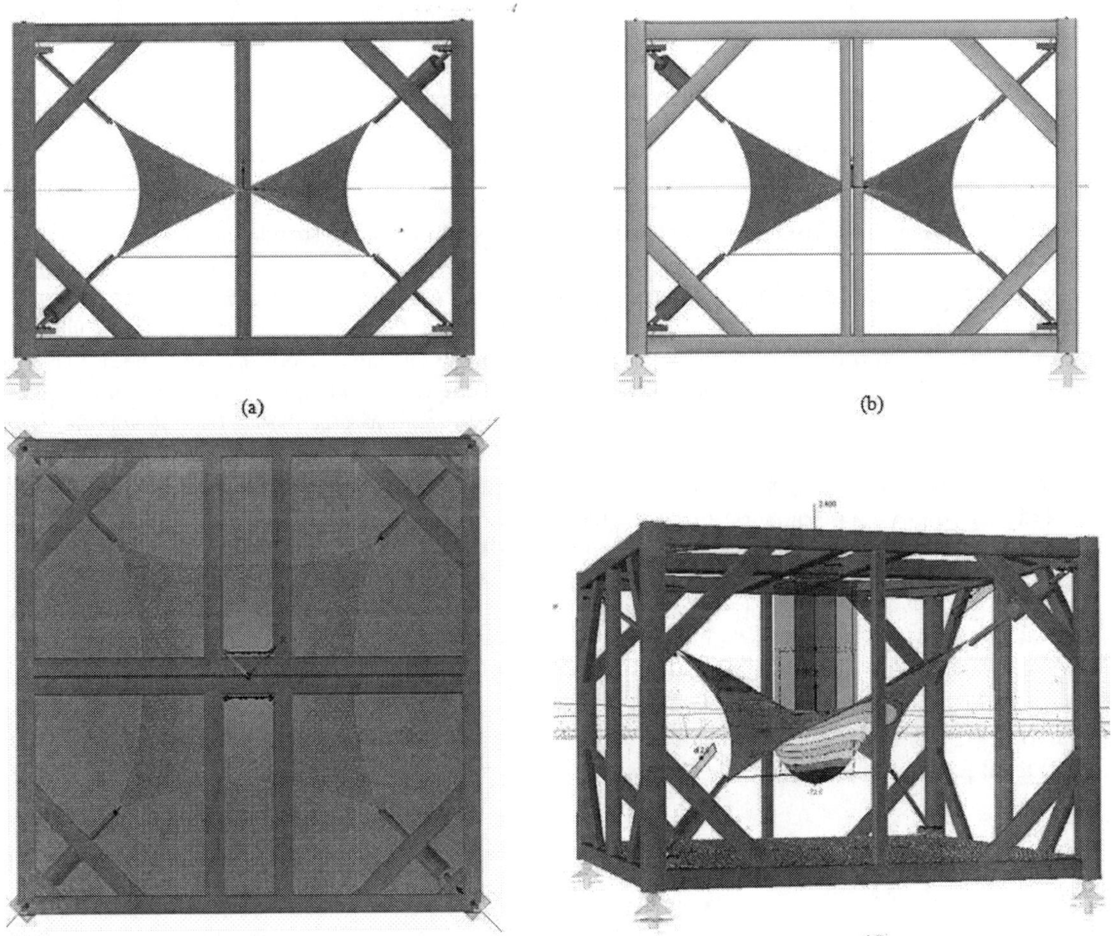

(a) (b) (c) (d)

Figure 1. Detailed computational model.
(a) left view, (b) right view, (c) top view, (d) axonometric view of model

4. Analysis of the results

Due to the large number of combinations, only the most relevant data were included in the graphical comparison of the results, i.e. limit values in terms of the maximum and minimum shortening/extension of the action members and also the combination with no shortening/extension with the load.

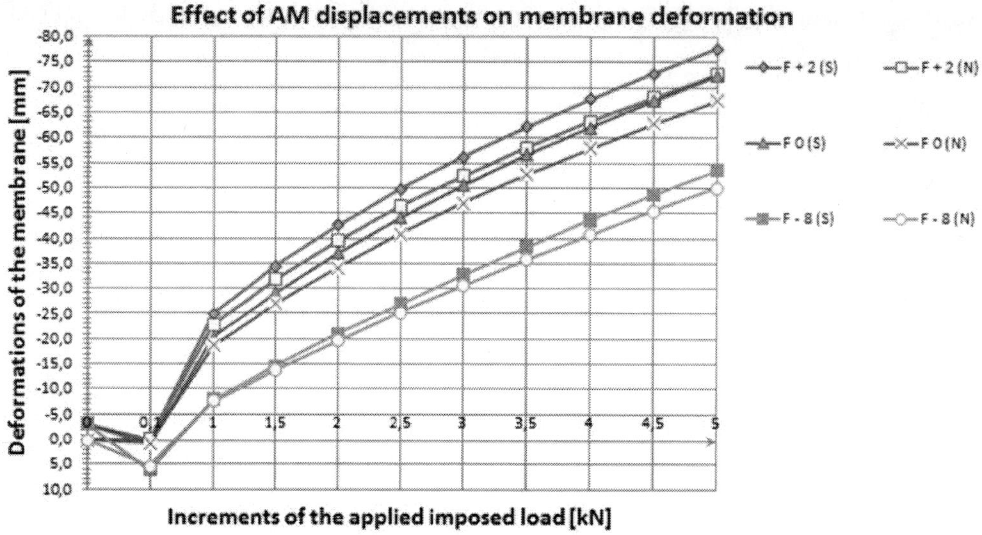

Figure 2. The influence of action members displacement to membrane deformation, the comparison of pre-analysis results (S) and new-analysis results (N).

As can be possible to see at figure 2, apart from the relative differences in the form-finding process and the combinations exclusively with its own weight, it can be stated that the maximum absolute difference in the deformation of the membrane surface is the 5.00 mm, from the combination (F + 2) which represents the incrementally loaded structure by the imposed load and the extension of the action members by 2 mm. The maximum relative difference of deformation is the 9.14% from the combination (F 0), the incrementally loaded structure by the imposed load without changing the length of the action members. At the figure 3, the maximum absolute difference in the axial force in the action members is the 2.00 kN from the combination of (F 0), the incrementally loaded structure by the imposed load without changing the length of the action members, which is also the maximum relative difference established at value 9.15%. At the figure 4, the maximum absolute difference in axial forces in the warp surface direction is the 0.28 kN/m width, for the minimal value of axial forces, combination (F 0), but the maximum relative difference is the 21.02% for the minimal value of the axial forces in the warp surface direction from the combination (F + 2). Similarly at the figure 5, the maximum absolute difference in axial forces in the weft surface direction is the 0.31 kN/m width, for the minimal value of axial forces, combination (F 0), but the maximum relative difference is the 30.13% for the minimal value of the axial forces in the weft surface direction from the combination (F + 2).

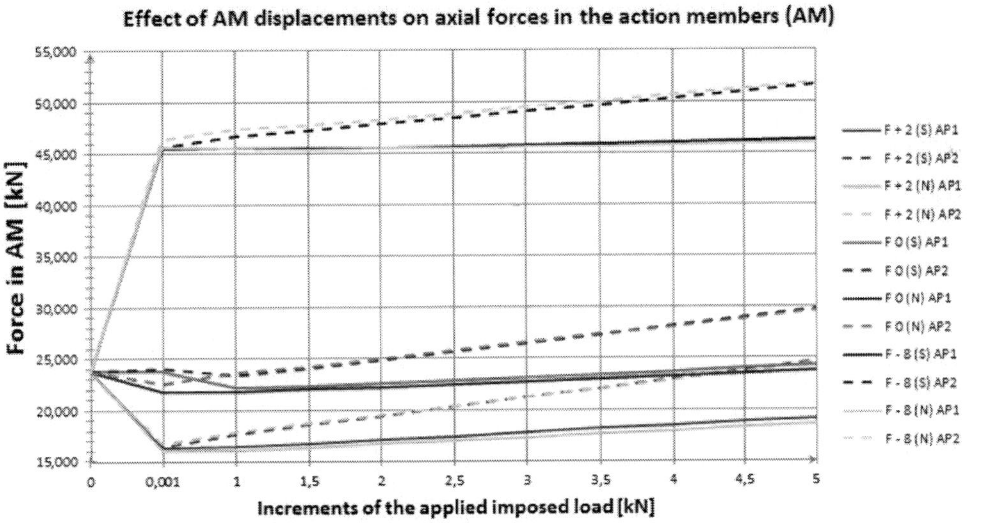

Figure 3. The influence of action members displacement to axial forces in action members, the comparison of pre-analysis results (S) and new-analysis results (N).

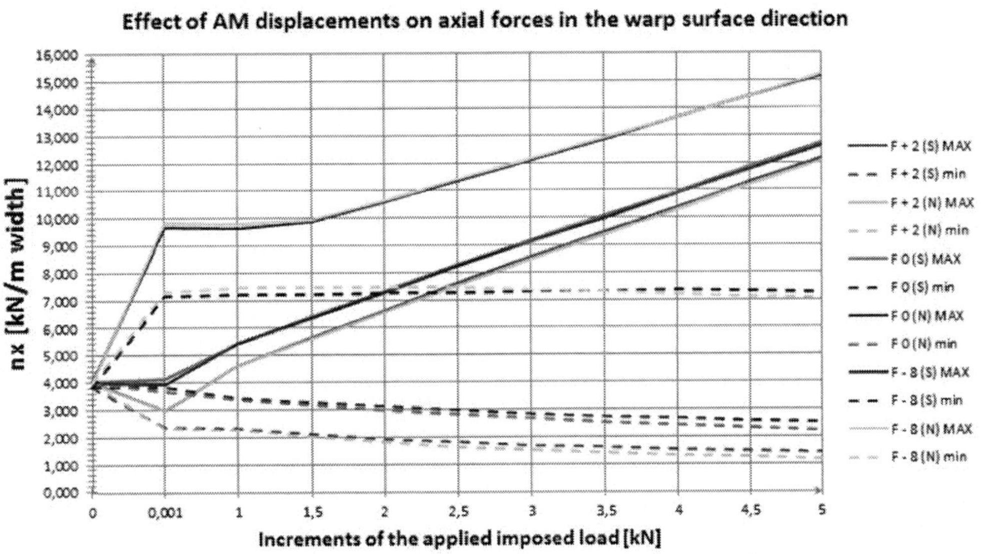

Figure 4. The influence of action members displacement to axial forces in the warp surface direction, the comparison of pre-analysis results (S) and new-analysis results (N).

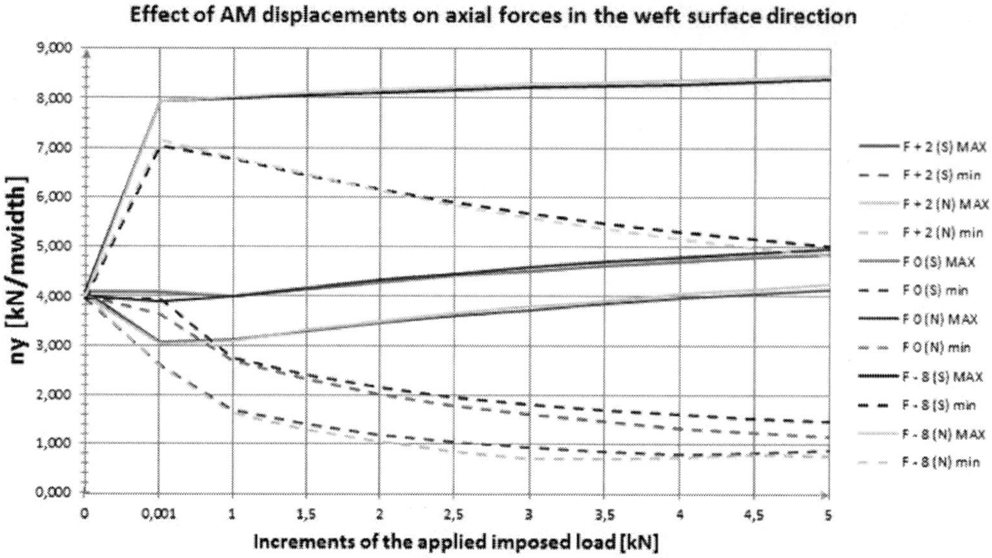

Figure 5. The influence of action members displacement to axial forces in the weft surface direction, the comparison of pre-analysis results (S) and new-analysis results (N).

5. Conclusion

From the results of the detailed computational model, it can be seen that there is no wrinkling at the membrane surface. The magnitude of the relative differences between the results of the pre-analysis and the results of that model are inversely proportional to the stiffness of the membrane surface (the smaller axial forces in the surface means the more significant relative difference). Due to the fact that the experimental device is limited by the force in the action members to the value 50 kN, the creation of the exact model is needed. These results will be compared in the near future with the results of the Ansys software and the results obtained from the experimental device.

6. References

[1] Kapolka L 2018 Static analysis of an adaptive model of membrane structure *Young Scientist 10th international scientific conference on civil and environmental engineering* (Košice: Technická univerzita v Košiciach) p 22

[2] Dlubal RFEM 5.17, student version of the finite elements method software

Acknowledgments

This work was supported by The Slovak Research and Development Agency under the contract No. APVV-15-0777 and also by The Slovak Scientific Grant Agency under the contract No. VEGA 1/0302/16.

IOP Conf. Series: Materials Science and Engineering **566** (2019) 012015 doi:10.1088/1757-899X/566/1/012015

Update on Research on Transparent Wood

V Karľa[1]

[1] Technical University of Košice, Slovakia, Faculty of Civil Engineering, Department of Architectural Engineering

Abstract. This article focuses on various methods of creation of transparent wood, which are derived of the original invention of transparent wood fabrication and describes the advantages of each method, newly obtained properties of the composite materials and suggestions of its possible use in building industry and architecture.

1. Introduction

Transparent wood – an invention first introduced by Siegfried Fink in 1992, who turned wood transparent to be able to observe its specific structures [1]. His work was rediscovered between years 2015 and 2016 by teams at Swedish KTH Royal Institute of Technology [2] and at University of Maryland [3]. Both teams were inspired by Fink's process of fabrication of transparent wood, which they modified to suit their needs and unlike Fink, they were examining properties of transparent wood itself. Both the processes were different, but the concept of creation of transparent wood was the same. The first step was to delignify the wood sample, as lignin gives wood its colour (at KTH, they used 1 wt% $NaClO_2$ with acetate buffer solution with pH 4.6 at 80°C; at Maryland, they used boiling solution, which contained NaOH and Na_2SO_3 and then, the samples were transferred into a solution of H_2O_2 to remove the remaining lignin). Delignified wood samples were then white. After, the second step was undertaken, which was to fill the sample with transparent polymers having matching refractive indexes under repeated vacuum and de-vacuum (PMMA at KTH and epoxy resin at Maryland). After the second step, transparent wood sample was obtained. Both teams then researched properties of their transparent wood samples and found that the thicker the sample was, the lower was its optical transmittance (the highest value of optical transmittance of 90% was measured for the 0.7mm thick TW sample, while for the sample of 3.7mm thickness, its optical transmittance dropped significantly at about 40%). Another finding was that transparent wood has high values of optical haze between 90% and 100%. The very important discovery, considering use of transparent wood in building industry, were its mechanical properties that were improved compared to both, the original wood and the original polymer. Researches then suggested its possible applications for lightweight low-cost structures in light-transmitting buildings, transparent solar cell windows, furniture, or as structural material in automobiles and optoelectronics. They were claiming the better cost and improved ecology compared to glass. However, there was some criticism about those claims, mainly caused by use of chemicals needed to delignify the wood and because of limited thickness, which can still become transparent and be fabricated. Having this criticism in mind, scientists at KTH created derivations of the original transparent wood and examined their properties. These innovations are the main theme of this article and suggestions of their use in building industry are presented.

IOP Conf. Series: Materials Science and Engineering **566** (2019) 012015 doi:10.1088/1757-899X/566/1/012015

2. Lignin-Retaining Transparent Wood and Its Suggested Applications for Various Types of Transparent Wood Beams

In 2017, researchers at KTH have invented a process of creation of transparent wood, which is leaving the lignin inside the wood cells by modifying its structure and only removing chromophoric structures. The process started by exposing balsa wood to alkaline H_2O_2 treatment. After 0.5 hour, the brightness of wood samples increased to 77% and after an hour it reached 79%, which was acceptable for creation of transparent wood. Longer process did not increase brightness more than to 80%. With delignification process, in contrast, it took 6 hours to get 80% brightness. Samples of delignified and lignin modified wood appeared similar. The large sample of balsa wood with dimension of 100x100x3mm was prepared in 5 hours using a new method, while it took 24 hours with the delignification process. Leaving the lignin in wood has also another benefit. Absence of lignin weakens the wood structure, so handling and fabrication of large samples becomes challenging and unpractical and some species, pine, for example, breaks into pieces when delignified. The polymer infiltration was the same process as the previous one using PMMA. [4]

Mechanical properties of the lignin-modified wood proved to be better compared to delignified samples. Measured wet strengths for various samples are shown in Table 1. [4]

Table 1. Comparisons of two methods (delignification method and lignin modification method) to prepare the wood templates from various wood species for transparent wood fabrication [4].

Wood Species	Treatment Method(s)	Time (h)	Weight Loss (wt%)	Lignin Content (wt%)	Wet Strength(a) ∥ Fiber (MPa)	Wet Strength(a) ⊥ Fiber (MPa)
Balsa	Lignin modification	2	12	21.3	7.9±1.2	0.2±0.09
	Delignification	6	26.4	2.5	6.9±1.3	0.2±0.04
Birch	Lignin modification	2	10.6	20.1	14.4±3.3	0.8±0.2
	Delignification	12	25.3	3.3	1.4±0.4	0.07±0.03
Pine	Lignin modification	8	25.0	22.3	14.4±2.2	0.1±0.02
	Delignification	18	40.9	5.2	(b)	(b)
Ash	Lignin modification	4	15.5	22.4	13.9±1.4	0.2±0.05
	Delignification	18	31.1	5.3	0.8±0.3	(b)
Basswood Thickness ∥ fibre	Delignification and Bleach	12	-	≈ 1.5	-	-
Basswood Thickness ⊥ fibre	Delignification and Bleach	6	-	≈ 1.5	-	-
Cathay Poplar	Delignification and Bleach	16	-	-	-	-

(a) For mechanical test, the samples are cut into dimension of 50 mm×10 mm×1.5 mm before chemical treatment. (b) The samples are too weak to keep the shape for the test. Lignin content of the original wood: pine: 32.5wt %, birch: 24.2wt %, balsa: 23.5wt %, ash: 27.1wt %.

Lignin-retaining transparent wood demonstrated high optical performance. The evaluated sample was of 1.5mm thickness and its optical transmittance was 83%. Its value of haze was 75%. Values for delignified sample were also measured for comparison. Optical transmittance of delignified sample was

86% and its haze was around 68%. Scientists at KTH have speculated that the increased haze values in lignin-retaining transparent wood may be caused by the comparably large refractive index mismatch between lignin (1.61) and PMMA (1.49). [4]

At KTH they also ran a 3-point bending test to assess whether transparent wood could replace glass in structural applications. The stress at breakage was comparable for both. For transparent wood it was at 100.7±8.7MPa and for glass it was 116.3±12.5MPa. Strain to failure went even much better for transparent wood (2.18±0.14%) than for glass (0.19±0.02%). It is thanks to reinforcing wood template skeleton in transparent wood. [4]

These properties of lignin retaining transparent wood can be useful in applications of transparent wood beams. Various types of such beams can be created. To use transparency at the highest level, the optimal option is to create some transparent wood "I" beams. The web of the beams would be made of thin but relatively tall transparent wood with retaining lignin. Flanges would be made of thick and short lignin retaining TW. The beam is illustrated in figure 1. [7]

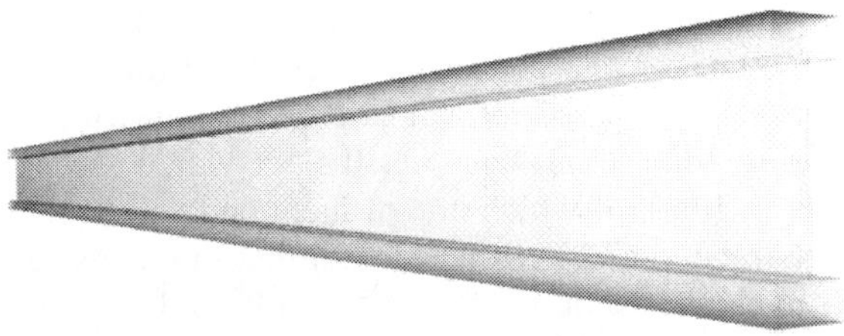

Figure 1. Transparent wood "I" beam rendering [7].

Another variation is creating some box beams. Such beams would be relatively similar to "I" beams, but they would contain two parallel webs, making the beams less transparent, but stronger. To reinforce the beams even more, while sacrificing some more transparency, Vierendeel box beam is the next option. Such beam would be created as Vierendeel truss set up with thick lignin retaining TW with webs of thin TW fixed at its side. The Vierendeel box beam is illustrated in figure 2. [7]

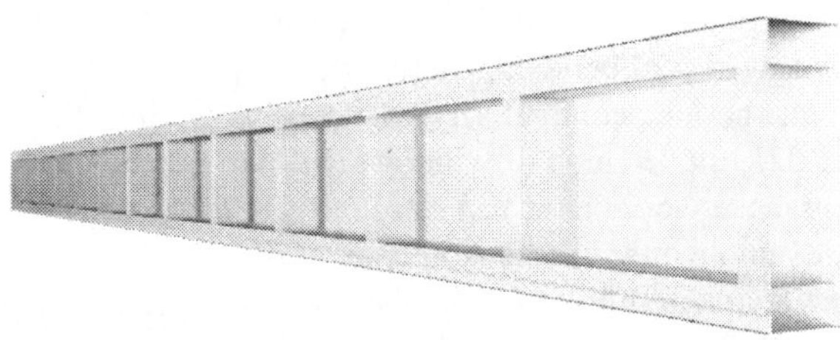

Figure 2. Transparent wood Vierendeel box beam rendering [7].

Yet another option is to create trusses of lignin retaining transparent wood. For this case there are multiple options, of course, but the lightest version is to only use one flange made of TW and the flange would be stretched by a steel cable. The heavier versions would be all TW trusses using thick members of lignin retaining TW as is illustrated in Figure 3. [7]

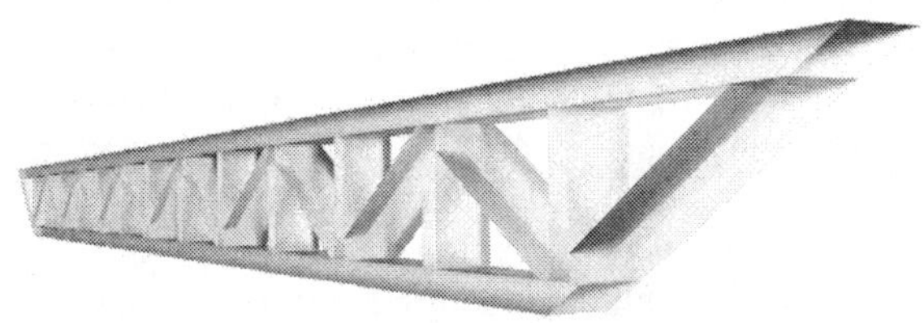

Figure 3. Transparent wood truss rendering [7].

3. Highly Transparent Wood and Its Possible Applications as Façade Material

In 2018, at KTH, they have tried to improve optical transmittance of transparent wood. For this case they acetylated delignified wood samples, using acetic anhydride, for better compatibility with PMMA. It was because wood, especially delignified, is hydrophilic in nature, while polymers as PMMA or epoxy are hydrophobic, which leads to compatibility issues. To obtain the optimized parameters for the acetylation, a systematic study was conducted. Samples with dimensions of 25x25x1.5mm were used for the study.

After acetylation, optical properties of transparent wood improved. For the 1.5mm thick template, optical transmittance reached 92 ± 1%, while the same thickness non-acetylated transparent wood had its optical transmittance of 83 ± 2%. For the neat PMMA it is 95%. Also, wood texture was not obviously visible in the acetylated transparent wood sample. Improvement in optical transmittance was even more visible in thicker samples. While the value of optical transmittance for 3mm thick acetylated transparent wood sample was 89%, it was only 60% for non-acetylated sample of the same thickness. Acetylation also lowered optical haze of transparent wood. For 1.5mm thick samples, optical haze decreased drastically from 70% to 50% and decrease was even larger for 3mm thick samples from 78% to 53%. The haze was still much higher compared to neat PMMA. Comparing two sets of transparent wood samples with 1.5mm thickness (one set always contains acetylated and non-acetylated TW sample of the same cellulose volume fraction), of which one of them had cellulose volume fraction of 30% and other had cellulose volume fraction of 5%, it was found that the increase of optical transmittance for sample with 30% cellulose volume fraction is from 64% to 90%. For the 5% cellulose volume fraction sample it was the increase from 83% to 92%. [5]

Another tests considering highly transparent wood were done on samples with thicknesses of 1.5mm, 3mm, 7mm and 10mm. Optical tests showed that with higher thickness, optical transmittance decreases and haze increases, but compared to non-acetylated samples it is significantly improved. The transmittance of 10mm thick acetylated sample was 40%, similar to 3mm thick non-acetylated sample. Haze increased from 50% in 1.5mm thick acetylated sample to 70 % for the 10mm thick sample. [5]
This higher transparency, which was achieved can be very well used to create transparent wood façades. The factor that is to be considered here, aside from transparency, is thermal conductivity of the material. It can be speculated, that if we used beech to create transparent wood with cellulose volume fraction of 30%, its thermal conductivity would be somewhere around 0.17W/mK (as the thermal conductivity of PMMA is 0.18W/mK and thermal conductivity of beech is around 0.15W/mK). If it is taken into consideration, that after 2021, the U value of windows will have to be 0.6W/m^{2}K, then the thickness of a solid transparent wood façade would have to be around 285mm, which is impossible thickness to achieve, as a 10mm thick sample only has transparency of 40%.

To create TW façades, an inspiration must be taken from double and triple glazing systems in order to make them valid as a construction material. To make the façade system as transparent as possible, 1.5mm thick transparent wood panes should be used. Double pane transparent wood façade with a usual 16mm thick layer of argon would have a U value of 1.1W/m^{2}K. To satisfy the current standard 18mm thick layer should be used, but to reach the standard after 2021, a 30mm thick layer of argon would have to be used. It is different with krypton, as only 17mm thick layer would satisfy the standard after 2021. To satisfy the current standard, only 10mm thick layer of krypton would be necessary. The usual thickness of krypton layer in glazing systems is 12mm, which makes TW façade's U value of 0.82W/m^{2}K.

Triple pane transparent wood façade with usual 16mm thickness of argon layers would have its U value of 0.55 W/m^{2}K. This is even more efficient than the standard after 2021 requires. To satisfy this standard, only 15mm thick layers of argon are necessary. For the current standard it would only be 9mm thick layers of argon. With the usual 12mm thick layers of krypton in a triple pane transparent wood façade system, the system would have its U value of 0.41 W/m^{2}K. For the satisfaction of the standard valid after 2021, only 9mm thick layers of krypton would be necessary and to satisfy the current standard only 5mm thick layers of krypton would be required. All the options of façade set ups mentioned above are summarized in table 2.

Table 2. Summary of U values of different transparent wood façade set ups.

Facade type	Thickness of layer(s) (mm)			U (W/m^{2}K)
	Transparent wood	Ar	Kr	
Solid wall	170	-	-	1
	285	-	-	0.6
Double pane	2x1.5	16	-	1.1
	2x1.5	18	-	0.98
	2x1.5	30	-	0.59
	2x1.5	-	10	0.98
	2x1.5	-	12	0.82
	2x1.5	-	17	0.58
Triple pane	3x1.5	2x9	-	0.97
	3x1.5	2x15	-	0.59
	3x1.5	2x16	-	0.55
	3x1.5	-	2x5	0.97
	3x1.5	-	2x9	0.55
	3x1.5	-	2x12	0.41

(shaded) satisfies current EN standard
(shaded) satisfies EN standard after 2021
Thermal conductivity of transparent wood was considered as 0.17W/mK

4. Transparent Plywood and Solid Building Systems

It was created at KTH using delignified veneers, which were compressed by the force of 75kN for 25 minutes under 25°C after delignification. These delignified compressed veneers were then infiltrated by PMMA and turned transparent. Then they assembled five layers of transparent veneers in perpendicular direction to each other (0/90/0/-90/0, cp-TPW) and also assembled other five layers in a quasi-isotropic structure by increased 45° (0/45/0/-45/0, qi-TPW). These layers were then laminated, again using PMMA. [6]

Both, optical and mechanical properties of transparent plywood were evaluated. The findings are as follows. Unlike regular transparent wood, transparent plywood displays isotropic optical properties. Optical transmittance for cp-TPW with the thickness of 3.0 – 3.5mm is 83%. For qi-TPW with the same thickness it is 75%. Single ply TW with the thickness of 0.8mm has the same optical transmittance as a pure PMMA with the thickness of 3mm having the value of 90% – 95%. Optical haze is for both types of TPW at a value of 80%. Single ply TW has the haze of around 50% - 60%. For the PMMA, the haze is lower than 5%. [6]

A setup was designed and built for the measurement of transmitted light intensity to study the scattering behaviour of TPW. The scattering patterns and light intensities of single ply TW and qi-TPW at different angles were observed. It was found that the scattered light distributions of single ply TW are different between L and T directions. This is caused by anisotropic structure of wood. In contrast, the scattered light distribution in qi-TPW is almost homogenous. [6]

Mechanical tests of TPW revealed that the ultimate strength for single ply TW increases linearly with cellulose volume fraction. Table 3 summarizes the mechanical properties of single ply TW and TPW. [6]

Table 3. Summary of the mechanical properties of single ply and TPW, and the Young's modulus predictions from laminate theory [6].

Wood Species	Tensile Test Direction	Cellulose Volume Fraction (%)	Size	Mechanical Properties			
			l x w x t (mm)	E (GPa)	E^a (GPa)	σ (MPa)	$\varepsilon_c \approx$ (%)
Single Ply	L	12	60x5x0.8	4.3 ± 0.4	-	62.5 ± 2.7	1.5
TW	T	12		2.4 ± 0.2	-	14.6 ± 1.2	0.7
Cp-TPW	L	10	60x5x3.5	4.1 ± 0.3	3.7	50.1 ± 2.6	1.2
	T	10		3.9 ± 0.1	3.3	44.9 ± 1.3	1.3
Qi-TPW	L	10	60x5x3.5	3.9 ± 0.2	3.4	45.4 ± 2.9	1.2
	T	10		3.5 ± 0.3	3.0	42 ± 1.8	1.4

'-': No data available. 'l × w × t': Length, width and thickness, respectively. 'E': The Young's modulus from the experimental data. E^a: The predicted Young's modulus is determined based on laminate plate theory. 'σ': Ultimate strength. 'ε_c': Failure at break. '*': T The mechanical performance values were calculated from tensile test with AVE.

Transparent plywood could find its purpose in building industry as semi-transparent non-bearing partition walls or as load bearing solid beams for short bridges or for stairs, which would also act as railings (figure 4). If the thicker samples were created, they could be arranged similarly to cross laminated timber and would be able to span even further distances.

Figure 4. Rendering of transparent wood solid beams/railings used for ramps at Tathra Beacon (designed for MNPG Arch competition).

Also, panel building systems can be considered using transparent plywood (or transparent cross laminated timber), where TW panels would be at the outer surfaces of the panel and inside, the panel would be insulated with aerogel and reinforced with transparent wood grid made of thick TW members, so the structure would be all transparent with sufficient thermal resistance. Such system is illustrated in figure 5.

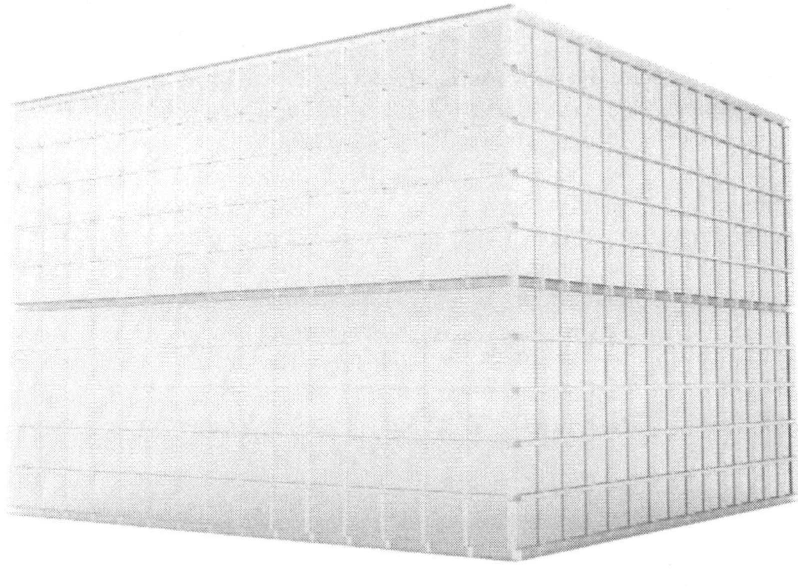

Figure 5. Rendering of transparent wood + aerogel panel system

5. Conclusions

The research on transparent wood has only begun recently, but it is already advancing rapidly, as was shown in this article. With improvements in technology of fabrication of transparent wood, it is becoming more ecologically friendly and more easily obtained material with lignin-retaining transparent wood. With acetylation, transparent wood optical properties become even more attractive and thicker samples can be created as the research on highly transparent wood has shown. Finally, transparent plywood made properties of otherwise highly anisotropic material isotropic and transparent plywood showed even better structural properties than, before also outstanding, original transparent wood composite. One can speculate about its future possible applications as façade material, probably mostly in situations where there is required sunlight just as much as privacy, as its high optical transmittance together with high optical haze predetermine it. As was shown, compared to glass systems, transparent wood facades would require less material and lesser amount of inert gases, therefore would also save more money. Another speculations considering this material are to use it structurally as a load bearing material due to its mechanical properties in various types of beams, namely "I" beams, box beams and various forms of trusses, or even as a glued solid beam. What the future holds for transparent wood is uncertain now, but the possibilities of future architectures are exciting for sure.

6. References

[1] Fink S 2009 Transparent Wood – A New Approach in the Functional Study of Wood Structure *Holzforschung - International Journal of the Biology, Chemistry, Physics and Technology of Wood* 46(5) pp 403-408

[2] Yuanyuan L, Qiliang F, Shun Y, Min Y and Lars B Optically Transparent Wood from a Nanoporous Cellulosic Template: Combining Functional and Structural Performance *Biomacromolecules* 2016 17 (4) 1358-1364

[3] Mingwei Z, Jianwei S, Tian L, Amy G, Yanbin W, Jiaqi D, Yonggang Y, Wei L, Doug H and Liangbing H 2016 Highly anisotropic, highly transparent wood composites *Advanced Materials* 28(26) 5181–5187

[4] Li Y, Fu Q, Rojas R, Yan M, Lawoko M and Berglund L. 2017 A new perspective on transparent wood: Lignin-retaining transparent wood *ChemSusChem*

[5] Li Y, Yang, X, Fu Q, Rojas R, Yan M and Berglund L 201. Towards centimeter thick transparnt wood through interface manipulation *Journal of Materials Chemistry A* 6(3) 1094-1101

[6] Fu Q, Yan M, Jungstedt E, Yang X, Li Y and Berglund L A 2018 Transparent plywood as a load-bearing and luminescent biocomposite *Composites Science and Technology*

IOP Conf. Series: Materials Science and Engineering **566** (2019) 012015 doi:10.1088/1757-899X/566/1/012015

[7] Katunský D, Kanócz J, Karľa V 2018 Structural elements with transparent wood in architecture *International Review of Applied Sciences and Engineering = IRASE* Budapest (Hungary: Akadémiai Kiadó) ISSN 2062-0810 9 (2) 101-106

Acknowledgments
This contribution came (resulted) from the solution of research project VEGA 1/0674/18 supported by the Scientific Grant Agency, project KEGA 046TUKE-4/2019 supported by the Cultural Grant Agency of the Ministry of Education, Science, Research and Sport of the Slovak Republic and the Slovak Academy of Sciences. This paper was elaborated with financial support projectswith ITMS code: 26220220182 (TECHNICOM), ITMS: 26220220064 (VUKONZE).

Dynamic influence of a swinging bell on a traditional wooden structure of the Lemko orthodox church belfry

T Kochański[1,2]

[1] Institute of Building Materials and Engineering Structures L-1, Faculty of Civil Engineering, Cracow University of Technology, Cracow, Poland
[2] Member of the Polish Chamber of Civil Engineers

Abstract. Paper treats about dynamic influence of a swinging bell on a traditional wooden structure of the orthodox church belfry (example calculations made for a bell and belfry of the Lemko's orthodox church). On the basis of the conclusions of the analysed belfry, there were made modifications aimed to improve its dynamic parameters. The brief summary provides comparison between the effects of such modifications and existing wooden structure.

1. Introduction

1.1. Lemko orthodox church

While settling in Carpathian region, Lemkos started to build specific type of the orthodox churches, very typical for the north Carpathian region (southern east Poland, north east Slovakia, west Ukraine and north Romania). Almost all of the Lemko orthodox churches structures were timber (most available resource), with traditional types of joints which are very common. Need to mention that before Austro-Hungarian regency orthodox churches did not included timber belfry, which nowadays is the most recognised part of the building. In most buildings, there could be found signs that belfry was built up to the existing building (such as old roof structures elements and claddings in belfry interior), so due to that fact they have independent structure – built as the column-and-struts structures [2][3].

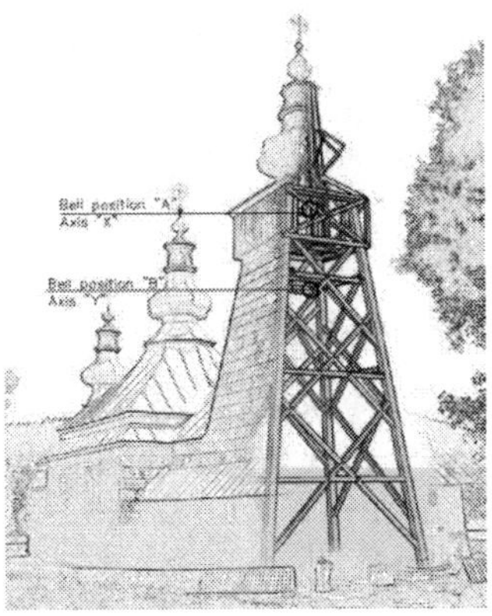

Figure 1. Analysed belfry structure.

Figure 2. Main belfry structure (columns, struts and head-piece).

IOP Conf. Series: Materials Science and Engineering **566** (2019) 012016 doi:10.1088/1757-899X/566/1/012016

1.2. Belfry structure

Main structure of the belfry is built as 4 columns (cross sections about 30x30cm to max 60x60cm) supported on the sandstone simple foundations. Columns are horizontally supported by struts and beams, so the main structure may looks like a truss, but due to the fact that the elements are not symmetrically connected, it works similarly to the frame – providing also momentum of force. On top of it, the belfry has mostly only one axis of symmetry (side walls are symmetrical, interior and exterior walls are asymmetrical to each other, due to main church door and existing main building beam structure in belfry interior) and has old traditional joints which in some aspects are rigid types, but with limited work angle (dependent on connection loose) and limited resistance to tension. The Highest part of the belfry is built as a turret, with convergent columns connected among themselves by struts, ended with a strut structure supporting cantilever (with a cross) [4].

1.3. The bell

Bell mostly is hanged on the higher interior part of the main belfry structure (around a level of the main columns' endings and turret) and by the special beam, which is supported by the main structure struts or top beams. Hanging axis varies, but mostly, there are two directions: perpendicular to the building and parallel with the building, same as the various types of bells (mass of bells and dynamical parameters are varying and should be approx. calculated in every case).

1.4. The problem

Presented calculations were provided for the timber belfry of the Lemko's Orthodox Church in Berest (Lesser Poland, Poland). The structure was chosen due to problems with usage of the belfry in the past (problems with instability of the main timber structure), so there is the question "what is the influence of a bell on a timber belfry structure?". To answer that question, adequate calculations must be performed.

2. Assumptions, FEM model building

The belfry structure was measured, considering cross sections of the belfry elements, levels of the connections with struts and beams, characteristics of the connections, dead loads (timber cladding, roof steel panels and timber structure). Then, based on measurements was built calculation (FEM) model. Then the bell was measured, its diameters, thickness and approx. geometry to make assumptions of its dynamic characteristic (considered only low frequencies, high frequencies was not considered due to fact they have negligible dynamic influence on the belfry). The bell was hanged in two various positions, first on the highest part of the belfry (swinging parallel to the building main axis), then moved on a lower level (swinging perpendicular to the main building axis) and in the end taken out of a belfry (hanged on a tree, which broke under the bell). Mentioned hanging positions in the belfry were taken for consideration in calculations.

2.1. The bell

Parameters of the bell were determined by geometry measurements and based on [6].

Table 1. Approximated bell parameters, according to [6].

Diameter [mm]	930
Mass [kg]	400
Strike frequency [Hz]	1
Swinging angle (strike) [°]	65
Tone	gis' as'

According to the following bell parameters and [1], dynamic parameters and related loads have been calculated. Horizontal harmonic components are considered as the ones responsible for potential instability and/or resonance. Vertical harmonic components (i=2.4)are integrated as a bell gravity load (dead), considered in FEM model.

IOP Conf. Series: Materials Science and Engineering **566** (2019) 012016 doi:10.1088/1757-899X/566/1/012016

Table 2. Characteristics of the main harmonic horizontal components (i = 1,3,5).

Component	Period [s]	Frequency [Hz]	Component force [kN]	Proportion of components [%]	Summary horizontal force /static/ [kN]	Oscilation of the resultant horizontal force (x=[s]; y=[kN])
1	2	**0.5**	**1.66**	43.6	3.8	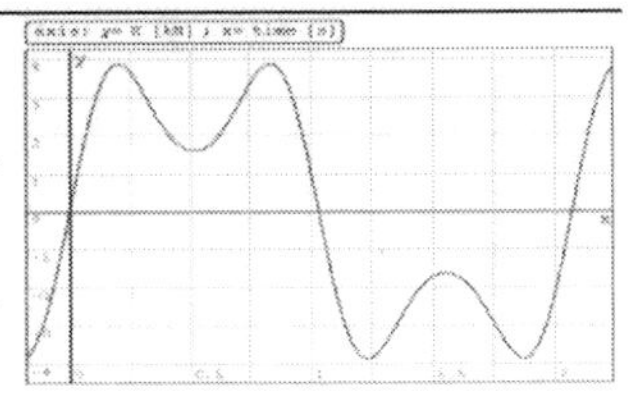
3	0.67	**1.5**	**2**	52.7	3,8	
5	0.4	2.5	0.42	3.7	3,8	

Values presented in Table 2 shows dominating horizontal components with proportion, parameters and forces values related to each component. Mentioned components have major impact on belfry structure and their frequencies are close to belfry natural frequencies. Most dominating (for analysed bell characteristic) are i=1 and i=3 responsible for about 96% of total horizontal force, and those characteristic components frequencies must be compared with belfry natural frequencies while searching potential resonance.

2.2. The belfry

Belfry geometry, nodes type and elements relevance was converted to 3D model which is basis for FEM calculation. Some assumptions has been simplified (every node as pinned connection – various loose and rotation possibility, pinned support on footing, not included rigidity of cladding).

Table 3. Characteristic dimensions of the belfry and structure layouts.

North and south side (side walls) [m]	West side (front side with main door) [m]	East side (church side) [m]

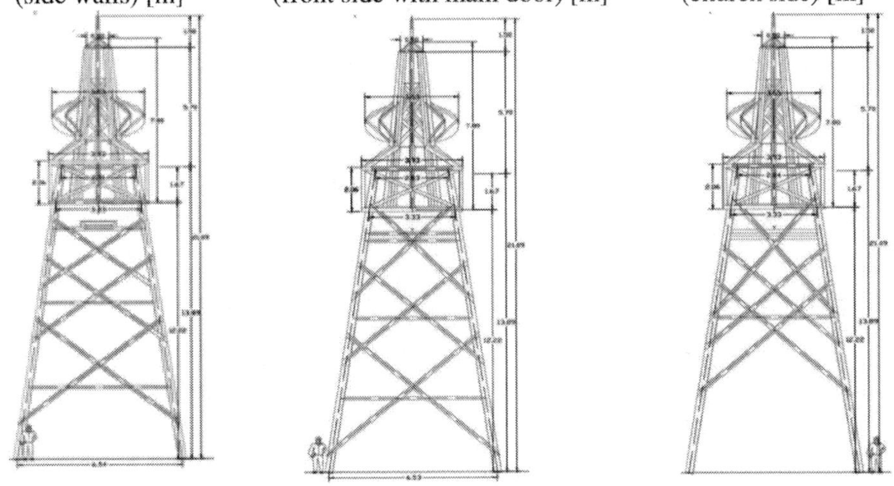

IOP Conf. Series: Materials Science and Engineering **566** (2019) 012016 doi:10.1088/1757-899X/566/1/012016

Table 4. Joints – nodes. Presented following traditional connections for the timber belfry.

Main structure columns and struts connection	Belfry interior	Head-piece of main structure

Belfry main column support	Support beams of the higher turret	Higher turret columns and struts

Table 3 shows dimensions of the belfry and structure layout according to measurements, Table 4 presents typical traditional connection solutions for presented structure. Traditional belfry's structures are complicated, so measuring of the structure and creating structural layout require time to acquire and analyse necessary data. Next step is to build model representing real structure as close as it's possible. Model main structure input data should be possibly accurate due the fact that every element (stiffness) and every node (released and blocked reactions, displacements) has considerable impact on model output data (especially natural frequencies). It took few attempts to build possibly accurate model, shown as Figure 3 and Figure 4.

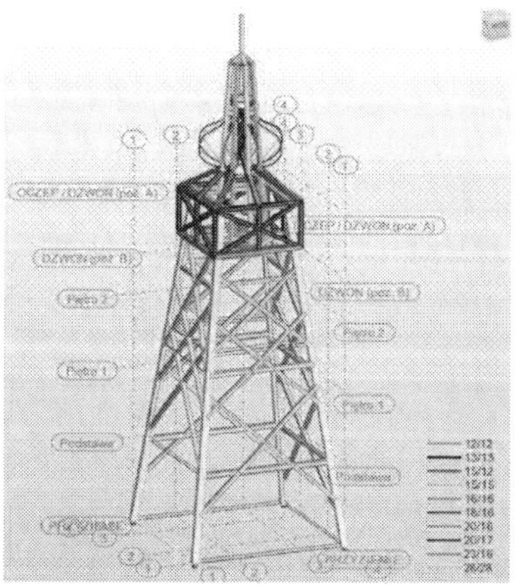

Figure 3. Completed model including eccentricity of connections.

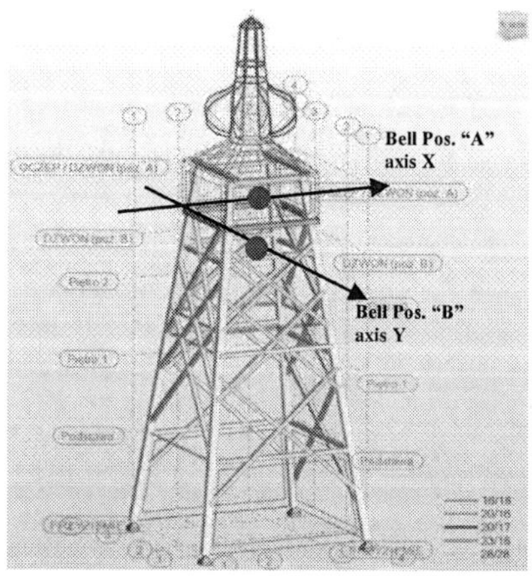

Figure 4. Main belfry structure (columns, struts and head-piece).

3. FEM model analysis

Model including dead loads (structure load, cladding load) and static bell loads was calculated for static forces and then converted to model with distributed masses for modal calculations.

Table 5. Belfry Eigenmodes.

1 eigenmode	2 eigenmode	3 eigenmode	4 eigenmode	5 eigenmode

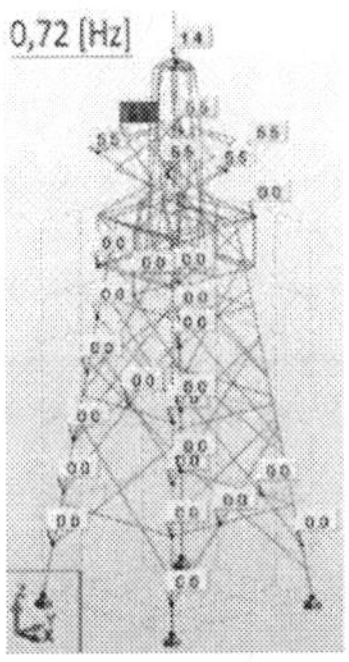

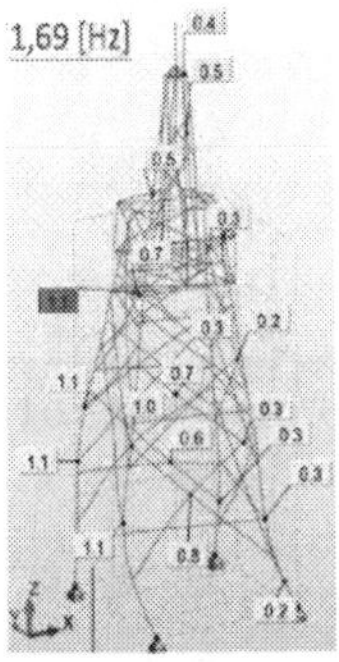

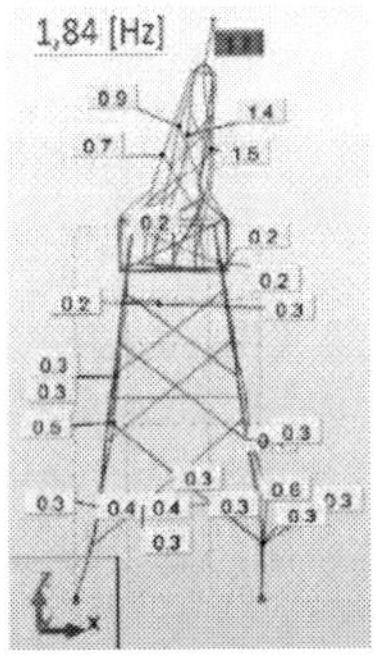

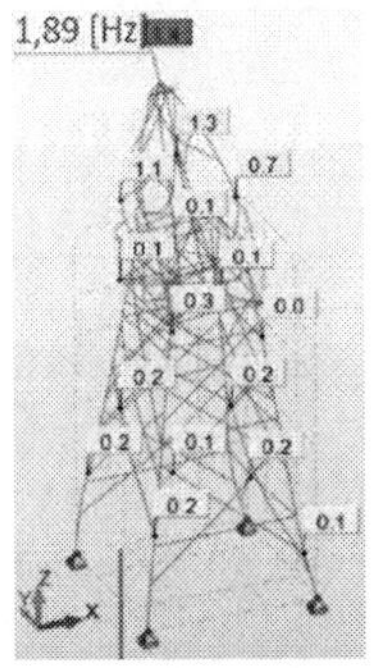

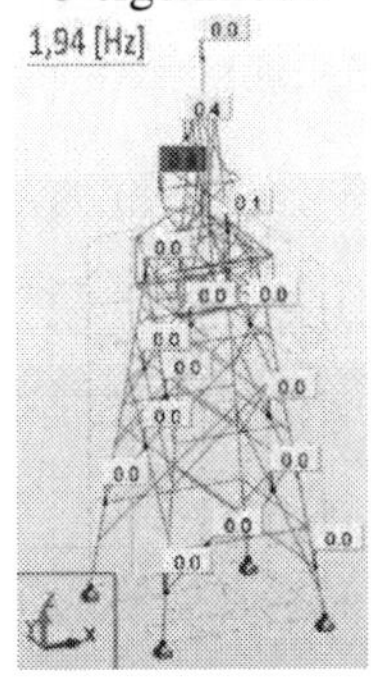

Table 6. Natural frequencies oscillating mass (related to axis).

Eigenmode	Frequency [Hz]	Related mass X axis [%]	Related mass Y axis [%]	Total mass [kg]
1	0.72	0	0.01	
2	**1.69**	0.07	**61.4**	
3	1.84	9.3	0.33	23406
4	1.89	0.35	1.1	
5	1.94	0	0	

Due to monosymetry of the structure natural frequencies has not exactly one directed character. The easiest way to describe the character of the frequency is to analyse related to each axis oscillating mass. Analysing masses one can deduce that dangerous is frequency 2 on the axis "Y" (due to mass taking action).

4. Resonance possibilities

Comparing bell components frequencies and belfry natural frequencies it's possible to find dangerous resonance situation and related equivalent static horizontal force generated while resonance.

Table 7. Frequencies comparison, convergence of frequencies natural-harmonic.

Comparison	Natural belfry		Related mass Y axis	Bell pos. "B" harmonic (axis Y)		Convergence
	Eigenmode	[Hz]	[%]	Comp.	[Hz]	[%]
1	1	0.72	0.01	1	0.5	69 (<80)
/Resonance/ 2	**2**	**1.69**	**61.4**	**3**	**1.5**	**89 (>80)**
3	3	1.84	0.33	3	1.5	81.5 (>80)
4	4	1.89	1.1	3	1.5	79 (<80)
5	5	1.94	0	5	2.5	78 (<80)

According to Table 7 resonance may occur for impacting third harmonic component (related mass oscillating on the axis same as the bell and convergence of those frequencies), and it should be considered in calculations using dynamic factor for static forces. Convergence more than 80% is considered as 100% resonance due to uncertainty of calculation assumptions. In that fact dynamic factor for the third component is set as maximum: η=10, according to [1].

5. Resonance impact on the main belfry structure

Convergence of natural frequencies and harmonic vibrations for position "B" is critical for the belfry. Table 8 shows values of the forces in elements and impact descriptions when the belfry is resonating with the bell.

Table 8. Critical resonance forces impacting belfry with the bell situated on position "B" (axis "Y").

Force type	Value		Impact on the belfry
Rz (N) – main column support	(Max) 81.5 kN	(Min) -4.95 kN	**Instability of the main structure. Disconnection of the support.** Traditional support on the ground beam cannot afford tension. **Destruction of joints connections in main columns** (tensioned).
R (V) – main column support	Axis X (max) 12.9 kN	Axis Y (max) 14.7 kN	Shearing of the support pivot. **May lead to pivot destruction and change of the static scheme** leading to significant increasing of forces impacting belfry (destruction of sensitive tensioned joints).
N – struts, connections among elements	(Min) -8.1kN	(Max) 13.7 kN	**Destruction of joints sensitive for tension**, disconnection of the elements. **Compressed slim elements may lose stability.**

6. Modifications of the structure. Improving dynamic parameters of the belfry

Resonance situation according to Table 7 must be considered as unwanted and dangerous for the belfry. Need to mention that resonance can occurs while a standard swinging of the bell, when the bell has the same swinging angle as strike angle and vibrations are harmonic (bell ringer is actively preventing damping). Figures 5 and 6 shows two, separated kinds of structural approaches. Figure 5 presenting approach with inserting additional elements to make belfry structure close to bisymmetric. Figure 6 presenting reinforcing and modification of the main structure joints (rigid connections). Two mentioned approaches include also inserting small columns between turret and main structure struts.

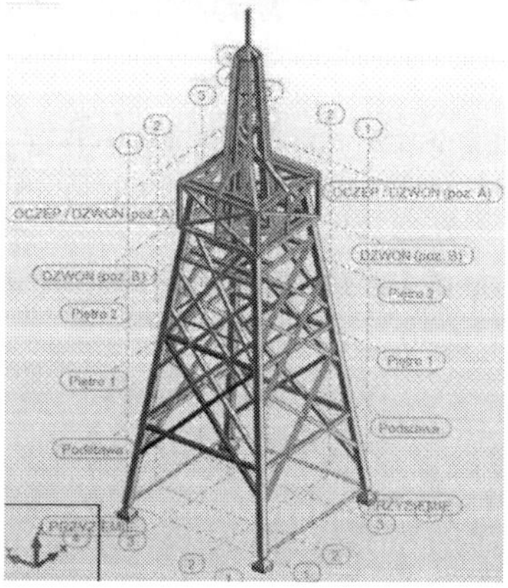

Figure 5. Bisymmetrisation of the structure (additional struts and beams in east side wall).

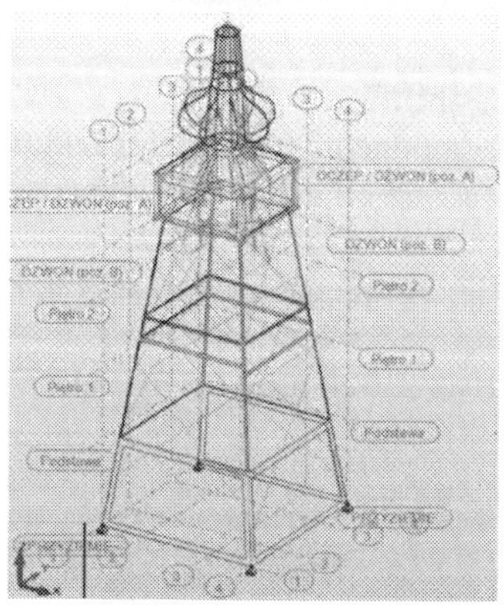

Figure 6. Main structure joints modification (rigid connections. Red appointed connections left as joints).

Modifications are aimed to improve dynamic parameters of the belfry (mainly to increase natural frequency of the belfry). Results of such modifications shows Figure 7.

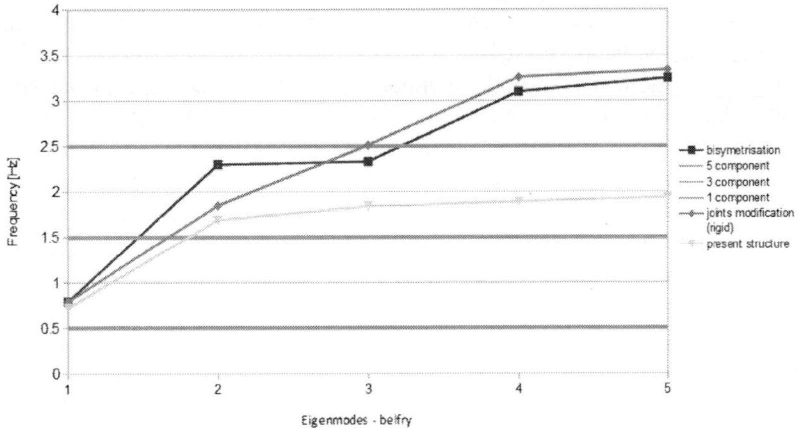

Figure 7. Comparison of eigenmodes frequencies changes.

Not only frequencies has changed due to modification, Table 9 show level of forces impacting the belfry structure. Both structural changes in the belfry structure led to a better forces redistribution among elements, and that is considered as very desirable result for a safety of the structure. Considering redistribution of forces, joints modification gave limited results, on the contrary bisymmetrisation gave as suspected the best results, allowing structure equable redistribution of forces (bolded in Tab.9).

Table 9. Comparison of characteristic forces impacting structure during considered hanging.

Force type	Present resonance		**Bisymmetrisation**		Joints modification (rigid)	
Rz (N) – main column support	(Max) 81.5 kN	(Min) -4.95 kN	**(Max) 45 kN**	**(Min) 31kN**	(Max) 50.7 kN	(Min) 25kN
R (V) – main column support	Axis X (max) 12.9 kN	Axis Y (max) 14.7 kN	**Axis X (max) 6.6 kN**	**Axis Y (max) 7.5 kN**	Axis X (max) 7.9 kN	Axis Y (max) 9.2 kN
N (tension) – struts, connections	(Min) -13.7 kN		**(Min) -3.8kN**		(Min) -4.7kN	

6.1. Modification summary

Both presented belfry modifications are increasing natural frequencies of the structure to the safe level (considered as non resonating with the bell 1 and 3 components) and more (bisymmetrisation) or less (joints modification – rigid) are ensuring equable redistribution of forces among elements. For 2 eigenmode resonance may occure with the component 5, but due to very low component impact on the total horizontal force (<4%) it's not considered.

7. Summary

Swinging bell has significant dynamic impact on considered timber belfry structure in the field of low frequencies, leading to the resonance, structural instability and potential structural failure. Conducted calculations gave results explaining problems with the belfry usage in the past and shows that direction of the bell swinging may have more impact on the structure than level of hanging due to resonance. Proposed structural modification by bisymmetrisation of the structure gave most

IOP Conf. Series: Materials Science and Engineering **566** (2019) 012016 doi:10.1088/1757-899X/566/1/012016

satisfactory results, eliminating possibility of the resonance and ensuring equable redistribution of forces among belfry elements. Both modifications could be performed on a timber belfry with the column-and-struts structures struggling with an instability due to resonance impact and in general will lead to a significant improvement of the belfry dynamic parameters, force balance and ultimately possibility of restoring usability of the belfry.

For engineering (practical) use both methods should be undertaken on the structure at the same time (reinforcing of the main belfry joints (for rigid) and Bisymmetrisation of the structure) to ensure belfry stability and structural safety. It is essential due to disparity of the technical conditions among belfry elements and joints. From the author personal experience the highest parts of the belfry (turret structure) and main columns around foundation, including ground beam must be carefully revised due to possibility of leaking and various humidity conditions leading to a progressive biological destruction of the timber sections and joints.

8. References:

[1] Rapp P 2014 Historial Bells – Mechanics and drive problems *J. Heritage Conserv.* **40** 41-57

[2] Brykowski R 1986 *Wooden Orthodox Church Architecture of the Lemkov People in Poland, Slovakia and Carpathian Russia* (Wrocław: Ossolineum)

[3] Brykowski R 1995 Drewniana architektura cerkiewna na koronnych ziemiach Rzeczypospolitej (Warszawa: Towarzystwo Opieki Nad Zabytkami)

[4] Kopkowicz F 2009 Ciesielstwo polskie (Warszawa: Arkady)

[5] Gromysz K 2016 Dynamika budowli. Obliczenia układów prętowych i o masach skupionych (Warszawa: Wydawnictwo Naukowe PWN)

[6] DIN 4178:2005-04 Glockentürme. Berechnung und Ausführung. Table A.1

IOP Conf. Series: Materials Science and Engineering **566** (2019) 012017 doi:10.1088/1757-899X/566/1/012017

Experimental house and stratification preheating tank

J Kosican[1]

[1] Technical University of Košice, Department of civil engineering, Institute of Architectural Engineering, Slovakia

Abstract. The aim of this study is to simulate two types of usage of solar panel systems and their share of heating and hot water in the experimental house. Two differently designed solar systems are tested in an experimental house with stratification and the other without stratification. Both systems provide annual demand for heat and demand for hot water. The simulation program is T * SOL 2018. By using a variant comparison program, it allows you to examine which type is better in different standards and also compare both solar panel outputs. The results can be analysed in different environments, such as different weather conditions or different heat losses of a building. The following research deals only with heating and water heating[2][2][2][2][1].

1. Introduction – Variant with stratification

The first variant has 20 solar panels Thermoslar Žiar TS 300 with a surface of 2 m^2 with 45° inclination and 40.6 m^2 of total roof area. The solar preheating tank has a capacity of 2050 litres with a heat loss of 6.36 W / K, for hot water is designed tank with capacity of 200 litres, auxiliary power is supplied by a wooden boiler with a rated output of 15 kW. The experimental house is designed in a passive standard. The object has designed a 300-liter buffer tank for floor space heating as shown in Figure 1. The location of the experimental house is at a research institution at the Technical University in Košice in a climate chamber, that can simulate various external conditions. For the following research, thermal losses are 3.1 kW and calculated climatic temperature conditions for Košice with calculated external temperature (-14 ° C).

The principle of stratification is the storage of thermal energy, which is distinguished according to the chosen physical principles. To increase the use of energy from solar collectors and other heat sources, the so-called " temperature stratification, i.e., j. storaging by layers, e.g. in the stratification storage tank. In the storage tank is gravity stratification of water according to the volume of water depending on its temperature. The circulation of water in stratification storage tanks depends on the temperature of the heating circuits from the individual heat sources.

Thus, the circulation of water depends, for example, whether it is a low-temperature system (+40 to +60 ° C), or a high-temperature system (+60 to +90 ° C). Thermal stratification is the layering of the tank volume by temperature controlled by the storage of thermal energy in layers at the same or similar temperature. The cooler and heavier the water is held in the bottom of the tank, the hotter rises up, i.e. water with a density ρ is always supplied below a layer, which has a lower density. Individual temperature layers retain different temperatures.

In most cases vertical storage tanks are used to support the natural storage of temperature layers of the tank volume due to buoyancy forces.

For the correct function of the heat energy layering, it is necessary to create not only the layering according to the temperature, but also to ensure that the layering of thermal energy in the storage tank to not degrade. The quality of the temperature stratification of the volume in the stratification preheating

IOP Conf. Series: Materials Science and Engineering **566** (2019) 012017 doi:10.1088/1757-899X/566/1/012017

tank will affect the operating parameters of the whole system. The temperature layering is mainly influenced by the way in which the working substance is fed into the storage tank [2].

Table 1. Input data.

Thermal losses of the object	3,1kW
Average dayli DHW consumption	120 litres/day
Annual energy requirement for DWH	2031 kWh
Object area	53 m2
Calculated outside temperature	-14°C
Solar preheating tank	2050 litres
DWH tank	200 litres
Auxiliary heat source- wooden fire boiler	15kW
Space- heating buffer tank	300 litres

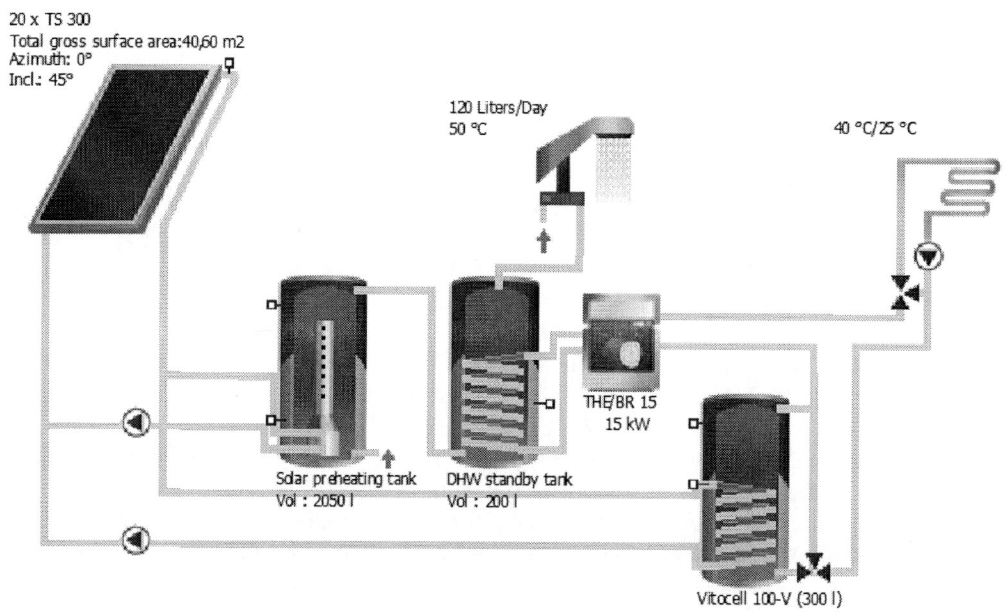

Figure 1. DWH system with stratification.

The results of the yearly simulation for the experimental house are shown in Figures 2 and 3. The simulation time was from the beginning of January to the end of December. The simulation for heating was throughout the year, but without a period from May to August, because at that time is no heating season. Figure 2 shows that the share of the solar panel for DHW is 94% - the red column, the share of the solar panel for heating - 20% - the blue column and the total solar energy is 37% - the green column.

IOP Conf. Series: Materials Science and Engineering **566** (2019) 012017 doi:10.1088/1757-899X/566/1/012017

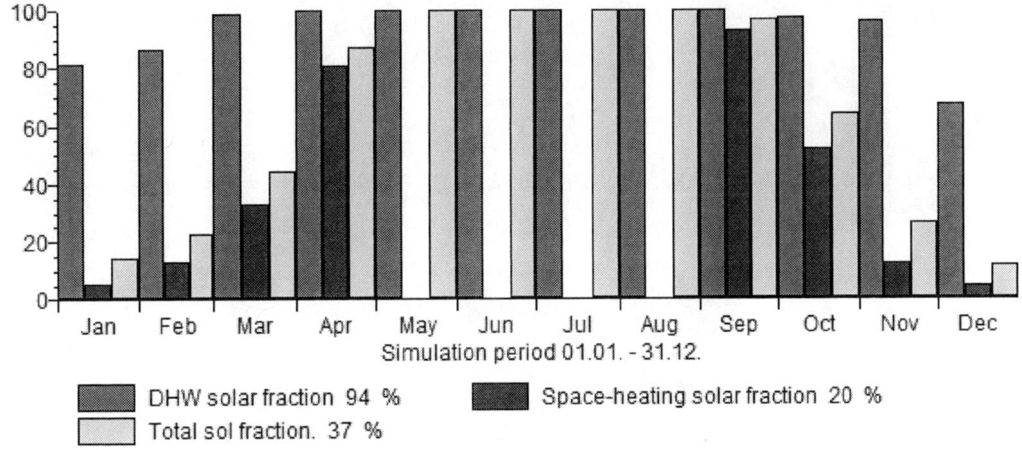

Figure 2. Share of solar energy for heating and DHW- variant 1.

Installed collector power:		28,42 kW
Installed solar surface area (gross):		40,6 m²
Irradiation on collector surface (active):	45 184,34 kWh	1 269,22 kWh/m²
Energy delivered by collectors:	9 219,53 kWh	258,98 kWh/m²
Energy delivered by collector loop:	7 480,93 kWh	210,14 kWh/m²
DHW heating energy supply:		2 040,65 kWh
Space-heating energy supply:		9 040,29 kWh
Solar energy contribution to DHW:		2 602,52 kWh
Solar contribution to heating:		1 801,61 kWh
Energy from auxiliary heating:		7 417,4 kWh
Wood chips - dry savings:		**1 343,2 kg**
DHW solar fraction:		**93,6 %**
Heating solar fraction:		**19,9 %**
Total solar fraction:		**37,3 %**
Relative savings of supplementary energy (DIN EN 12977):		**35,4 %**

Figure 3. Efficiency of solar energy per year – variant 1.

Figure 3 shows that the installed power of the collectors is 28.42 kW, the amount of energy delivered for DHW is 2040.65 kWh, the amount of energy for floor heating is 9040.29 kWh, the amount of solar energy input for DHW is 2602.52 kWh, the amount of solar energy for heating 1801.61 kWh and the amount of energy from an additional source - wooden boiler- 7417.4 kWh. The relative saving for additional energy is 35.4%.

2. Variant without stratification

The second variant for the experimental house is designed with the same number and area of solar panels with the same additional heat source, i.e., wood boiler with a rated output of 15 kW with a different 200 litres pre-heated solar tank and a heat loss of 2.14 W / K. External conditions were designed for the climatic region of Kosice with a calculated outside temperature of -14 ° C.

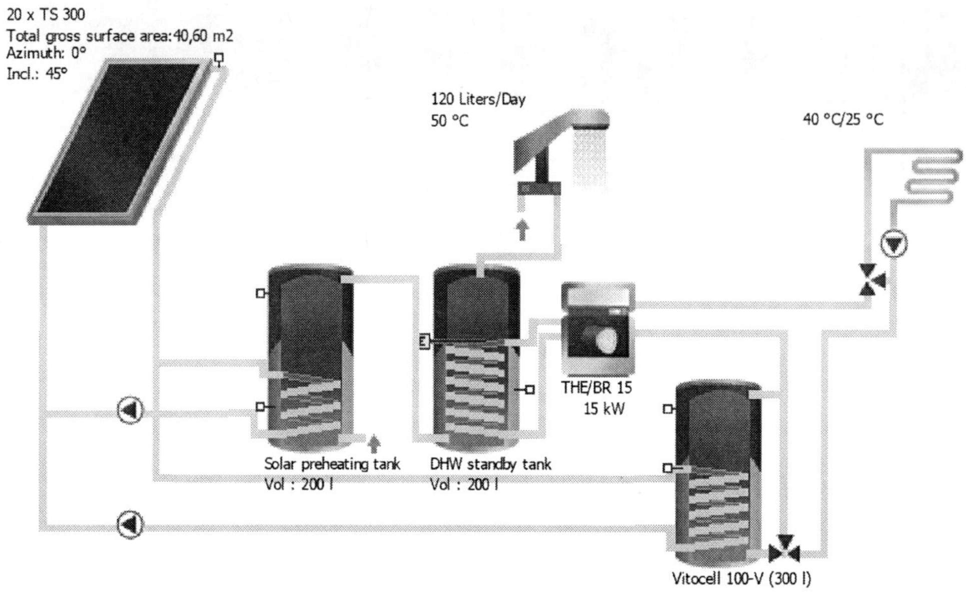

Figure 4. DWH system with heating buffer tank – variant 2.

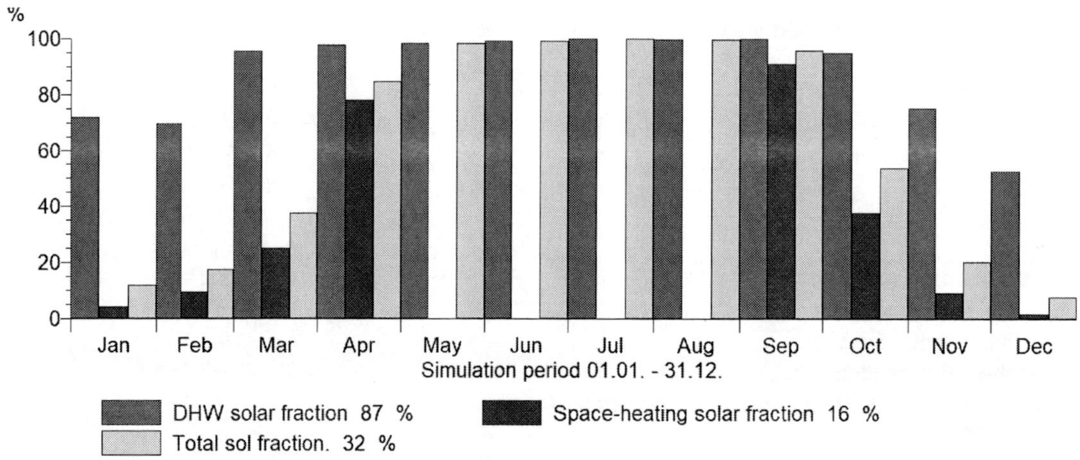

Figure 5. Share of solar energy for heating and DHW- variant 2.

Figure 5 shows that the share of the solar panel for DHW is 87% - the red column, the share of the solar panel for heating - 16% - the blue column and the total solar energy is 32% - the green column. The simulation for heating was year-round except for the months from May to August.

Figure 6 shows that the installed capacity of the collectors is the same as the first case of 28.42 kW, the amount of energy supplied for DHW is the same as the first variant - 2040.65 kWh, the amount of energy for floor heating has been reduced to 9039.78 kWh, the amount of solar energy for DHW has also been reduced to 2402.47 kWh, solar energy for heating also decreased to 1430.27 kWh, and the amount of energy from an additional source - the wood boiler has increased to - 7966.9 kWh. The relative saving for additional energy is 30.6%

Results of annual simulation

Installed collector power:		28,42 kW
Installed solar surface area (gross):		40,6 m²
Irradiation on collector surface (active):	45 184,34 kWh	1 269,22 kWh/m²
Energy delivered by collectors:	6 991,86 kWh	196,40 kWh/m²
Energy delivered by collector loop:	5 112,60 kWh	143,61 kWh/m²
DHW heating energy supply:		2 040,65 kWh
Space-heating energy supply:		9 039,78 kWh
Solar energy contribution to DHW:		2 402,47 kWh
Solar contribution to heating:		1 430,27 kWh
Energy from auxiliary heating:		7 966,9 kWh
Wood chips - dry savings:		**1 171,4 kg**
DHW solar fraction:		**87,0 %**
Heating solar fraction:		**15,8 %**
Total solar fraction:		**32,5 %**
Relative savings of supplementary energy (DIN EN 12977):		**30,6 %**

Figure 6. Efficiency of solar energy per year – variant 2.

3. Conclusion

The main difference between the classical accumulation preheating tank and stratification preheating tank is: The hot water in the classic tank is greatly mixed and the outlet water temperature does not reach the required collection point parameters, whereas the stratification tank has an internally directed thermal stratification of hot water by stratification elements (automatic check valves, plastic cone or plate control elements), which allows to extract hot water from a different temperature tank as required [2]. The comparison of the two variants suggests, that the stratification tank in combination with the solar panel is more efficient in terms of the amount of heating energy, the amount of solar energy for DHW and the reduced amount of energy from additional sources[2].

4. References

[1] Dr. Valentin Gerhard ,T*SOL basic version 5.0 Design and Simulation of thermal solar systems User Manual (2012)

[2] https://www.asb.sk/stavebnictvo/technicke-zariadenia-budov/vykurovanie/akumulacne-zasobniky-teplej-vody-na-principe-stratifikacie

Acknowledgements

This study was financially supported by the project VEGA No. 1/0307/16. Výskum kvality vnútorného prostredia budov pre školstvo a vzdelávanie zameraný na ochranu zdravia detí a mládeže and project ITMS Centrum výskumu účinnosti integrácie kombinovaných systémov obnoviteľných zdrojov energií - VUKONZE

IOP Conf. Series: Materials Science and Engineering **566** (2019) 012018 doi:10.1088/1757-899X/566/1/012018

Assessment of the possibility of using modern diagnostic methods to determine the changes in the properties of timber elements installed in a river bed

P. Kotásková[1], J. Fialová[1], J. Tippner[2] and J Marková[1]

[1] Mendel University in Brno, Faculty of Forestry and Wood Technology, Department of Landscape Management, Czech Republic

[2] Mendel University in Brno, Faculty of Forestry and Wood Technology, Department of Wood Science, Czech Republic

Abstract. The aim of the research was to find out which non-destructive methods can be used for determining the properties of wood placed in humid area and water. The research was carried out in order to propose suitable methods for detecting state of timber during regular inspection and maintenance of timber structures and elements in the watercourse. Wooden thresholds located in the watercourse belonging to the Divoká Orlice River Basin were selected for the measurement. The measured timber elements were already 9 years in the flow and the degree of damage to the built-in wood had to be assessed. The extent of biotic damage to water-soaked wood cannot only be determined by sensory methods. Resistance stop devices, a non-destructive ultrasonic device, a micro drill and an acoustic tomograph were selected for the evaluation. However, all commonly used modern diagnostic methods used for measurements in dry environment surveys cannot be used in the above terrain.

1. Introduction

It is recommended to use two natural materials - stone and wood - for buildings in the landscape environment. Wood from forests is the ideal building material that one can use in the natural environment. A significant advantage of wood is its impact on the environment. Wood is an environmentally friendly material with minimal energy costs for mining, processing and waste, a renewable resource. Compared to its weight, it exhibits high strength and rigidity. Small watercourses, as stated by Zuna [1], make it necessary to maintain acceptable living conditions even in low flows, which is particularly important for mountain streams and streams whose flow rates are very volatile. Thus, wood can be used for transverse gravity objects (thresholds, steps, dams) to reduce the inclination of the bottom level and to stabilize the channel profile. Wooden objects on watercourses are built where there is enough wood and it would be difficult to transport other building materials and also where different materials would be inappropriate in terms of integration into the landscape. The disadvantage is that the durability of the wood is reduced if the wood is not in a dry environment.

To find out what changes occur in wood that is installed in a damp environment or directly in the water, a watercourse was selected where wooden strips and sills were installed as part of the modifications. The reason for conducting research on wood elements was the detection of the occurrence of biological attack after 9 years. In addition, a fresh sample placed in the laboratory water was measured to compare the values.

Observed cross-threshold objects from debarked round timber are formed by the threshold body itself, 4 or 5 pieces of debarked round timber diameter 190 to 250 mm, lengths 5 and 6 m in the base part, damping round timber from 1 piece of debarked round timber of 100 to 140 mm, 6.0 m long, which

IOP Conf. Series: Materials Science and Engineering **566** (2019) 012018 doi:10.1088/1757-899X/566/1/012018

is mounted on the air side of the threshold between two upper logs 0.10 m lower than the upper edge of the threshold. A flow section with a width of 1.6 m in the bottom is created by means of the side wings, which consists of 1 piece of debarked round 200 mm diameter 2.3 m in length with a minimum binding of 1.6 m. Netex geotextile is attached to the upstream side of the wooden sill section.

It is obvious that the individual threshold elements are alternately in the wet or in the water all year round, some are only exposed to increased humidity. The ends of the elements are bound to the shore where humidity is increased. The largest differences in humidity are exposed to the side wings of the building, which may dry out in dry periods of the year.

Water absorption of wood is the ability of wood to absorb water into which it is immersed. The maximum amount of water the wood can absorb is called water capacity. The wood will reach full water capacity, ie it will not stop absorbing water until after a very long time. [2]

Water absorption is the ability of wood to absorb water in the form of a liquid due to the porous structure. Wood is fully soaked in water if it has the maximum amount of bound water (per hygroscopicity) and the greatest amount of water available. As a rule, the higher the density of wood, the lower its ability to absorb water. [3]

It follows that water absorption depends mainly on the pore volume of the wood. [2] States that water absorption depends on other factors in addition to bulk density, such as the anatomical and chemical composition of wood.

Wood with high humidity is attacked by wood-destroying and wood-staining fungi, which only cause deterioration of mechanical properties to wood degradation. It is clear from the information provided by the flow managers that the timber element life is about 10-15 years. The need to reconstruct the watercourse is currently based on visual inspection only. The extent of the damage determined by modern diagnostic methods would be important in the possibility of postponing the reconstruction, thus saving savings. Or, on the contrary, the early detection of the need for repairs and thus the prevention of damage by the sudden destruction of the object.

2. Materials and Methods

The determination of the condition and degree of damage to the wooden elements is done by visual evaluation, but the instrumental methods, which are used, for example, in the assessment of historical trusses, can be used to more accurately assess the condition of the elements.

The research was carried out in-situ on spruce wooden transverse thresholds of the height of 0.2 m, made of debarked round timber of 6 m. side wings of the object. The measurement points are shown in Figure 1. Several thresholds have been selected to compare the status of the individual wood elements. Furthermore, measurements were made in a laboratory on a 500 mm spruce log sample with a diameter of approximately 200 mm. The sample was measured before immersion in water, then immersed in water and the saturation state was measured by weighing on a KERN CB 12 K1N laboratory balance. Further measurements were made at the time of maximum saturation. The sample was then dried, first freely, followed by an oven.

Pilodyn 6J, a non-destructive Arborsonic Decay Detector, a Resistograph and a Fakopp 2D tomograph were selected for evaluation.

Increased water content can affect the results of non-destructive testing, so the moisture of wood elements was determined before each measurement. Hygrotest 6500 resistance hygrometer with stop probe was used to measure the humidity, a resistance humidity meter with integrated GMH 3810 measuring tips was used to measure the sample in the laboratory (see Figures 2 and 3). The principle of the method lies in the influence of moisture in the wood on the electrical resistance. Resistance hygrometers utilize electrical one-way resistance and wood conductivity.

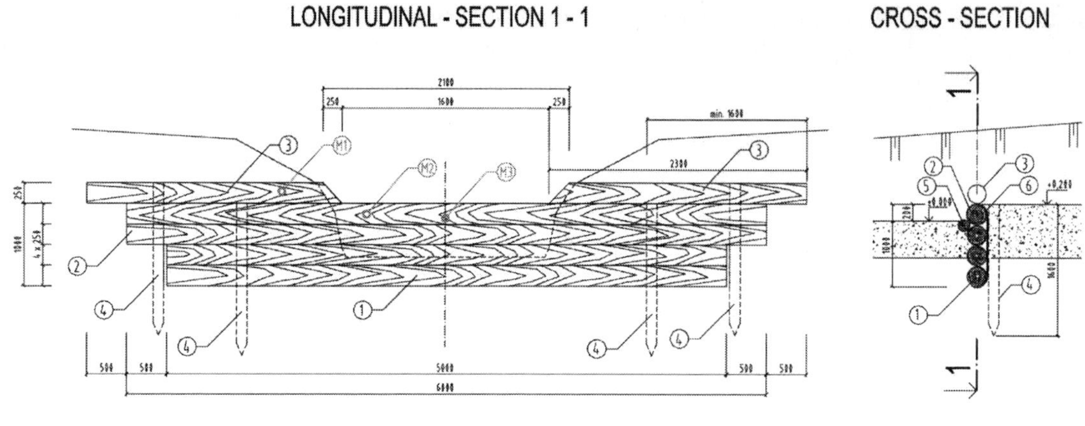

Figure 1. Measuring points on the wooden transverse threshold in the trough.
M1 - measurement on the side wing - wood in the wet, M2 - measurement on the overflow edge in the edge of the visible part - wood in the water, M3 - measurement on the overflow edge in the middle - wood in the water. 1 - foundation section of the building - debarked round Ø 250 mm, length 5 m; 2 - overflow section of the building - debarked logs Ø 250 mm, length 6 m; 3 - side wings of the building - debarked round Ø 250 mm, length 2.3 m min. commitment to shore 1.6 m; 4 - vertical stabilizing piles - debarked rod Ø 100 to 140 mm, length 1.6 m; 5 - damping logs - debarked logs Ø 150 mm, length 6 m; 6 - Netex geotextile on the upstream side of the wooden threshold

Figures 2 and 3. Hygrotest 6500 Resistance Hygrometer and Resistive Moisture Meter with GMH 3810 Integrated Probes.

Determination of moisture

Based on the determined wood sample weights in the wet, wet and absolutely dry laboratory after laboratory drying, moisture was determined and compared to the moisture measured. Subsequently, saturation and drying courses were evaluated by Excel.

Pilodyn

Measurement with the Pilodyn 6J Forest is a semi-destructive diagnostic method because the damage to the test material is very small and almost negligible. It is a simple mechanical device to measure the penetration depth of a 2.5 mm mandrel fired in wood at a constant driving force of 6 J (see Figures 4 and 5). The tip penetrates the surface of the material while the penetration depth is measured. The maximum penetration depth of the mandrel is limited to 40 mm by the device design. With Pilodyn, a low-invasive wood density estimate can be made. Density is a parameter closely related to the hardness of the wood, as Bonamini et al. [4] suggest; Kasal and Antony [5].

IOP Conf. Series: Materials Science and Engineering **566** (2019) 012018 doi:10.1088/1757-899X/566/1/012018

By assessing the depth of penetration of the tip, the degree of decomposition of the wood or its biotic damage can be estimated. In case of insect damage or decay caused by rotting, the tip will penetrate significantly deeper. However, this method provides us with a new way to measure the degree of damage to the surface affected by soft rot, which primarily affects the surface of the wood. The same is true for wood damage by insects, which causes the greatest damage in the surface layers.

A disadvantage is that during penetration of the mandrel only surface penetration into the material occurs and thus only the surface information of the state of the structural member is known. The measurements are mainly influenced by the moisture content. With higher moisture content Pilodyn easily penetrates into wood. Therefore, it is necessary to correct the moisture content for each measurement.

However, as Micko [6] states, the effect of moisture is only evident at a moisture content of 0-30% (fiber saturation point) above the fiber saturation point, the change in moisture content has no effect on the test result.

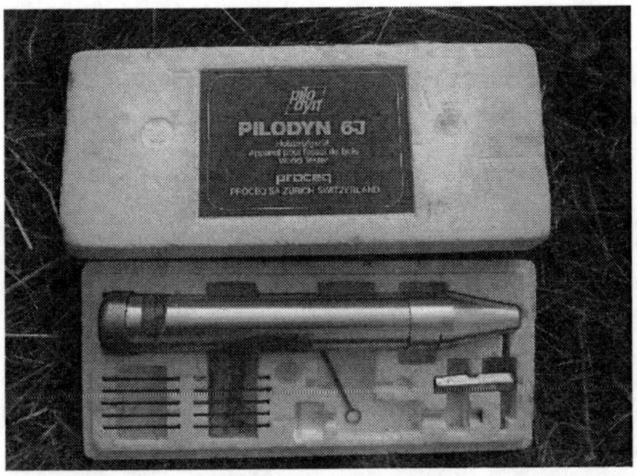

Figures 4 and 5. Pilodyn 6J Resistance Stroke Instrument - Measurement on Threshold Side Wing.

Measurement of ultrasonic wave velocity - Arborsonic Decay Detector

Arborsonic is a detector that uses ultrasound to detect and evaluate the internal state of the wood based on ultrasound rate (see Figures 6 and 7).

The instrument measures the speed of the 77 kHz ultrasonic signal passing through the measured material between the two sensors (transmitter and receiver). Since the signal propagates the fastest in the solid wood mass, the signal rate slows down in each decay or defect area. In damaged areas by rotting, the sound speed will be significantly lower than in intact wood. [7]

Figures 6 and 7. Non-destructive Arborsonic Decay Detector - Measurement on the sill side.

According to scientific publications [8], dynamic modulus of elasticity can be calculated using a relationship:

IOP Conf. Series: Materials Science and Engineering **566** (2019) 012018 doi:10.1088/1757-899X/566/1/012018

$$E = \varsigma \cdot c^2$$

where:
E is the modulus of elasticity of the material [MPa]
ϱ is the density of the material [kg.m^{-3}]
c is the speed of sound propagation through the wood in a given direction [m.s^{-1}].

The sound propagation rate decreases with increasing wood moisture, as the water fills its capillaries in which air was found. As a result, the resistance of the environment to the sound wave propagation increases [9].

Resistograph 4453 Resistance Micrometer

Resistance drilling is based on measuring the resistance to slow penetration of the small diameter drill by the material being investigated. The advantage is that the device offers an overview of the internal state of the element. Due to the minimal material damage, the semi-destructive methods of determining the condition of timber structures are among the methods. [10]

Thin drill senses resistance when penetrating wood. This implies that less drilling resistance, which requires less engine torque, is due to less wood density, cavities, damage, splintering, and cracks. Resistograph output is a graphical record (density profile, dendrogram).

Figure 8 and 9. Resistograph 4453 Resistance Micrometer - Measurement on Overflow Edge in the Middle (Wood in Water).

The drilling record corresponds to the energy consumed to maintain a constant drilling speed on the y-axis and the x-axis shows the distance from the surface of the measured element. The peaks in the graphical record correspond to higher energy, ie higher resistance and higher density, while lower points are associated with lower energy, ie lower resistance and density. [11]

The Fakopp 2D acoustic tomograph

The Fakopp is capable of reconstructing the internal state of the strain by measuring the speed of the sound pulse caused by hammer blows into 2 interconnected probes (see Figures 10).

Two sensors are connected to each other and are connected to the meter (see Figures 11). An acoustic signal is emitted by tapping the hammer on the probe and the instrument senses the time it takes to pass the signal from the currently activated probe to the opposite probe. The generated pulse spreads through the material for some time and reaches the second probe - the receiver. The ultrasonic pulse transmission time is displayed on the device display in microseconds. Based on the passage times and the geometry of the cross-section, the velocity is calculated and the state of the strain is reconstructed.

Figures 10 and 11. Acoustic tomograph Fakopp 2D.

The sound velocity in wood is proportional to the density and stiffness of the wood being measured. As density increases, it increases with increasing modulus of elasticity. Both properties are a sensitive indicator of the condition of the wood, disturbance by wood-destroying fungi causes a change in the rate of propagation and can be detected. If there is a cavity inside the trunk, the sound wave must bypass the longer path and decrease the signal speed again and can be detected. [12]

3. Results

The measurements in the laboratory show a large water intake 14 days after immersion of the dried wood in the water. For the next 3 months, the weight gain was slow. Upon withdrawal from the water, the weight loss graph shows a greater weight loss for 7 days, then a gradual drying of the sample until the sample is put into the oven for complete drying. After removal from the dryer, there was a slight increase in moisture due to air humidity.

The wood moisture on the sample in the laboratory determined by calculation was compared with the measured values of the hygrometer. It was found that at higher humidity values above 50% the measured values did not correspond to the values calculated. Due to the high humidity values of wet wood in the stream (measured 90 to 137%), it can be stated that the measured values of wet wood in the field measured on the sash side wings are high, but they do not fully correspond to reality. The hygrometers used can be used more accurately for a range of 6 - 30%, as noted by the manufacturer, with considerable variations outside this range.

Since it was assumed that the timber elements in the stream would be soaked with water, only the listed instruments were selected for the measurements. This assumption was confirmed. The measured values were above 90%, although, as verified, this figure cannot be considered relevant.

It was not possible to use Fakopp for water and water-washed elements from the selected devices. It was used only on the side wings of the object. The calculated velocity of elastic wave propagation, based on the measured values and the distance between the measured points, is some 640 m / s for some elements, but up to 1112 m / s for some elements. Sound waves propagate through the shortest path, but only through intact cell walls of wood. If there are wood-destroying fungi in the wood, for example, the sound waves will bypass these places, which can then be determined by comparing the measured propagation speeds with the reference speeds. A certain disadvantage in the interpretation of the results is that other factors, such as changes in humidity, may cause modification of the sound wave propagation rate in wood. [13]

The Arborsonic was also used only on the side wings of the object, which are not in water, but the humidity of these parts is high. When measuring some elements, the instrument showed an error of 999 - no measurement can be made. Cross-fiber measurements were performed. The ultrasonic wave propagation speed of some elements was about 1080 m / s. According to Kloiber [11], it would be a moderately damaged element, as it ranges from 920 to 1260 m / s. However, this estimation of the degree of damage applies to wood moisture of 12 - 16%. Since the sound propagation rate in wood depends on the moisture content of the wood, this method cannot be relied upon for such soaked samples. With increasing humidity, the speed of sound wave propagation decreases. With a speed value of, for example, 757 m / s, it is not clear how heavily the element is damaged and how much wood moisture plays.

From the graphical records (see figures 14 and 15) it was possible to observe on the first view higher resistances for wood in water than for wood only in wet. Some measurements have seen a decrease in recorded resistances, which is probably due to the effect of rot. The best overview of internal damage is obtained when drilling at multiple locations of one element. However, the Resistograph measurement is very demanding in the watercourse. Although wood elements are known to have internal damage without surface changes, this method cannot be considered suitable for assessing elements in the presented environment.

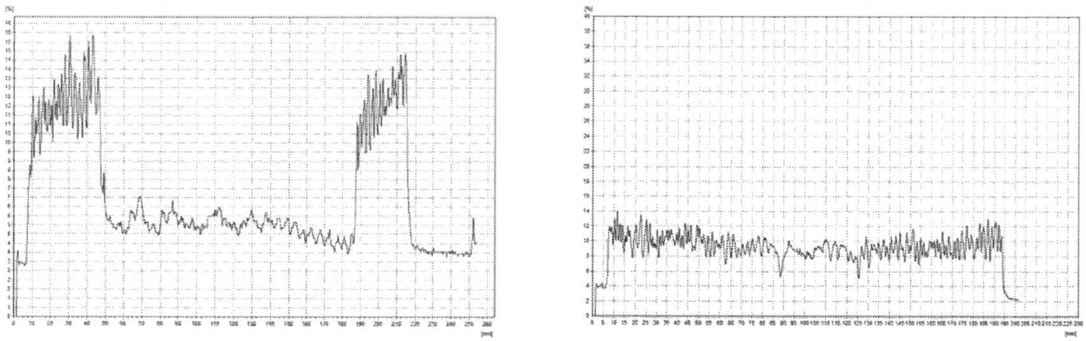

Figures 14 and 15. Resistograph output - dendrogram of rot affected wood and Resistograph output - dendrogram of healthy wood - measurement in laboratory.

Pilodynem measurement is simple, but we only determine the surface densities of the wood and thus only the surface damage to the wood. From the visual survey it was clear that some elements of the transverse objects in the flow show signs of surface damage, mostly wet elements in contact with the earth, affected by the ground moisture. Due to the high humidity, the increased penetration depth of these elements is also evident. It can be seen from the measured values that the elements in water show lower values (reaching a lower depth) comparable to the element measured in the laboratory.

4. Discussion and conclusion

Within the framework of the research of timber structures on watercourses, it was ascertained whether it is possible to supplement the visual check of the condition of wooden elements with a more exact method, which would specify their evaluation. The survey includes the detection of moisture that affects measurements on all instruments. However, it was found that at higher humidity values above 50% the measured values no longer corresponded to the values calculated.

Pilodyn 6J, a non-destructive Arborsonic Decay Detector, a Resistograph and a Fakopp 2D tomograph were selected for evaluation. From the measurements made, it can be stated that the elements that are in the water are less damaged than the side wings, which are in a humid environment. However, none of the selected methods can be described as reliable. When measuring not only in water, the Arborsonic ultrasonic machine often has error messages. Resistograph measurement is very challenging in watercourse conditions, even if the outputs reveal the best internal damage, eg by fungi. Facopp was used only on the side wings and the measured values were compared with the sample in the laboratory. The effect of increased humidity is evident and therefore the results cannot be considered relevant. The only suitable device seems to be Pilodyn, which only reveals sub-surface damage.

5. References

[1] Hanák K, Tlapák V, Šálek J, Kupčák V, Skoupil J, Zuna J 2008. *Stavby pro plnění funkcí lesa* ČKAIT Praha ISBN 978-80-87093-76-4

[2] Vanin S I 1955 *Nauka o dřevě* Praha SNTL 428

[3] Matovič A 1993 *Fyzikální a mechanické vlastnosti dřeva a materiálů na bázi dřeva* VŠZ Brno ISBN 80-7157-086-9

[4] Bonamini G 1995 Restoring timber structures – inspection and evaluation STEP – EUROFORTECH vol 2

[5] Kasal B, Anthony R 2004 Advances in in situ evaluation of timber structures *Progress in Structural Engineering and Materials* vol 6 (London: John Villey & Sons) pp 94-103

[6] Micko M et al. 1982 Determinaton of wood specific gravity in standing white spruce using a Pilodyn tester Forestry Chronicle 58[4]

[7] Špaček T 2007 Metodika stavebního technického průzkumu dřevěných prvků zabudovaných do památkových staveb Disertační práce MZLU v Brně

[8] Hoyle R J, Pellerin R F 1978 Stress wave inspection of wood structures, fourth international non-destructive testing of wood *Symposium p. 33. Pullman* (Washington, Washington state university)

[9] Požgaj A, Chovanec D, Kurjatko S, Babiak M 1993 Štruktúra a vlastnosti dreva 1. vyd. (Bratislava Príroda) 485 ISBN 80-07-00600-1

[10] Kloiber M, Kotlínová M 2008 Nondestructive defectoscopic devices used for build-technical surveys of historical wood construction *In: Stavební ročenka* (Praha JAGA) pp 36-41

[11] Kloiber M and Drdácký M 2015 Diagnostika dřevěných konstrukcí (Praha: ČKAIT. Technická knižnice) ISBN 978-80-87438-64-0

[12] Přístrojové měření kmenů [online]: http://www.vyspra.cz/rizikove-kaceni/pristrojove-mereni-kmenu

[13] Kolařík J 2006 Přehled přístrojových metod pro hodnocení stromů In Salaš P, and Litschmann T *Trendy ve veřejné zeleni sborník přednášek* (Lednice: Ústav šlechtění a množení zahradnických rostlin) ISBN 80-239-9051-9

IOP Conf. Series: Materials Science and Engineering **566** (2019) 012019 doi:10.1088/1757-899X/566/1/012019

Study of zinc removal from water solutions using hornbeam wooden sawdust

Z Kovacova[1]

[1] Technical University of Kosice, Faculty of Civil Engineering, Institute of Environmental Engineering, Vysokoskolska 4, 042 00 Kosice, Slovakia

Abstract. The polluted water from an industry, in particular with heavy metals, aromatic molecules and dyes, is a serious environmental problem because their potential environment toxicity. Therefore, there is a need to develop methods that can remove toxic contaminates from wastewaters. The sorption process is one of the most popular methods for the environment treatment. Adsorption on low-cost materials is studying with great interest by researchers in last years. Hornbeam sawdust is a natural resource and an inexpensive material which can be used for water treatment. In this study, the hornbeam sawdust was used as for the removal of zinc from an aqueous solution. Fourier transform infrared spectroscopy (FTIR) was used to determine functional groups of sawdust. The pH decreased from 6.3 to 5.8 during 2 h and in case of 24-hours experiment on the value of 5.4. Hornbeam sawdust had efficiency of zinc removal from aquatic model solutions height than 75 % which means that there is significant potential for sawdust as a sorbent material for heavy metals removal from water.

1. Introduction

The problems of polluted wastewater are rising during the past decades because of urbanization, industrialization, and agricultural practices. The major source of the aquatic pollution is untreated wastewaters from the industries. This water has due to the defend kind of industries various composition and the combination of pollutant. The most common contaminants are pesticides, dyes, aromatic hydrocarbons, oils, heavy metals etc. and represent a risk in most countries [1-4].

In recent years, heavy metals in wastewaters present a major problem in the environment, because of its flexibility, accumulation, non-biodegradable and endurance. They are very dangerous to the living organisms and human's health even at very low concentration [5, 6].

The zinc is applied in industrial activities like steel processing, mining and coal combustion. Though it is required by the humans in trace level zinc dose over 100 mg/day may lead to vomiting, skin inflammation, fever, anemia dysfunctions of respiratory track and brain. The main resources of zinc in water are electroplating industries, paper and pulp industries, steel making industries, and brass metal works [7, 8].

Despite the existence of many water treatment technologies as: chemical precipitation, ion exchange, different membrane techniques, coagulation, flocculation, ion exchange, flotation, chemical precipitation, electrochemical treatment, researches are still searching the unique methods of getting more economically and environmentally sustainable technology [9].

Over the last decade, bio-based wastewater treatment materials have received increased attention due to the global effort to increase environmental and cost effective purification processes. The biosorption is a cheap and fast separation method to heavy metals removal. The main sorption mechanism is ion exchange through the adhesion of positively charged metal ions on an adsorbent surface which consist of negatively charged polymeric functional groups.

IOP Conf. Series: Materials Science and Engineering **566** (2019) 012019 doi:10.1088/1757-899X/566/1/012019

The utilization of low-cost sorbents has been found as a perspective alternative to methods conventionally applied for metal removal. As biosorbents can be used lot of materials of natural origin: algae, microorganisms, coffee and tea grounds, yeasts, nut shells, citrus peels, rice husks, or sawdust [10, 8].

This paper presents the using of hornbeam wooden sawdust in removing zinc ions from aquatic solutions. Wooden sawdust is widely available and has a good mechanical stability and due to lignocellosic composition. Sorbent from hornbeam sawdust was analyzed by infrared spectrometry and efficiency of Zn(II) removal was measured by colorimetric method. The changes of pH were also determined.

2. Material and methods

2.1. Materials

The hornbeam wooden sawdust from local resources was sieved and in experiment only fraction less than 2 mm was used. The sawdust was analysed by infrared spectrometry by Bruker Alpha Platinum-ATR spectrometer (Bruker Optics, Ettingen, Germany). In the range of 4,000–400 cm^{-1} there were performed a total of 24 scans of hornbeam sawdust.

Aquatic solutions were prepared form appropriate amount of $ZnSO_4.7H_2O$ dissolved in distilled water. The concentration of Zn(II) in solution was 10 mg/L.

2.2. Experimental methods

To investigate the kinetic study 1 g of sawdust and 100 mL of model solutions was used. The sorption experiments last 5, 10, 15, 30, 45, 60 and 120 min, respectively. During this time the sorbent was intensively stirred in adsorbate.

In another experiment the same amount of sawdust and aquatic solutions was left during 24 h in laboratory temperature.

After established time was sample filtered and in filtrate was measured by pH meter (Mettler Toledo FG2, Schwerzenbach, Switzerland) and concentration of zinc by colorimetric method (Colorimeter DR890, Hach Lange, Germany).

The efficiency of zinc removal was calculated using the following equation:

$$\eta = \frac{(c_0 - c_e)}{c_0} \times 100\,, \tag{1}$$

where η is the efficiency (%), c_0 is the initial concentration of appropriate ions (mg/L) and c_e equilibrium concentration of ions (mg/L).

3. Result and discussion

3.1. Infrared spectra

The FTIR spectra was measured due to determine the potential functional groups that involve in biosorption process of zinc onto sawdust (Figure 1).

Sawdust of different trees is a by-product of the timber industry, mainly composed of cellulose (45-50%) and lignin (23-30%). In their structure it contains functional groups such a non-negligible portion of lipids and waxes, bearing functional groups like alcohol, ketone and carboxylic groups which provide its metal binding capacity. These groups can be involved in complexation reactions with zinc cation and can be viewed as natural ion-exchange materials on the surface of sorbent [11, 12].

The strong broad band between 4,000 and 3,300 cm^{-1} can be assigned to hydroxyl groups and O–H bands stretching. In the band between 3,000-2,800 cm^{-1} indicate symmetric or asymmetric C–H stretching of methyl and methylene groups. It was observed in peak 1,731 cm^{-1} presence of stretching vibration of C=O bonds due to groups of hemicellulose. Lignin bands were indicated at 1,503, 1,454, and 1,321 cm^{-1}. Cellulose can be seen in the band between 1,420 and 890 cm^{-1}. Aromatics, carboxylic acids, alkyl halides were determined at 830 cm^{-1} [13].

IOP Conf. Series: Materials Science and Engineering **566** (2019) 012019 doi:10.1088/1757-899X/566/1/012019

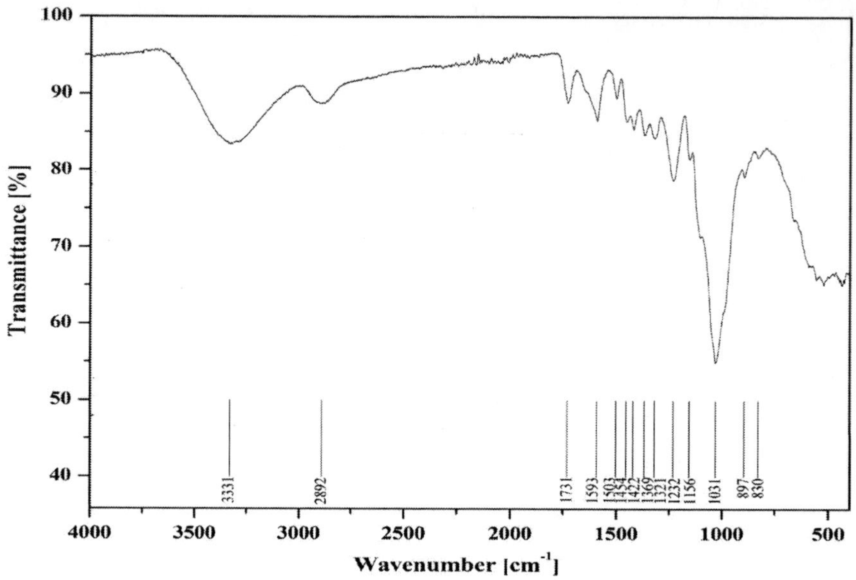

Figure 1. Infrared spectra of hornbeam wooden sawdust

3.2. Sorption experiments

The effect of contact time on the sorption of zinc at initial concentrations 10 mg/L by sawdust is shown in Figure 2. Bouziane et al. [14] determined that the most important factors to zinc sorption onto sawdust are initial concentration, sawdust mass and contact time.

Hornbeam shows high efficiency to zinc removal. In all experiment was more than 75% removal. The high efficiency was shown after 10 min.

The pH value is one of the most important factors influencing the solution chemistry of the heavy metals: hydrolysis, complexation by organic and/or inorganic ligands, redox reactions, precipitation, and in the adsorption processes the speciation and availability of heavy metals.

The effect of pH on the sorption of zinc on hornbeam sawdust can be determined with respect to the surface charge on the adsorbent material. Lower pH results in lower removal efficiency, due to high positive charge density on the surface sites adsorbent and electrostatic repulsion between metal ion and H^+ ion will be high. On the other hand increasing pH means an enhancement of metal sorption because the electrostatic repulsion decreases and that show reduction of positive charge density of proton in the adsorbent´s sites [15].

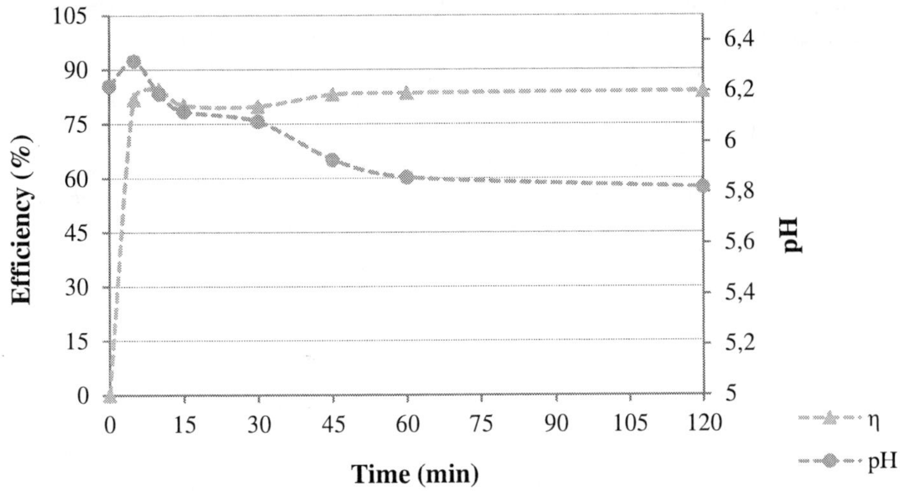

Figure 2. Dependence time from efficiency of Zn(II)

In 24-hours experiment was removal efficiency 90% at pH 5.4. Pragaty [16] published that 90% of zinc ion vas removed after 120 min of contact time at pH 5. As was mentioned before, pH and contact time are very important parameters which can affect sorption process.

4. Conclusion

The sawdust obtained from the local sources is an inexpensive material which can be used as effective adsorbent for removal of zinc from aqueous solution.

Fourier transform infrared spectroscopy (FTIR) of sawdust indicated the functional groups, like; C–O, C=O and C–N are responsible for metal binding and suggested the main mechanisms involved in the removal of heavy metals ions might be the ionic exchange and complexation. Efficiency of zinc removal was height than 75% during experiment. The highest efficiency (85 %) was after 10 min at pH 6.2. In 24-hour experiment was equilibrium concentration of ions 0.96 mg/L at pH 5.4. Changes of pH are due to adsorption and ion exchange. The hornbeam wooden sawdust could be used as an alternative economically and environmentally friendly sorbent for removal of Zn (II) from aqueous solutions.

5. References

[1] Renuka N, Sood A, Ratha S K, Prasanna R a Ahluwalia A 2013 Evaluation of microalgal consortia for treatment of primary treated sewage effluent and biomass production *J. Appl. Phycol.* **25** pp 1529-1537
[2] Balintova M, Demcak S, Pagacova B 2016 A study of sorption heavy metals by natural organic sorbent *Int. J. Energy Environ.* **10** pp 189-194
[3] Carstea E M, Bridgeman J, Baker A a Reynolds D M 2016 Fluorescence spectroscopy for wastewater monitoring: a review *Water Res.* **95** pp 205-219
[4] Xu, P el al 2012 Use of iron oxide nanomaterials in wastewater treatment: a review *Sci. Total Environ.* **424** pp 1-10
[5] Peligro F R, Pavlovic I, Rojas R and Barrig C 2016 Removal of heavy metals from simulated wastewater by in situ formation of layered double hydroxides *Chem. Eng. J.* **306** pp 1035-1040
[6] Demcak S, Balintova M, Hurakova M, Frontasyeva M V, Zinicovscaia I, Yushin N 2017 Utilization of poplar wood sawdust for heavy metals removal from model solutions *Nova Biotechnol Chim* **16** pp 26-31
[7] Petcu C, Purcar V, Radu A L, Ianchis R, Elvira A, Sarbu Ion-Ebrasu D, Miron A R, Modrogan C, Ciobotaru A I 2015 Removal of zinc ions from model wastewater system using bicopolymer membranes with fumed silica *J. Water Process Eng.* **8** pp 1-10

[8] Jakobik-Kolon A, Bok-Badura J, Karoń K, Mitko K a Milewski A 2017 Hybrid pectin-based biosorbents for zinc ions removal *Carbohyd. Poly.* **169** pp 213-219

[9] Božić D, Gorgievski M, Stankovič V, Štrbac N, Šerbula S a Petrović N 2013 Adsorption of heavy metal ions by beech sawdust–Kinetics, mechanism and equilibrium of the process *Ecol. Eng.* **58** pp 202-206

[10] Gogoi H, Leiviskä T, Heiderscheidt E, Postila H a Tanskanen J 2018 Removal of metals from industrial wastewater and urban runoff by mineral and bio-based sorbents *J. Environ. Spravovať.* **209** pp 316-327

[11] Šćiban M, Radetić B, Kevrešan Ž and Klašnja M 2007 Adsorption of heavy metals from electroplating wastewater by wood sawdust *Bioresour. Technol.* **98** pp 402-409

[12] Crini G 2005 Recent developments in polysaccharide-based materials used as adsorbents in wastewater treatment *Prog. Polym. Sci.* **30** pp 38-70

[13] Wahab M A, Jellali S and Jedidi N 2010 Ammonium biosorption onto sawdust: FTIR analysis, kinetics and adsorption isotherms modeling *Bioresource Technol.* **101** pp 5070-5075

[14] Bouziane L, Bendebane F, Ismail F and Delimi R 2012 Removal of zinc and cadmium from an aqueous solution using sawdust as a low-cost adsorbent: application of Plackett–Burman design *Desalin. Water Treat.* **49** pp 189-199

[15] Sulaiman O, Ghani N S, Rafatullah M, Hashim R and Ahmad A 2011 Sorption equilibrium and thermodynamic studies of zinc (II) ions from aqueous solutions by Bamboo Sawdust *J. Disper. Sci. Technol.* **32** pp 583-590

[16] Pragati P, Trivedia S S and Pandey L 2015 Removal of zinc from synthetic waste water by saw dust as an adsorbent *IJIRSET* **2** pp 307-354

Acknowledgement

This work has been supported by the Slovak Grant Agency for Science (Grant No. 1/0419/19)

Slab deformations caused by shrinkage – experiment vs. numerical calculation

D Kušnírová[1], **S Priganc**[1],

[1] Technical university of Kosice, Faculty of Civil Engineering, Institute of Civil Engineering, Slovakia

Abstract. This article gives an overview of the experiment, which aims to detect the effect of the shrinkage on the stress and the strain of the slabs reinforced at one surface by different types of reinforcement (steel and GFRP reinforcement). Some partial results from the strain measurements on steel reinforced slabs and their comparison with numerical calculation are also reported in the article. From publications found larger bending tensile creep towards compression creep can explain the difference between calculation and experiment.

1. Introduction

At the Faculty of Civil Engineering Technical University of Kosice, we investigate the effect of shrinkage on the strain and stress in reinforced concrete. Due to the implementation of new reinforcement types into practice, our goal is to determine the impact of these new reinforcement types, namely GFRP (Glass Fibre Reinforced Polymers) reinforcement. The main reason of different behaviour of GFRP reinforced elements is a significant difference in the modulus of elasticity in comparison to the conventional steel reinforcement [1, 2]. The results of the comparison of the proposed calculation model based on the theory of mechanics with the experiment for slabs reinforced by steel reinforcement are given in this article.

2. Description of experiment

2.1. Test samples

The slabs of dimensions 1800x600x120mm reinforced only at one side with different reinforcement ratio were used as test samples. The slabs were reinforced with a conventional steel reinforcement (3 specimens), GFRP reinforcement (5specimens) and a reference sample of plain concrete (1 specimen). In this article compared slabs were reinforced by different amount of steel (B500) bars: 4ϕ8mm, 7ϕ8mm and 7ϕ10mm. The concrete cover was 10mm. Material characteristics of concrete were investigated experimentally. In 28th day were measured values: *43MPa* for f_{cm}, *6,83MPa* for $f_{ctm,fl}$ and *30,7GPa* for modulus of elasticity.

2.2. Storage of samples

The slabs were demoulded on the seventh day and placed on the stands in a vertical position. This storage method has been chosen to maximally eliminate the influence of the self weight on the strains and deflection of the slabs (figure1). It is then possible in the other calculations to neglect generated stress and strain of the minimum values [3].

Figure 1. Storage of slabs.

3. Numerical calculation

The reinforcement as a rheologically stable material prevents free concrete shrinkage, resulting in tensile stresses in the surrounding concrete and the pressure force F_{cs} in the reinforcement [4].

$$F_{cs} = -A_s \varepsilon_{cs}(t, t_s) E_s \tag{1}$$

The stress created in concrete by this force:

$$\sigma_c = F_{cs}\left(\frac{1}{A_i} + \frac{e_s}{I_i} z\right) \tag{2}$$

Then the relative strain of the concrete is:

$$\varepsilon_c(t) = \sigma_c / E_{cm} + \varepsilon_{cs}(t, t_s) \tag{3}$$

The values of strains ε_{cs} measured on the surfaces of the unreinforced slab (reference sample) were used for calculation. Different values on both surfaces of the unreinforced plate are caused by non-homogeneity of concrete through the high of cross-section as a result of vibrating.

4. Comparison of experimental results with numerical calculation

In this manner calculated relative strains of concrete on both sides (A-side with reinforcement, B-without reinforcement) of slabs reinforced by steel reinforcement are compared with experimentally obtained values in figures 2-4 (ε meas =measured strain, ε cal = calculated strain, ε unr = strain of unreinforced slab).

On the graphs below, we can see that strain values calculated according to the theoretical assumptions of the mechanics do not correspond to the values measured on the samples during the experiment (difference 20-63%). These values are closing with the decreasing reinforcement ratio.

IOP Conf. Series: Materials Science and Engineering **566** (2019) 012020 doi:10.1088/1757-899X/566/1/012020

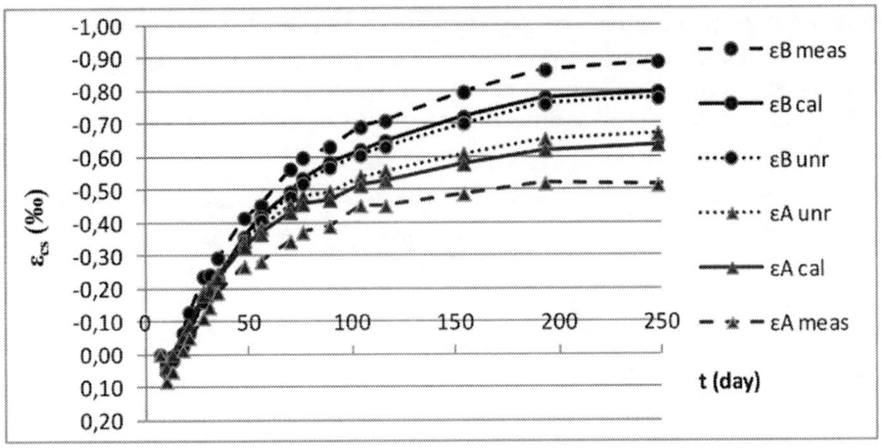

Figure 2. Comparison of calculation and experiment for RS1 slab (4ϕ8mm).

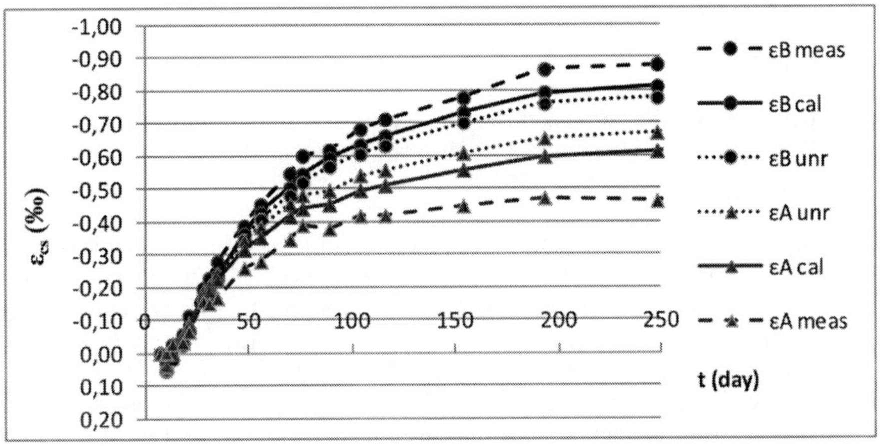

Figure 3. Comparison of calculation and experiment for RS2 slab (7ϕ8mm).

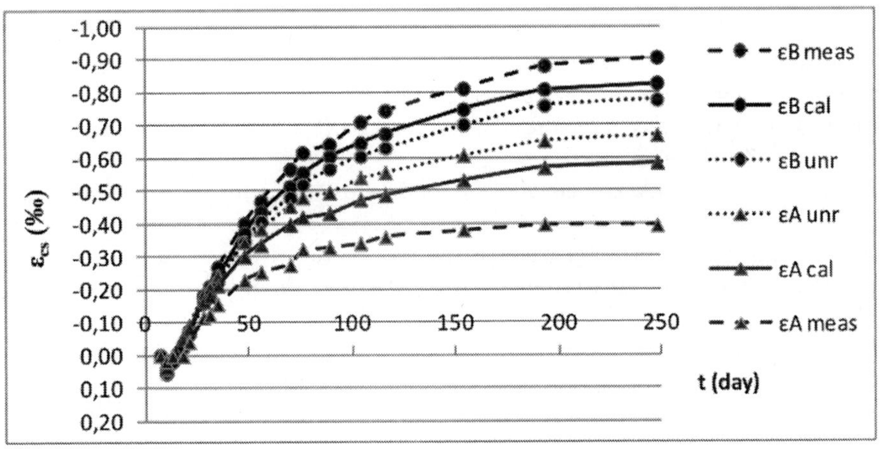

Figure 4. Comparison of calculation and experiment for RS3 slab (7ϕ10mm).

Such numerical model doesn´t correspond with measured values, so it was modified. The influence of creep was included:

$$\varepsilon_c(t) = \sigma_c / E_{c,eff} + \varepsilon_{cs}(t, t_s) \tag{4}$$

The resulting values changed only by 8-30%, which is not satisfactory enough. According to [5,6] the coefficient for the bending tensile creep of concrete is larger than the compression creep by a factor of 2-4 depending on load level and class of concrete. If we presuppose 2.5-times greater value of creep coefficient, we get comparable results (difference from measuring 1-4%) (figure 5).

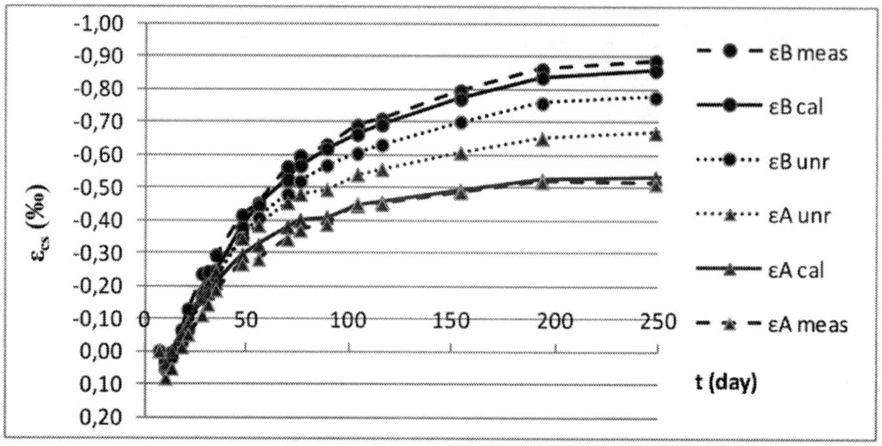

Figure 5. Comparison of calculation and experiment for RS1 slab (4ϕ8mm) using 2.5-time greater value of creep coefficient for calculation.

5. Conclusion

From the performed calculations and their comparison with the experimentally measured values, it is clear that the chosen numerical model based on the theory of mechanics without creep influence is not suitable for determining the shrinkage strains of one-side reinforced slabs. The same model with creep influence gives better values, but not enough. The most appropriate results are given by model in which the creep coefficient was multiplied by factor 2.5. The finding of [5, 6] about larger bending tensile creep towards compression creep can explain the difference between calculation and experiment. But it is necessary find out how to predict the exact value of bending creep coefficient depending on many possible variables.

6. References

[1] SchöckComBAR 2013*SchöckComBARTechnical Information* (Baden-Baden) p 44
[2] Armastek 2018 Online: http://www.armastek-csp.sk/
[3] Fecko L 1986 Sledovanie pretvorení železobetónových dosiek od zmrašťovania *Stavebnícky časopis 34* **8** 615-630
[4] Bilčík J, Fillo Ľ, Benko V and Halvoník J 2008 *Betónové konštrukcie* vol 2 (Bratislava: STU) p 374
[5] Meloun V and Adámek J 1998 Comparison of concrete creep deformations determined from specimen bending and pressuring tests *Concrete and Concrete Structures* (Košice: TU Košice) 109-113
[6] Adámek J, Meloun V, Juránková V and Juránek P 2003 Concrete creep deformations caused by bend and pressure stress specimen test *International Symposium (Non-Destructive Testing in Civil Engineering 2003) Proceedings BB 85-CD ISBN 3-931381-49-8* (Berlin: DGZfP) p 028

Acknowledgments
The article was supported by the scientific grant agency of the MŠVVaŠ SR and SAV project VEGA 1/0661/16 The behaviour of load bearing elements from ordinary and lightweight concrete influenced by temperature.

IOP Conf. Series: Materials Science and Engineering **566** (2019) 012021 doi:10.1088/1757-899X/566/1/012021

The Effect of Openings on Stress Distribution in Glass Beams – Dependence on Diameter of Openings

M Lavko[1], V Kvočák[1], Z Sokol[2]

[1] Technical university of Kosice, Faculty of Civil Engineering, Department of Steel and Timber Structures, Slovakia
[2] Czech Technical University in Prague, Faculty of Civil Engineering, Department of Steel and Timber Structures, Czech Republic

Abstract. Presented paper describes the behaviour of simply supported single ply glass beam with discontinuities in its geometry. Discontinuity is introduced by four openings symmetrically placed over the beam. This type of simplified structure represents cut out of real constructions where similar glass components are used as primary load bearing structures supporting glazing of roofs or decks of pedestrian bridges. The main focus of this article is in research of stress distribution around openings and how different diameter of a hole influences stresses in its close proximity. As Eurocode standards for glass structures are still not codified and most of the nowadays literature only provide suggested diameters of holes based on the thickness of the glass pane, this paper will provide the future designers of glass load bearing structures with closer look at stress distribution around mentioned area. In order to bring those informations to theirs seekers, both experimental analysis and numerical analysis together with parametric study were used.

1. Introduction

Among traditional materials used in constructions such as concrete, steel, timber and masonry, more often materials like glass emerged in last decades. Glass, substance used for centuries only as a fill of openings serving for bringing light into a building and creating transparent barrier between exterior and interior is nowadays, thanks to its unique properties used for constructions like slabs, beams and columns [1]. When used as a load bearing structure it not only transfers self-weight to other load bearing structures, but actively transfers loads like wind, snow or forces acting on structure during fire, to foundations and then to soil.

Glass, compared to other common materials used in building industry, has a unique mechanical and optical properties which open whole new range of construction details where it can be applied. Its high level of flatness allow engineers to design slender columns, beams and slabs, which also fulfil strict strength, reliability and safety requirements for buildings [4]. Relative to timber, glass is isotropic, which means, it has the same material properties in each point of a specimen. Moreover, it has equal tensile and compressive strength and strength of tempered glass exceeds compressive strength of commonly used concretes.

The aim of this work is to better understand stress distribution in glass beams with two or more vertical openings and help designers by providing them with information about stress distribution around holes depending on mutual distances between holes and on the diameter of openings.

2. Experimental analysis

The experiments were held indoor with room climate conditions. For the experimental analysis 3 specimens made out of heat-strengthened glass were used, numbered N01, N02 and N03. All three specimens ware made out of heat-strengthened glass of thickness 12 mm with suggested minimal characteristic strength of glass 70 MPa. The beams were asked and fabricated in dimension 1000 x 200 mm. In the distance of 350 mm perpendicular to shorter edge two holes of diameter 20 mm were drilled above each other. From the longer edge the holes are 50 mm far and the distance between the centres of the openings is 80 mm. The specimens are symmetric, which means, that there are 4 openings in total. The centre to centre distance of supports is 900 mm. Cylindrical supports are placed 50 mm from the edge of the beam from both sides. The tested beams represent the cut of real structures. Beams of similar geometry as the tested once can be found in structures of roof glazing or pedestrian bridges. Four point flexural bending test was used, which leads to constant moment in between the forces. The consequences of moment distribution as described is that the crack will appear in the weakest point of the specimen. Then, from the crack pattern of the broken specimen is obvious in which point stress exceeded the load bearing capacity of the specimen the first. The tests were running until the failure of a structure.

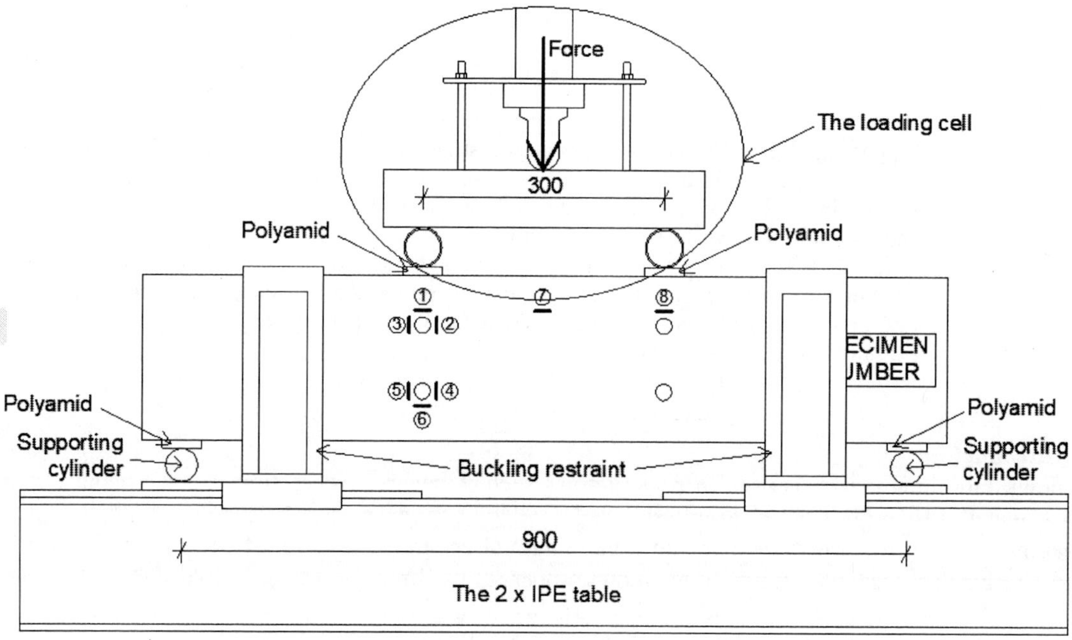

Figure 1. Scheme of the loading assembly (distances in mm).
⑦ – Number of strain gage

The vertical deflection was measured by 4 rod extensometers. Two of them were placed in the middle of span from each side of the beam. Rod extensometers were placed on the horizontal plate of the two L-shaped steel components. Measuring the deflection from both sides allowed us to observe, if the beam buckles. Other 2 rod extensometers were placed on the top edge of the specimen in the point above the supports. As there were plastic washers from polyamide inserted between the steel supporting cylinder and the glass beam, the deformation of the polyamide showed as significant. To obtain the real deflection of the beam in the middle of its span, vertical deformation measured in the centre must be subtracted by the vertical deformation measured by extensometers placed in the points above the supports.

The failure of glass structures is brittle and occurs immediately after a maximum load bearing capacity of an element is reached. Shards of heat-strengthened glass are similar to those of float glass,

IOP Conf. Series: Materials Science and Engineering **566** (2019) 012021 doi:10.1088/1757-899X/566/1/012021

which allow us to fold the broken pieces of each specimen and determine where the crack was initiated, furthermore where the maximum stress was developed during the test. Ultimate limit state was reached for each beam at different force and deflection aligned to that load. Variety in maximum load applied to each specimen is presumably caused by the process of how heat-strengthened glass is produced. Comparison of values obtained from the tests and average value of force and deflection reached at breakage is shown in the table below.

Table 1. Maximum force and deflection reached for each specimen.

	Max. force [kN]	**Max. deflection** [mm]	**Stiffness** [kN/mm]
N01	39,06	1,259	31,017
N02	32,07	1,081	29,677
N03	36,38	1,146	31,746
Avarage	35,84	1,162	30,813

As shown in the diagram of comparison of force-deflection lines of all 3 specimens, the inclination of line of specimen N02 is different. The angle between the line and horizontal axis is lower which indicates lower stiffness. As all 3 tested beams have the same cross-section and length, Young's modulus is lower too, compared to specimens N01 and N03 which stiffnesses are similar.

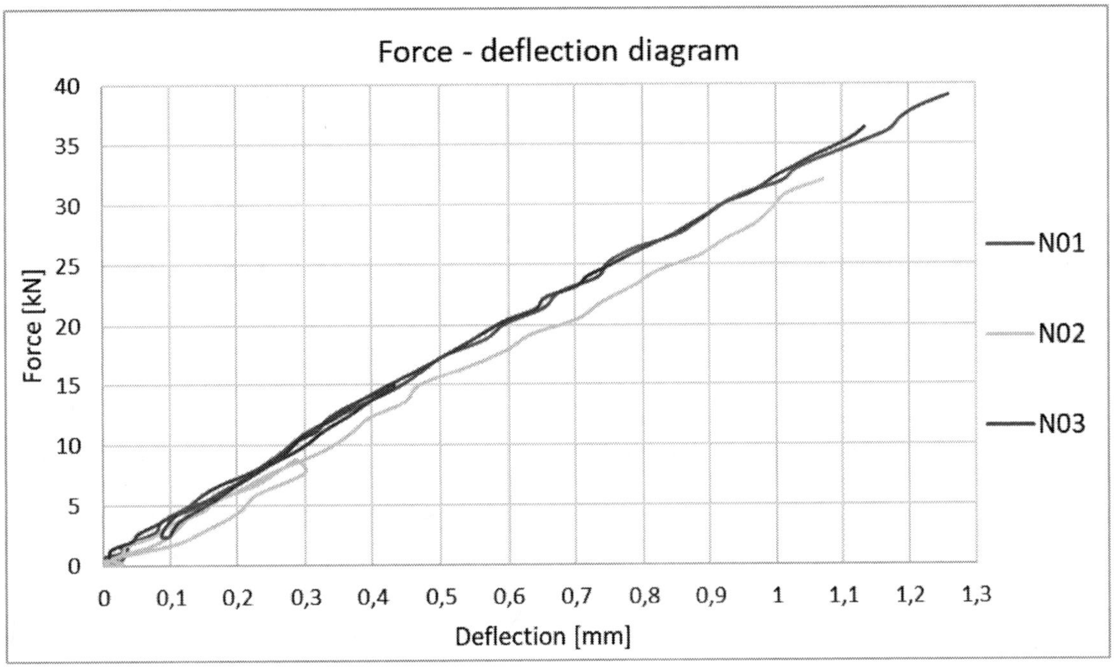

Figure 2. Force-deflection diagram of all 3 specimens.

As observed from the crack pattern of specimens N02 and N03, the initiation of crack was found at the edge of low left opening which signals that during the tests maximum stresses occurred at the described point.

3. Numerical analysis

For numerical modelling was used ANSYS Workbench 19.0 (student license) software. Mentioned software as one of many uses finite elements method (FEM) for analysing the structures. The Workbench interface provide easy orientation alongside conserving the high standard of ANSYS software. A three-dimensional model was created which will represent tested beams. To define

IOP Conf. Series: Materials Science and Engineering **566** (2019) 012021 doi:10.1088/1757-899X/566/1/012021

geometry of same dimensions as specimens, the SpaceClaim interface was used. The shape of component allowed us to model only half of beam and use the symmetry of tested specimens.

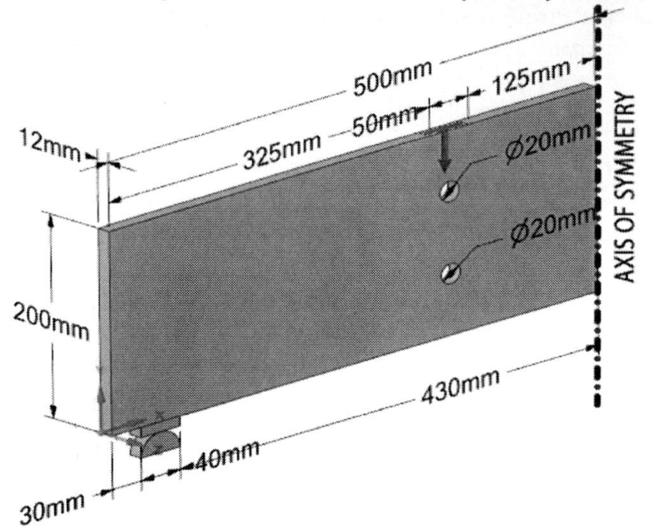

Figure 3. Geometry of used model.

Numerical model reached during loading by force F = 17,92 kN which is half of the total average force applied by loading arm in experiments, maximum vertical deflection of 1,155 mm. Estimated stiffness as well as deflection corresponds with the average values stated in table 1. Calculated stiffness of numerical model is K = 31,041 kN/mm which is almost identical as average stiffness of tested specimens, which is K = 30,813 kN/mm. In numerical model the maximum stress appears at the bottom edge of lower opening, at the point, where initial crack happened during the tests. As glass is elastic material, the force – deflection diagram of tested beam in ideal conditions should be linear as shown in the graph below.

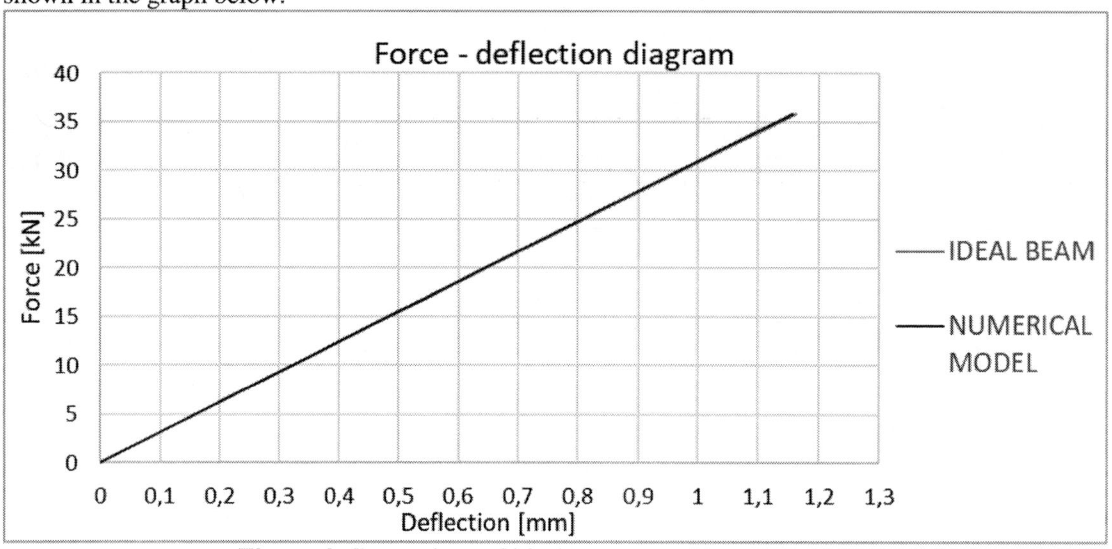

Figure 4. Comparison of ideal beam and numerical model.

3.1. Parametric study

Validated numerical model with the same material characteristics and with unchanged way of support and load application, will be used in parametric study. Essential to this parametric study is to monitor changes in stress distribution around the openings, while using different radius of holes. The radiuses used for this parametric study are shown in Table 2 together with maximum stress recorded at the bottom edge of lower opening and stiffness of modelled beam. Radiuses of both openings were changed simultaneously and centre of rotation of holes remained unchanged.

Table 2. Input and output parametric quantities for variable radius of opening.

r [mm]	5	8	9	10	11	12	13	15	20	25	30
Max σ [MPa]	74,73	88,01	86,92	90,13	83,89	88,71	92,44	97,24	108,14	122,52	143,51
K [kN/mm]	31,36	31,19	31,12	31,04	30,95	30,85	30,73	30,48	29,66	28,48	26,82

Influence of radius of the openings on stress distribution is shown in the diagram below. In presented diagram output quantity is stress at the bottom edge of lower opening. As centre of rotation of openings remained in the same position for all used diameters, the effect of increase respectively decrease in stress caused by linear distribution of stress alongside the height of the beam starting with zero at neutral axis and reaching its maximum at the edges is not involved in the following diagram.

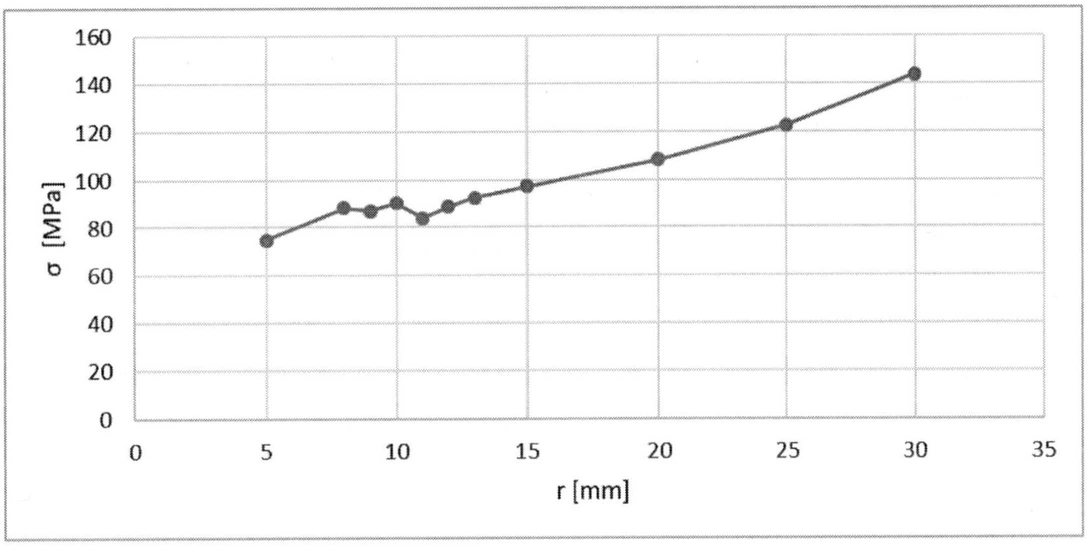

Figure 5. Correlation between the radius of the opening and maximum stress at the edge of hole.

4. Conclusion

Experiments consisted of 3 specimens made out of heat-strengthened glass which were tested in four point flexural test. When the strength of specimens was exceeded, brittle failure characteristic for glass structures appeared immediately. Then, the crack pattern and shards were studied carefully and the bottom edge of lower opening was marked as the point of initiation of crack which spreaded promptly across the whole beam. The average stiffness of all 3 specimens was K = 30,813 kN/mm and maximum average vertical deflection was δ = 1,162 mm. Parametric study provided the data which will be compared to stated suggested diameters of openings in literature. As suggested by [1] diameter of a hole in a in-plane loaded glass pane should fulfil the formula $D \geq S$, where D is diameter and S is thickness of glass pane. But from the perspective of stress quantity the most effective radius for openings in glass is around 11 mm (diameter 22 mm) as shown in chapter 3.1. Thickness of glass panes used for this research study is 12 mm. The parametric study not only confirms proposed formula, moreover adds new condition $D \approx 2S$ in order to reach economic design with respect to maximum stress.

IOP Conf. Series: Materials Science and Engineering **566** (2019) 012021 doi:10.1088/1757-899X/566/1/012021

5. References

[1] Wurm J *Glass structures: design and construction of self-supporting skins* Basel: Birkhäuser, c2007 ISBN 978-3-7643-7607-9

[2] O'Regan Ch and col. *Structural use of glass in buildings (Second edition)* London: The Institution of Structural Engineers, c2015 ISBN 978-1-906335-25-0

[3] Wald F, Macháček J and Eliášová M, et al. *Novinky v navrhování ocelových a dřevěných konstrukcí se zaměřením na skleněné konstrukce*. Praha: České vysoké učení technické v Praze, 2015. ISBN 978-80-01-05780-3

[4] Eliášová M *Lectures of Glass structures*, Department of Steel and Timber Structures, Faculty of Civil Engineering, CTU in Prague. 2015

[5] Vencl R (2011) *Analýza chování nepředepnutých šroubovaných spojů konstrukcí ze skla.* Dizertačná práca

[6] Haldimann M, Luible A and Overend M *Structural Use of Glass*, Zürich: ETH Zürich, c2008 ISBN 978-3-85748-119-2

[7] Weller B and Härth K *Experimental Research on Glass-Polycarbonate Beams* Challenging Glass 2 – Conference on Architectural and Structural Applications of Glass TU Delft, The Netherlands 2010 ISBN 978-90-8570-524-6

[8] Weller B, Meier A and Weimar T *Glass-Steel Beams as Structural Members of Facades* Challenging Glass 2 – Conference on Architectural and Structural Applications of Glass TU Delft, The Netherlands 2010 ISBN 978-90-8570-524-6

[9] DIN 18008-3 *Glas im Bauwesen – Bemessungs- und Konstruktionsregeln – Teil 3: Punktförmig gelagerte Verglasungen (Glass in Building – Design and construction rules – Part 3:Point fixed glazing)* 2013

[10] Feldmann M, Kasper R and col. *Guidance for European structural design of glass components: support to the implementation, harmonization and further development of the Eurocodes.* Luxembourg: Publications Office for the European Union, 2014 EUR ISBN 978-92-79-35093-1

Acknowledgments

The presented paper was supported by the project: VEGA 1/0188/16 "Static and Fatigue Resistance of Joints and Members of Steel and Composite Structures" of the Scientific Grant Agency of the Ministry of Education, science research and sport of the Slovak Republic and the Slovak Academy of Sciences.

IOP Conf. Series: Materials Science and Engineering **566** (2019) 012022 doi:10.1088/1757-899X/566/1/012022

The fire characteristics of the Municipal noise absorbing screen

Z Mašková[1] and O Bret[1]

[1] Czech Technical University in Prague, Faculty of Civil Engineering, Department of Railway Structures, Czech Republic

Abstract. This article deals with the fire resistance of an innovative noise protection – so called the *Municipal noise absorbing screen* and its resistance in other dangerous situations with the occurrence of fire. The article briefly describes the characteristics of the noise absorbing screen's element which is made mostly of the cemented rubber granulate. Due to its geometrical, spatial and material solution the Municipal noise absorbing screen innovatively extends technical opportunities in the field of noise reduction from the rail transport, especially in the urbanized areas. This article also covers the problems of possible causes of fire considering the real installation of the element into the tramway track. The main focus of the article lies in the selection of possible dangerous fire situations. In connection with this, within the experimental part there were carried out reaction to fire tests according to the arranged fire scenarios. There were realized three basic tests in total and their several modifications. In order to that the article describes the effects of the flame on the noise absorbing screen's element and its damage. Final part is dedicated to the summary of all results from the previous experiments and their comparison. Last but not least there are designed future construction improvements of the element.

1. Introduction

Traffic is an integral part of our everyday life whether it is road or rail traffic. According to the increasing requirements for the quality of life as well as health of population we have to deal with constant tightening of limits for the noise emission from traffic, especially in built-up areas.

Noise reduction of railroads focuses mostly on tram tracks which are unlike the rapid transit (the underground/subway) usually run on the surface and often directly through the residential area. Higher traffic intensities can lead to the noise level's maximum or worse – they can be exceeded. [1][2] Currently, an effort is made in developing various low noise protection screens which are, in cities, much better alternative than classic noise barriers limiting view from vehicles. [2][3] As the low noise protections are considered constructions with their height above the terrain not exceeding 1.5 m [1].

An example of the new noise protections of this kind is so called Municipal noise absorbing screen developed by the Department of Railway Structures, Faculty of Civil Engineering, CTU in cooperation with MONTSTAV CZ Ltd. From the acoustic point of view the choice of the screen's material is crucial as well as its shape and the future location. The Municipal noise absorbing screen uses a rubber granulate as a material with great noise absorption. [1][2][4] The following text is dealing with the element of the municipal noise absorbing screen, the description of possible fire hazard and the design of appropriate tests.

IOP Conf. Series: Materials Science and Engineering **566** (2019) 012022 doi:10.1088/1757-899X/566/1/012022

2. The Municipal noise absorbing screen

The Municipal noise absorbing screen ("MNAS", in Czech "Městská protihluková clona – MPHC") is a passive noise protection for the noise attenuation particularly from tram traffic. The screen is designed for the urban setting although not to interfere with the view of a driver or passengers and not to disturb the city panorama. "MNAS" is typical for its low height – only 300 mm above the top of a rail. [1][2] In compliance with technical regulations and rolling-stock, the screen is placed as close as possible to the structure gauge where it effectively absorbs noise coming from the wheel-rail contact. The principal of the attenuation is visible below (Figure 1).

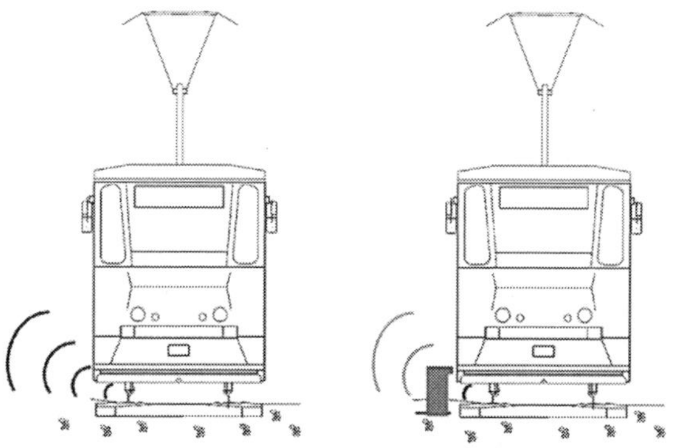

Figure 1. The principal of the noise attenuation with "MNAS“ [4].

Figure 2. The Municipal noise absorbing screen in Prague – Braník.

Thanks to its shape and material used the screen acts like an absorptive element and intercepts the noise at the source. On top of that, thanks to the screen's small dimensions and the whole construction it is easily dismountable. That is a great benefit beside the concrete elements in the case of an accident or in the need of intervention of the integrated rescue system. [1]

The main material component of the Municipal noise absorbing screen is a cemented rubber granulate emerging from recycling of eliminated tyres. Within the recycling process the tyres are freed from bead wires, other metals and textile. [1] [4] The final product – granulate of a given grain size is then bonded with high-molecular-weight polyurethane adhesive and placed into individual manufacturing forms. These forms are adjustable so various surface finishes can be achieved (e.g. waves, pyramids, etc.). These surface finishes help to a better acoustic absorption of the screen. [4] Figure 2 shows the design used in the test section in Prague – Hloubětín which was realized in 2018.

In connection with the construction of the second test section of the Municipal noise absorbing screen in Prague - Hloubětín (June 2018) there was made a request to verify other qualities of this element. In terms of the overall safety emerged the need to verify its reaction to fire and the risk of fire itself.

These real requirements inspired performing tests described in this article. The aim of this article is to experimentally assess the extent of fire risks in several considered situations which can occur during the screen's lifetime.

3. Experimental part

3.1. Initiation sources

Within the realization of the second Municipal noise absorbing screen test section in Prague – Hloubětín it was decided to approach the verification of the screen's reaction to fire and its behaviour during the contact with fire or another initiation source. [5] Critical situation with emergence of fire can occur near the tram track due to several causes – natural causes, technical problems and failures or intentional damages by vandals.

In the category of natural causes the risk of fire is mostly related to **the fire of adjacent greenery**. In case of fire in the track surroundings the fundamental test is the Fire resistance of shrubs described in ČSN EN 1794-2 [6] and ČSN EN ISO 11925-2 [7]. Considering the placement of the screen between both tracks and considering the already taken test on the similar element [8], this test with burning shrubs wasn't performed. The results are assumed as very similar.

A big problem lies in the screen's fire resistance to a **burning vehicle or its part**. The Faculty of Civil Engineering, CTU, isn't at this moment capable of providing the right conditions for the experiment simulating this situation. That is why this test isn't included here. Even though fires of vehicles, especially in the chassis area, are very rare, it is crucial to be concerned with this danger in the case of future boom of the screen.

Among others, risk of fire can occur within construction or maintenance work, specifically in the process of **rail welding** or **rail grinding**. During rail welding there is a prominent local increase of rail's temperature. This operation is done by qualified workers, the work takes place under a technical supervision and the workers have ordinarily a fire extinguisher at hand, so the fire risk is to a considerable extent eliminated. Similar conditions apply to the rail grinding as well.

By a long shot the most problematic situations are caused by **vandalism**. The main risk lies in early (on time) detection of hazards or arising danger and control over the area. There are several fire scenarios that come to consideration and are described in section 3.2.

3.2. Proposal of tests

For the purposes of this article there were considered tests simulating rail grinding and vandalism. In total there were realized three types of tests which are described in detail in the following subsections.

Trial sample – one segment of the Municipal noise absorbing screen consists of two separate parts. The lower rectangular part is 600 mm long, 400 mm high and 250 mm wide. The side adjacent to the track has a wave-like surface but the others have a plain surface. The upper part of the segment is also rectangular but with a quadrant cut out. It is 100 mm high, 250 mm wide at its bottom and 330 mm wide on the top. The segment was kept for more than 48 hours before the start of testing in a space with constant temperature 20°C ± 3°C and humidity 50% ± 10%.

The experiment was carried out in exterior. The segment was placed in the lee in the distance of 1 m in front of a barrier. Placement of the segment didn't prevented wind flowing. During the tests every now and then occurred a light breeze reaching the speed of max. 2 m/s. Temperature of air in the testing area moved around 19°C. Air humidity was around 60%.

3.2.1. Test 1 – Effect of fly away sparks simulating rail grinding. The aim of this test was a determination of segment's reaction on effect of sparks as a simulation of rail grinding. There were two takes within this test where a tube from cast iron was placed on a metal frame in two different distances from the segment. The first distance represented the actual distance from the screen's face to the track, or more precisely to the outer rail. The second configuration should provide bigger amount of fly away sparks coming to a more concentrated area on the face of the screen. The entire construction was secured against shifting. The tube was cut with an angle grinder so that the sparks flew away to the face of the screen (Figure 3). These sparks should simulate the state during rail grinding. In both cases, the segment was exposed to the sparks for 30 s. Results from this test can be seen in Table 1.

Table 1. Results of the Test 1.

Measurement	Distance of the tube from the segment (mm)	Temperature before (°C)	Temperature after (°C)
1.	670	18.2	21.5
2.	280	18.8	30.1

Figure 3. Test 1 - simulating rail grinding.

Conclusion: The risk of a flare from the fly away sparks is negligible. The effect of the fly away sparks caused a local increase of temperature by approximately 10°C. During rail grinding the screen is under supervision when the workers can immediately intervene in the case of potentially problematic situation (e.g. occurrence of fire or its expansion).

3.2.2. Test 2 – Effect of a burning match and cigarette. The aim of this test was to expose the screen's segment to possible vandal activity. The meaning lied in placing a burning cigarette on the segment and observing its behaviour as well as behaviour of the cigarette (3rd measurement). The same way the test with a match was run (4th measurement). Furthermore, it was necessary to measure the time using the stopwatches and to observe changes in behaviour.

Within the third measurement the cigarette was lit and dragged for a good incandescence Then it was placed on top of the segment and after that it wasn't manipulated with anymore. During the test there wasn't noticeable any reaction of the segment on the burning cigarette. Cigarette itself extinguished after 12 minutes and 28 seconds after burning about 2/3 of its length (Figure 4).

The fourth measurement went similar to the previous one. After lighting a fireplace match, it was immediately placed on top of the segment and observed. Likewise the take with the cigarette, there wasn't spotted any reaction of the segment to the burning match. The match itself extinguished after 55 seconds in about a half of its length (Figure 4). Results from this test can be seen in the Table 2.

Table 2. Results of the Test 2.

	Temperature before (°C)	Temperature after (°C)	Time of burning (min)
Cigarette	19.7	29.8	12:28
Match	19.9	49.8	00:55

Conclusion: The risk of a flare from both - the burning cigarette and the match is very low. As a result of the burning of the cigarette local temperature rose and there was a slight disruption of the surface appearance. However, the flash point wasn't reached in this fire scenario.

In the case of the burning match the local temperature was risen approximately about 30°C, but still the temperature was below the flash point. In addition, in the experiment was used a large fireplace match which has in comparison to an ordinary sized match bigger energy.

Figure 4. Test 2 - exposing the segment to vandalism – cigarette (left), match (right).

3.2.3. *Test 3 – Small fire attack.* The last test consisted of an intentional ignition of the screen's segment. The segment was exposed to a small fire attack using a kitchen lighter with 45 mm high flame for three intervals: 15 s, 30 s and 60 s. The ignition was performed on two different areas – the narrowest edge of the segment's upper part (5th, 7th and 8th measurement) and on the flat back side of the segment (6th measurement).

As the first testing area was chosen the narrowest front edge of the upper part of the segment. The lighter was placed under the edge into the cut out section and there the small fire attacked the segment for 60 s (Figure 5). After about 8 s a flare of the segment occurred and during the next 22 s the flame spread significantly. After 60 s the lighter was switched off and then the burning progress and its extent were observed. After another 5 min and 5 s the fire spontaneously extinguished.

The second testing area was the back side of the segment where the small fire attacked a flat surface, not an actual edge. Again the flame attacked the segment for 60 s 50 mm below the top edge of the segment. After 8 s a flare occurred and then the flame gradually spread towards the upper edge (Figure 5). After 60 s the lighter was switched off and then the burning progress and its extent were observed. After another 5 min and 26 s the fire completely extinguished.

Within the last two measurements the segment was exposed to the small fire attack for shorter time periods. In the first case (7th measurement) the small fire attacked the front thin edge for 15 s, however, no flare was observed. Within the very last - 8th measurement the same edge was exposed to the small fire attack, this time for 30 s. After about 6 s a flare of the segment occurred and after another 3 min and 44 s the flame completely extinguished.

Table 3. Results of the Test 3.

Measurement	Time of the small fire attack (s)	Flare time (min)	Extinguishing (min)
5.	60	0:08	5:05
6.	60	0:08	5:26
7.	15	-	-
8.	30	0:06	3:44

Conclusion: The risk of flare of the segment when using the kitchen lighter is needed to take seriously. Even with the use of an ordinary pocket lighter there is a possibility of a local igniting of the screen. The flame doesn't spread rapidly over the screen's surface and it has self-extinguishing properties. The problem can be a long-term continuation of this vandal activity which could lead to a larger fire of the element. The risk of fire rises with longer transport intervals (in the daytime – about 8 min, at night – about 30 min) when there can develop a time gap between starting a fire and its spotting. However, it is possible to say that even after the continuous small fire attack for 60 s, the element spontaneously extinguish within about 6 min.

IOP Conf. Series: Materials Science and Engineering **566** (2019) 012022 doi:10.1088/1757-899X/566/1/012022

Figure 5. Test 3 – small fire attack.

4. Overall conclusion

Based on the performed experiments it is possible to state that the element of the Municipal noise absorbing screen has a good resistance against the considered initiation sources in the fire scenarios set above. Several types of tests have been performed as to simulate real situations which the screen may be exposed to. All tests were carried out on dried element. Under operation though, the element will be affected by weather so it is assumed to have higher moisture content.

Within the test simulating construction or maintenance work (such as rail grinding) there was only an insignificant local increase in temperature so the risk of flare of the element is minimal. A similarly low risk arises when putting down a burning cigarette or a match. Potentially most risky appears to be a targeted vandal activity with the aim of damaging, or more precisely igniting an element of the MNAS- e.g. with a lighter. In these scenarios a local flare of the segment occurred and the flame then spread to the near vicinity of the segment's surface. However, in all cases, the flame spontaneously extinguished within a few minutes and the damage of the element was only local. The risk of an intentional vandalism rises especially with longer transport intervals when there can arise a time gap between starting and spotting the fire and thus also endangering the traffic.

Needed to say, this series of tests serves only as a primary insight into this issue and the identification of the greatest risks that can be used to design and perform other fire tests in the future.

5. References

[1] Bret O and Čížková P. 2017 Development of the Municipal Noise-Absorbing Screen and test section construction technology. Acta Polytecnica CTU Proceedings. p: 6-11. ISBN: 978-80-01-06297-5

[2] Čížková P and Bret O 2017 Městská protihluková clona v Praze Modřanech Akustika 28 p 3-12. ISSN: 1801-9064

[3] Čížková P and Lomoz L 2014 Comparison of efficacy of low and high acoustic screens Akustika 22 p 28-36. ISSN: 1801-9064

[4] Lidmila M, Krejčiříková H et al. Protihlukové prvky z drcené gumy Týden výzkumu a inovací pro praxi Hustopeče 2014 ISBN 978-80-85990-25-6

[5] ČSN EN 13501-1+A1: Fire classification of construction products and building elements – Part 1: Classification using test data from reaction to fire tests ÚNMZ 02/2010

[6] ČSN EN 1794-2: Road traffic noise reducing devices – Non acoustic performance – Part 2: General safety and environmental requirements ÚNMZ 08/2011

[7] ČSN EN ISO 11925-2: Reaction to fire tests – Ignitability of building products subjected to direct impingement of flame – Part 2: Single-flame source test ÚNMZ 07/2004

[8] Burning shrubs test of multifunctional gabion element record. Provided by Monstav CZ Ltd.

Acknowledgement
The published results were achieved with the support of the project: Concepts of the Faculty of Civil Engineering, CTU for Prague 2017 (Operational Programme Prague – Growth Pole of the Czech Republic, id. No. CZ.07.1.02/0.0/0.0/16_040/0000377)

IOP Conf. Series: Materials Science and Engineering **566** (2019) 012023 doi:10.1088/1757-899X/566/1/012023

The forced ventilation analysis of panel block building pilot project

D Miček[1]

[1] Brno University of Technology BUT, Faculty of Civil Engineering, Department of building services, Czech Republic

Abstract. The 21th century is characterized as a hectic age, which point out the inability of people to correctly understand and evaluate the physical parameter of the internal environment and microclimate. This misunderstanding often leads to the life endangerment in the way of carbon dioxide high levels and inappropriate living temperature. Latest technological development helps us to improve this situation by computational capabilities, which includes the possibility of measurement and collection data from long-term point of view. The contribution consists of long-term real-time measurement at panel block building pilot project in Brno, Czech Republic. The measurements are analysed and evaluated with respect to interior environment and microclimate requirements stated in Czech National Standards. The result comprise the proposals and simulations of internal microclimate, focusing on the carbon dioxide levels suggesting the most efficient forced ventilation unit operation, based on evaluation of measurement and simulations. The main asset is opportunity to regulate and optimize the forced ventilation units based on the simulation of the carbon dioxide level in the room, resulting in financial savings by proper regulation of the forced ventilation units and improving the microclimate of buildings.

1. Introduction

Low-carbon economy belongs to the main goals of European Union. The constant increasing of the requirements for the microclimate levels leads to the development of new technologies for its control and optimization. The contribute "*The forced ventilation analysis of the panel block pilot project*" point out the fact that in these days hectic age the people are not able to correctly evaluate the carbon dioxide levels and concentration. Therefore the long-term stay in such rooms leads to increased health risks and reduced working performance of individual residents, the main objective is to analyse selected panel block building room from temperature, relative humidity and carbon dioxide concentration point of view with the help of experimental measurements and the forced ventilation optimization based on computer simulation [4].

2. Carbon dioxide – measurement and concentration

The feeling of heavy and stifling air is often caused by carbon dioxide higher levels. Especially for air-tightly closed new building, where there is no automatic exchange of air and regular air exchanger is required. The concentration of carbon dioxide in poorly ventilated areas often exceeds 1 600 ppm (parts per million). In smaller and enclose, non-ventilated dormitories, the carbon dioxide level during the night may exceed 2500 ppm. According to these levels, it is not possible to rest during the sleep and in the morning rise with the sense freshness. The impact of carbon dioxide concentrations per living room inhabitant is shown in Figure 1[3].

CO CONCENTRATION IN AIR	INHALATION TIME AND SYMPTOMS
9 ppm	ASHRAE maximum allowable concentration for short exposure in a living area.
50 ppm	Maximum allowable concentration for continuous exposure in any 8-hour period.
200 ppm	Headache, tiredness, dizziness and nausea after 2 to 3 hours.
400 ppm	Frontal headache within 1 to 2 hours and life threatening after 3 hours. Maximum allowable amount (air-free) in flue gases.
800 ppm	Dizziness, nausea and convulsions within 45 minutes. Unconsciousness within 2 hours. Death within 2 to 3 hours.
1,600 ppm	Headache, dizziness and nausea within 20 minutes. Death within 1 hour.
3,200 ppm	Headache, dizziness and nausea within 5 to 10 minutes. Death within 30 minutes.
6,400 ppm	Headache, dizziness and nausea within 1 to 2 minutes. Death within 10 to 15 minutes.
12,800 ppm	Death within 1 to 3 minutes.

Figure 1. Carbon dioxide levels [3].

The main source of carbon dioxide in the interior is human breathing, where the concentration is although affected by cooking, the presence of animals (large dog release the same amount of carbon dioxide as adult human) and room plants. A sense of well-being is only possible with a sufficient oxygen supply. The concentration of carbon dioxide is measured in units ppm (parts per million), which express the number of carbon dioxide particles per one millions air particles.

According to statistical evaluations most people spend up to 90% of their day time in enclosed areas such as building interiors. Many times, the indoor air quality is worse than outdoor air quality. The evaluation of the air levels based on subjective feelings of cold, heat, exhaust air and odor gives us a signal to open windows and ventilation. An objective assessment of air quality is not enough. At present, it is becoming more and more popular to measure carbon dioxide concentration by data loggers [3].

3. Economic and technical attributes of the panel block building

The selected and afterwards analysed housing unit is located in panel block building situated in Brno, Nový Lískovec region on cadastral parcel with the area 434 m^2 , 9 over ground floors and 32 housing units. The type of vertical structure is panel blocks. Figure 1 shows the location of pilot project and simplified selected housing unit floor plan.

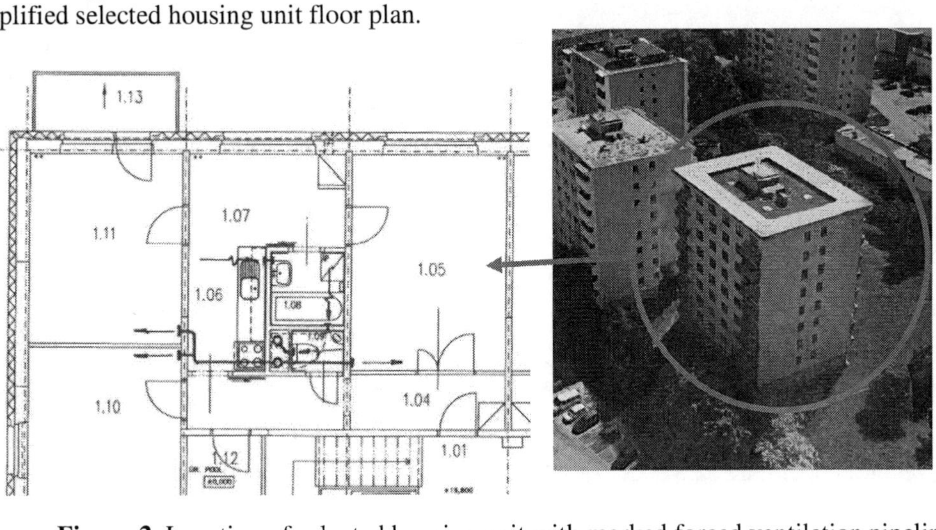

Figure 2. Location of selected housing unit with marked forced ventilation pipelines.

IOP Conf. Series: Materials Science and Engineering **566** (2019) 012023 doi:10.1088/1757-899X/566/1/012023

The residential unit is after a general reconstruction with the extra thermal insulation of the peripheral facades and the original wooden window replacement for the new triple-glass windows. The housing unit is inhabited by three persons, two of them aged 55-60 and one 28 years old. The time schedule of the inhabitants presence is show in Figure 2 [2].

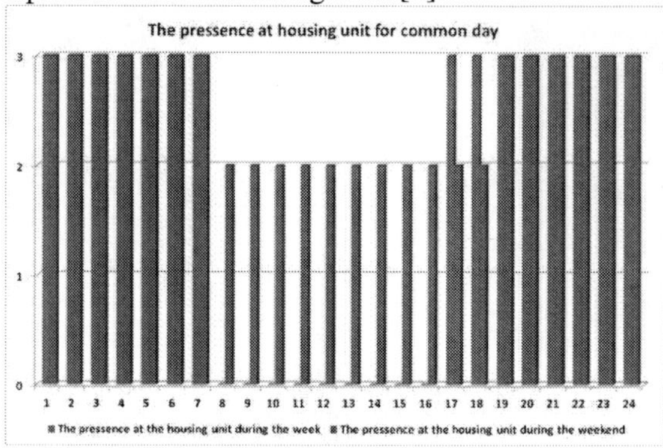

Figure 3. The presence at the housing unit.

The presence is based on individual interview with the inhabitants and is covering they common daily routine during the 24 hours, their presence. Based on the interview it is possible to state that inhabitants at the age 55-60 spend most of their time in unit by light physical activities, like watching television. The inhabitant at the age 28 spends most of its time in unit by working on computer.

4. The residential unit analysis

For the purpose of the residential unit analysis, there was chosen representative room, namely bed room based on which the analysis is divided into two part, where the first parts contains real-time measurements of relative humidity, carbon dioxide concentration in ppm units and indoor temperature provided by data loggers Comet. The second part contains the residential unit analysis and optimization of forced ventilation point of view [1].

4.1. The residential unit analysis – part one

The real-time measurement is based on data collection from four Comet sensors located in reference rooms of residential unit – two bed room, kitchen and living room. Following the statement in Chapter 4 as the reference room for the analysis was chosen the bedroom, where inhabitants aged 55-60 are sleeping.

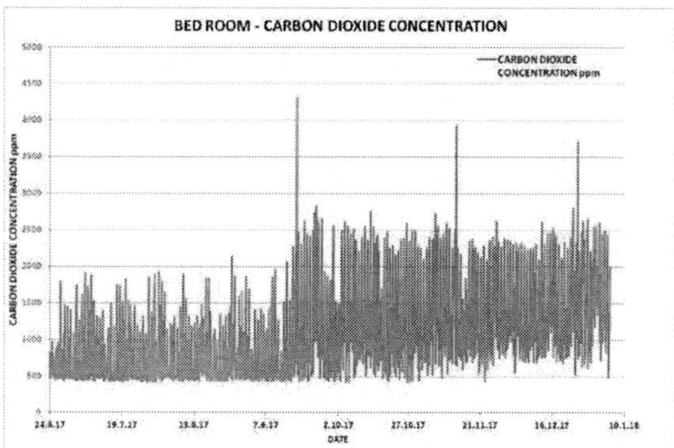

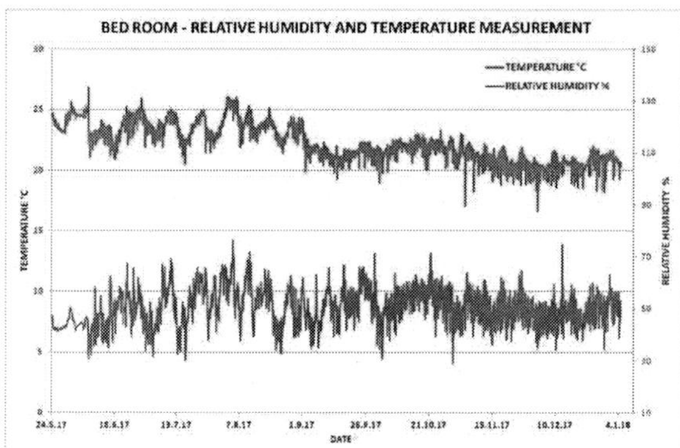

Figure 4. CO_2 concentration. **Figure 5.** Temperature and relative humidity.

The highest temperature during the observed period was 26,8 °C, the lowest temperature was 16,7 °C and the average temperature 22,34 °C – shown at Figure 3 and Figure 4. Refers to the Czech National Standard – ČSN EN 12831 has the temperature at living spaces not overcome the boarder 20 °C for rooms such as bedroom due to health risks. During the measurement the temperature in in bedroom overcome this border in 97,5 % of measured values, which from long-term point of view will leads to health problems. The data underlay the same proceedings as in article [3].

The highest relative humidity during the experimental measurement was 78,3 %, the lowest relative humidity was 29,2 % and the average relative humidity was 50,93 % - shown at Figure 3 and Figure 4. Refers to Czech National Standard – ČSN 060210, the relative humidity should be 60% due to health risks. During the experimental measurement was the relative humidity at the level 60 % for the 10 % of measured values which will from long-term point of view leads to health problems.

Based on provided experimental measurement it is possible to state, that the concentration of carbon dioxide at living area – bedroom was in intervals:

- < 1000 ppm 58,21 %
- 1000 – 2000 ppm 31,76 %
- 2000 – 5000 ppm 9,77 %

From health point of view 40% of measured values exceed border 1000 ppm, which resulted in concentration reduction, fatigue, approximately 10 % of measured values was in interval 2000 – 5000 ppm, which leads to headaches, fatigue and concentration reduction. Therefore, spending time in this room will from long-term point of view leads to health problems.

4.2. The residential unit analysis – part two

Based on experimental analysis for forced ventilation optimization was chosen as a reference day the Sunday 15.10.2017 at which measurement of the average carbon dioxide concentration values is shown in Figure 6. The physical activity during this day was set based in personal interview to the 150 W/m^2.

With respect to the boundary conditions and initial data for minimum carbon dioxide concentration 600 ppm – which refers to the average exterior carbon dioxide concentration, there was with help of MS Excel created carbon dioxide simulation during the reference day with the air exchange values 0,25; 0,5; 0,75; 1 h^{-1}. Figure 6,7,8,9 shows the graphical interpretation of carbon dioxide levels simulation, where after evaluation is possible to state the conclusion in form of recommendation for interior air exchanger value set to 0,25h-1, due to the fact that by this value the concentration level will stabilize at the approximate value 600 ppm, which is from medical and economic point of view the most suitable technical solution.

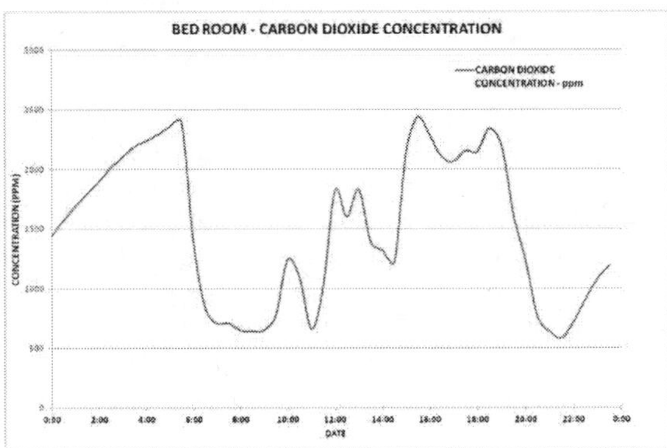

Figure 6. CO_2 concentration level.

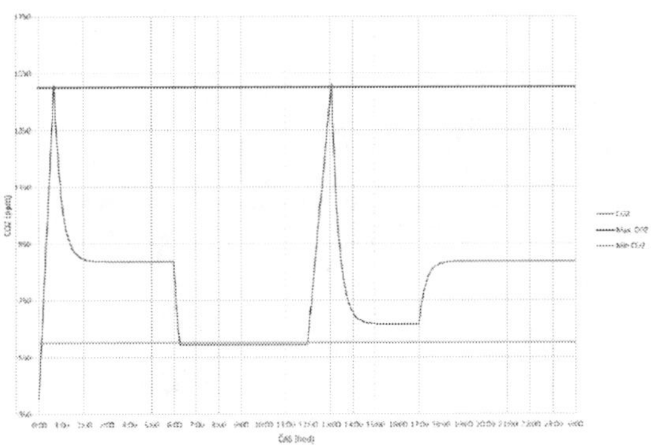

Figure 7. Air exchange: 0,25 h^{-1}.

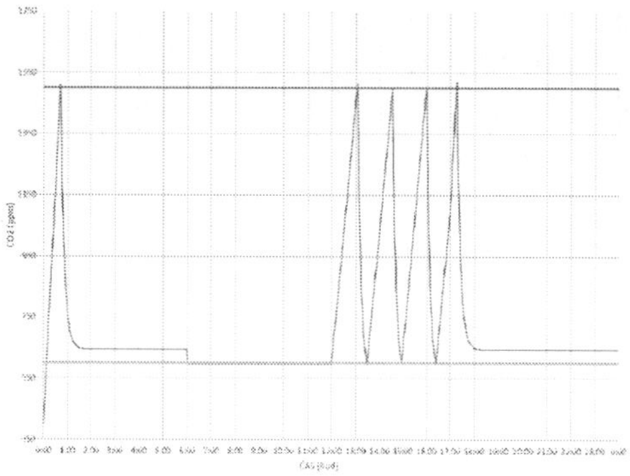

Figure 8. Air exchange: 0,5 h^{-1}.

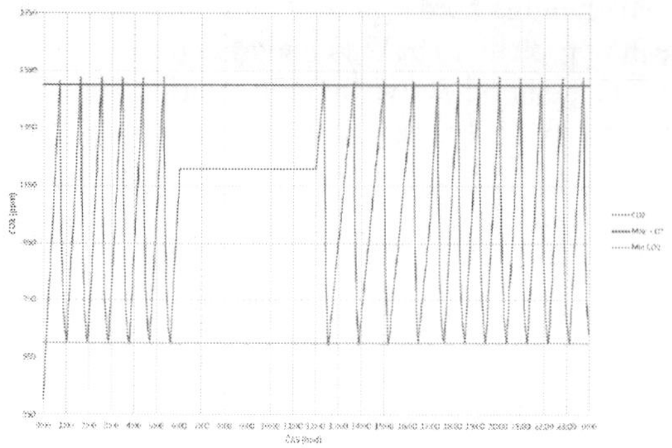

Figure 9. Air exchange: 0,75 h^{-1}.

4.3. The residential unit analysis – forced ventilation unit analysis

The forced ventilation unit at analysed panel block building is placed on roof top constructed by ATREA s.r.o. company with fabric name DUPLEX 2400 air condition unit with air recuperation. The composition consist of two ventilators AIRFLOW type 102GTL/4, recuperation heat exchanger hPs-K-B Duplex, two filters ATREAtype G4 520 520 x 2050 mm, bypass, controlling and safety components. The air condition unit on the roof top is currently out of order due to the contamination of both filters with dust and other particles. After visual exploration it is possible to say that both recuperation heat exchanger and bypass are in good condition. The air handling unit is undamped, with construction connected directly to the roof construction. Air condition system is currently out of control due to swirl pad fully opened (Figure 10 and Figure 11).

Due the fact that air condition unit is out of order, the panel block building inhabitants were complying about the smell coming out of air distribution pipelines and solved this problem by connection their own supporting ventilator into the air condition system.

Winter operation is based on recuperation heat exchanger usage for partial incoming air heating with two ventilators operation. Summer operation is based on bypass usage and rejection of recuperation heat exchanger with one ventilator enhancing the outcoming air – the incoming air is naturally flowing.

Figure 10. Detailed filter photo.

Figure 11. Air unit overview.

5. Conclusion

Based on the experimental analysis elaborated during the time period from 24.05.2017 to 05.01.2018 it is possible to clearly see, that housing unit will not satisfy Czech National Standard ČSN EN 12 831 and ČSN 060210 in the values of temperature in more than 93 % and the values of relative humidity in more than 93,5 % of total time. The concentration of carbon dioxide in housing unit overcome the border 1000 ppm at which people will start suffer from health problems in more than 31,2 % of total time. To be able to resist this dissatisfaction and to avoid health problems caused by spending time in such as area, where the current state is inappropriate and insufficient it is necessary to design forced air circulation respecting the economy, environmental and energy point of view. The article "The forced ventilation analysis of panel block building pilot project "provide complex analysis of reference room and gives recommendations about optimization of the forced ventilation unit presenting. The most important conclusion is that the ventilation unit is currently out of order due to the heavy dust sedimentation at filters and possible electric wires failure, where the recommendation is revitalization of current air condition unit, in the way of replacing damaged components and presenting the air exchange to the value specified in building simulations.

6. References

[1] Hojova A 2017 *Dýchame v miestnosti zdravý vzduch ?* (Bratislava: Festival vedy a techniky amavet) p 13
[2] Vidim J 2013 Monitoring the Indoor Air Quality and the Carbon Dioxide Content *Tzb haustechnik 3/2013* **21** 46 ISSN 1210-356X
[3] Hirs J, Dermekova S, Wawerka R, Volarik T and Mohelnikova J 2016 Information modelling process based on integrated data acquisition *Energy and buildings* **130** p 122-130 ISSN: 0378-7788.
[4] Mcgill G, Oyedele L and Mcallister K. 2015 Case study Investigation of Indoor Air Quality in Mechanically Ventilated and Naturally Ventilated UK Social Housing *International Journal of Sustainable Built Environment* **15** 10

Acknowledgement

This article "*The forced ventilation analysis of panel block building pilot project*" was elaborated within the Smart Regions - Buildings and Settlements Information Modelling, Technology and Infrastructure for Sustainable Development Designation: TE02000077 and project TAČR solved within AdMaS Research Center at the Faculty of Civil Engineering, Brno University of technology.

Semi-rigid operation of connection in timber engineering

M Neusch[1], J Sandanus[1], K Sógel[1]

[1] Slovak University of Technology in Bratislava, Faculty of Civil Engineering, Department of Steel and Timber Structures, Slovakia

Abstract. Consideration the semi-rigid behaviour of connections is an essential part of the modern way of the design of timber load-bearing structures. The computational models using the finite element method allow us to easily replace the use hinge operation with the appropriate stiffnesses of connections in the design. Taking the stiffness of the joint into consideration can have a significant influence on the load distribution and on the stiffness of the object as a whole. The paper contains the calculation process of the stiffness of selected connections. The mentioned stiffnesses could be used in modelling of timber load-bearing structures.

1. Introduction

One of the important indicators in the building industry is the economic aspect. The sound design of a load-bearing structure and structural connections play an important role in wooden constructions. A high-quality load-bearing structure of a timber building consists predominantly well-designed and well-manufactured construction connections. In the engineering practice connections with the hinge effect are designed and utilized most frequently. The joint connection allows for free rotation. However, almost all connections have a certain degree of rotational stiffness. The real behaviour is somewhere in between the hinge behaviour and the rigid behaviour. The modern connections allow for semi-rigid behaviour. In the case of atypical connection where the structural elements are not connected under the right angle it is possible to use welded T-pieces made of steel plates. This paper deals with the welded T-connector that is used for connecting the adjoining beam to the main beam. The semi-rigid behaviour of connections is demonstrated based on this connection.

2. Input data

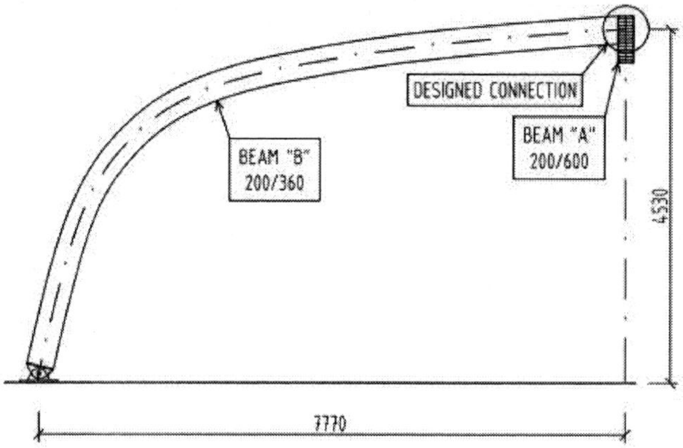

Figure 1. Structure.

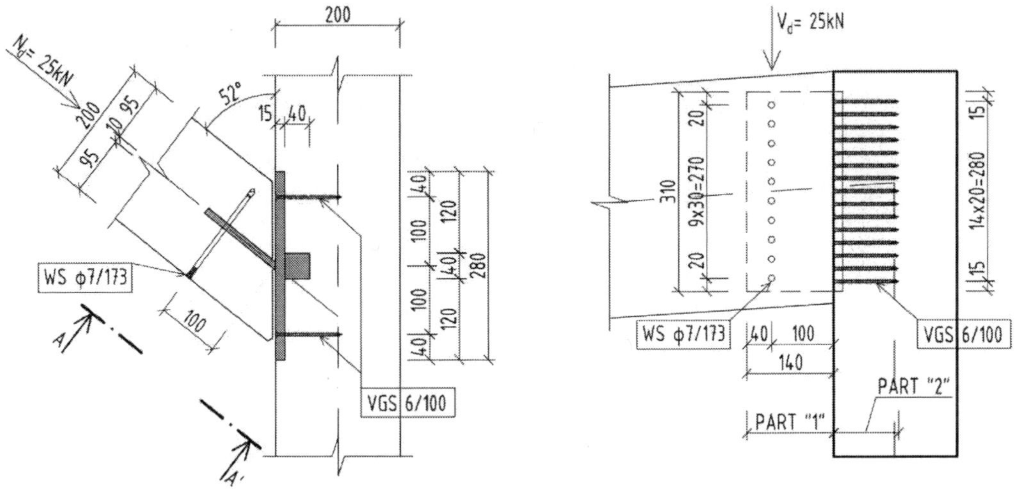

(a) (b)

Figure 2. Geometry of connection: (a) top view, (b) side view A-A'.

Table 1. Structural elements.

Beam	Measure [mm]	Class of wood
A	200/600	GL 24h
B	200/360	GL 24h

Secondary beam „B“ is connected to T-connector through dowels with (self perorating dowel) SFS WS 7/133 (part „1“). T-connector is connected to primary beam „A“ using screws VGS 6/100 (part „2“). Axial force is attached using notch. The aim of the calculation is the design and verification of the behaviour of the dowels and screws on shear force V_d= 25kN.

Table 2. Connection material.

Type	Diameter [mm]	Length [mm]	$F_{v,Rd}$ [kN]	$F_{v,Rd,30}$ [kN]	$F_{ax,Rd}$ [kN]
WS	7	173	2,92	6,80	-
VGS	6	100	1,64	-	2,74

Fv,Rd Design load-carrying capacity per shear plane per fastener.
Fv,Rd,30 Design load-carrying capacity per shear plane per fastener at angle of 30 degrees.
Fax,Rd Design value of axial withdrawal capacity of the fastener.

3. Rotation stiffness calculation according to [2]

To begin with, it is important to divide the connection into two parts – part 1 and part 2. Two different values of the rotational stiffness for the two parts will be obtained from the caltulation. At the end, the FEM model will contain the stiffness values.

3.1 Part „1"

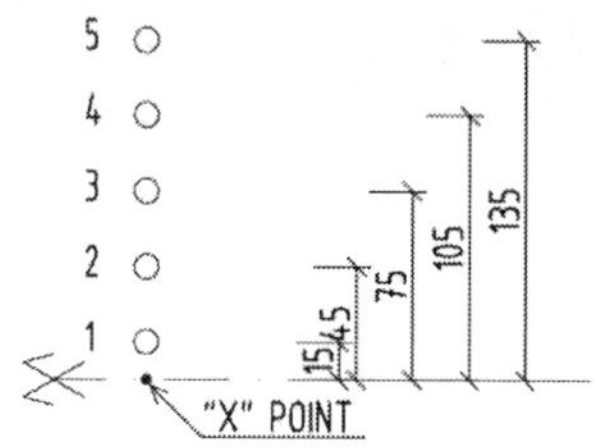

Figure 3. Distance of dowels.

Calculation of polar moment of inertia according to rotation point "X":

$$I_p=\sum r_i^2 \tag{1}$$

$$I_p=2\left(15^2+45^2+75^2+105^2+135^2\right)=74{,}25*10^3\ mm^2$$

Calculation of middle value of slip modulus K_{ser} according to [1]:

$$K_{ser}=\frac{\rho_m^{1,5}*d}{23} \tag{2}$$

$$K_{ser,1}=\frac{380^{1,5}*7}{23}*2=4509\ Nmm^{-1}$$

The K_{ser} is multiplied by two because this connection is of the wood + steel type. The variable ρ_m is the density for the GL 24h class. The variable *d* is the diameter of the dowel WS.

Calculation of stiffness for serviceability limit state according to [2]:

$$C_{\varphi,MSP}=K_{ser,1}*n_{sr}*I_p \tag{3}$$

$$C_{\varphi,MSP}=4509*2*74{,}25*10^3\ mm^2=0{,}67\frac{MNm}{rad}$$

n_{sr} Number of shear planes.

Calculation of stiffness for ultimate limit state:

$$C_{\varphi,MS\acute{U}}=\frac{2}{3}*C_{\varphi,MSP} \tag{4}$$

$$C_{\varphi,MS\acute{U}}=\frac{0{,}67*2}{3}=0{,}45\frac{MNm}{rad}$$

The verification of the calculation of stiffness through deformation of dowel:

$$F_M=\frac{M*r_{max}}{\sum r^2} \tag{5}$$

If $M=1Nmm$ then $F_M=1{,}818*10^{-3}\ N$

Deformation of dowel:

$$\delta = \frac{F_M}{K_{ser,1} * n_{sr}} = \frac{1{,}818 * 10^{-3}}{4509 * 2} = 2{,}016 * 10^{-7} mm$$

The deformation of the dowel is expressed using the angle of rotation, $\varphi = 1{,}49257 * 10^{-9} rad$. 1 Nmm caused angle φ. Using a simple calculation, we will obtain the stiffness of the connection, which is equal to 0,669 $\frac{MNm}{rad}$. We have verified the calculation of stiffness of part „1“.

3.2 Part „2“

Part 2 consists of two rows of screws VGS 6/100. The T-connector is attached to the main beam using the screws.

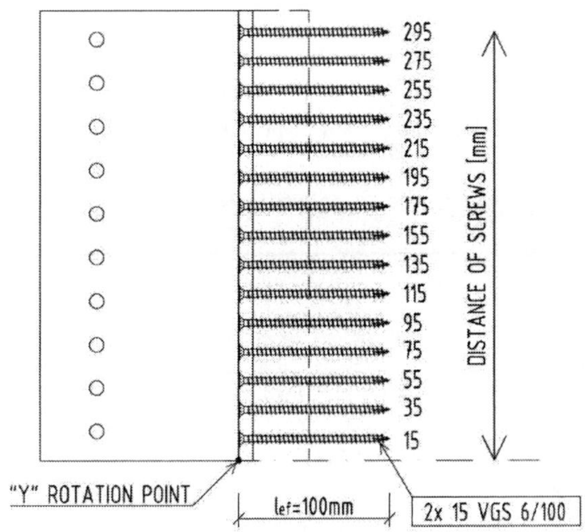

Figure 4. Distance of screws.

Calculation of polar moment of inertia according to rotation point "Y":

$$I_p = 2\left(15^2 + 35^2 + 55^2 + \ldots + 275^2 + 295^2\right) = 944{,}75 * 10^3\ mm^2$$

Calculation of middle value of slip modulus K_{ser} according to [1]:

$$K_{ser} = 100 * l_{ef} \tag{6}$$

$$K_{ser,2} = 100 * 100 = 10\ 000\ \frac{N}{mm}$$

Calculation of stiffness for serviceability limit state according to [2]:

$$C_{\varphi,MSP} = K_{ser,2} * n_{sr} * I_p$$

$$C_{\varphi,MSP} = 10000 * 1 * 944{,}75 * 10^3\ mm^2 = 9{,}45 \frac{MNm}{rad}$$

Calculation of stiffness for ultimate limit state according to (4):

$$C_{\varphi,MSÚ}=\frac{9,45*2}{3}=6,3\frac{MNm}{rad}$$

The verification of the calculation of stiffness through deformation of screw according to (5):

If $M= 1Nmm$ then $F_M= 3,1217*10^{-4}\ N$

Deformation of screw:

$$\delta=\frac{F_M}{K_{ser,1}*n_{sr}}=\frac{1,818*10^{-3}}{10\ 000}=3,12*10^{-8}mm$$

The deformation of the screw is expressed using the angle of rotation, $\varphi = 1{,}0571 * 10^{-10} rad$. 1 Nmm caused angle φ. Using a simple calculation, we will obtain the stiffness of the connection, which is equal to $9{,}459\ \frac{MNm}{rad}$. We have verified the calculation of stiffness of part „2".

Table 3. Rotational stiffness of connection.

Part	$C_{\varphi,MSÚ}$ $[\frac{MNm}{rad}]$	$C_{\varphi,MSP}$ $[\frac{MNm}{rad}]$
„1"	0,45	0,67
„2"	6,3	9,45

4. FEM model

To examine the connection, we will model the construction. The model will contain the calculated stiffness values of the connection from parts „1"and „2".

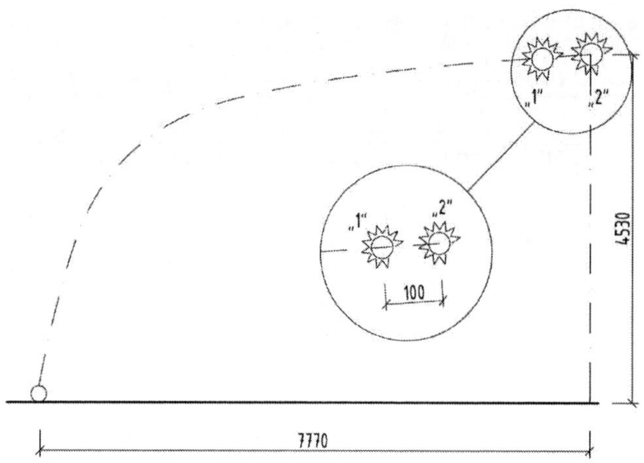

Figure 5. FEM model.

When designing a similar connection, the bending moment at the point of the connection is usually not taken into consideration. After a detailed analysis and using the semi-rigid impact we have obtained the bending moment which is equal to $M_{,max}= 3kNm$.

5. Report

Table 4. Final forces for the most stressed dowel or screw.

Part	F_M [kN]	F_V [kN]
„1“	5,50	2,50
„2“	0,94	0,83

5.1 Examination of part „1“

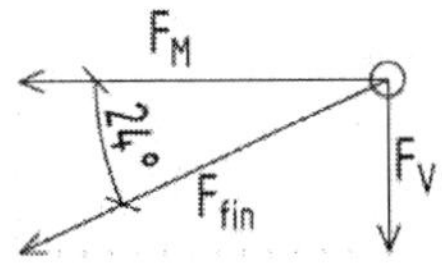

Figure 6. Final force

$$F_{fin}=\sqrt{F_M^2+F_V^2} \tag{7}$$

$$F_{fin}=\sqrt{5,5^2+2,5^2}=6,04kN$$

$$\frac{F_{fin}}{F_{v,Rd,30}} \leq 1,0$$

$$\mathbf{0,88 < 1,0}$$

5.2 Examination of part „2“

$$\left(\frac{F_M}{F_{ax,Rd}}\right)^2+\left(\frac{F_V}{F_{V,Rd}}\right)^2 \leq 1,0 \tag{8}$$

$$\left(\frac{0,94}{2,74}\right)^2+\left(\frac{0,83}{1,64}\right)^2 \leq 1,0$$

$$\mathbf{0,38 < 1,0}$$

6. Conclusion

When modelling the structure, we have utilized the real stiffness values of the connection in question. The detailed calculation better explains the behaviour of the structure connection and the overall allocation of the forces in the load-bearing structure. The results of considering the connection as a semi-rigid will be become evident not only in the area of safety but using more connection will also be more economical. Hinge behaviour is a less safe because this way doesn´t consider additional bending moment. By effect of bending moment the dowels are more stresses, and they should be designed properly.

7. References

[1] STN EN 1995-1-1 + A1: Eurocode 5 – Design of timber structures – Part 1-1: General – Common rules and rules for buildings, SÚTN, 2008

[2] Schickhofer G 2006 *Holzbau Nachweisführen für Konstruktionen aus Holz* (Graz: Institut für Holzbau & Holztechnologie Technische Universität Graz)

[3] Porteous J and Kermani A 2007 Structural timber design to Eurocode 5 (UK: Blackwell Publishing Ltd)

[4] Rothoblaas catalogue, available 21.2.2019

Acknowledgements
This paper has been supported by the Scientific Grant Agency of the Ministry of Education, science, research and sport of the Slovak Republic and the Slovak Academy of Sciences – grant VEGA 1/0773/18.

IOP Conf. Series: Materials Science and Engineering **566** (2019) 012025 doi:10.1088/1757-899X/566/1/012025

Usage of true stress-strain curve for FE simulation and the influencing parameters

A Petrík[1], and R Ároch[1]

[1] Slovak University of Technology in Bratislava, Faculty of Civil Engineering, Department of Steel and Timber Structures, Slovakia

Abstract. Computer simulations require appropriate input data to match the relevant calculated results with those of experiments. The true stress-strain curve suitable for finite element analysis is describing the non-linear behaviour of the steel material with a proper pre-peak as well as necking stress-strain relationship. The pre-peak part of true stress-strain curve can be easily covered by an analytical method (i.e. logarithmic equations), but the necking part of the curve cannot be described by such a relationship between the normalized and the true stress-strain curves. Simple tensile test of steel specimen was performed to establish the true stress-strain curve in ANSYS software. Different sizes of finite elements were applied for the non-linear FE analysis to find an appropriate size function while creating the necking part of true stress-strain curve using an iterative method. Stress and strain from the computer analyses were compared with the results of the tensile test. The dependence of size of finite elements and/or element type was analysed to achieve the most precise results.

1. Introduction

Nowadays, increasingly complicated and complex simulations require appropriate input to create and verify computer models with experiments. Especially the steel plastic behaviour be considered by various methods and material characteristics. For simple simulations, the usage of bilinear stress-strain curve with or without hardening is sufficient to obtain ordinary usable results. However, the most precise results are obtained by considering the engineering stress-strain curve with its elastic and plastic parts. The engineering stress-strain curve can be easily gained form the tensile coupon test.

In ANSYS software, as in other programs using the finite element method, the stress-strain data should be input in a correct way, i.e. to use the true stress-strain characteristics of the steel. The conversion between engineering and true stress-strain curves can be simply made by the logarithmical equations. This procedure can be used in elastic region and partially in plastic region from the yield point to ultimate strength point, where the strain of test specimen is uniform within the gauge length. The problem can be reached in the necking region (from ultimate strength point to fracture point) of the stress-strain curve where the conversion of curves cannot be done by those equations and other methods must be used. One of the methods is an iterative method based on continuous determining the necking part of the true stress-strain curve until the fracture point of the curve.

2. Theoretical background

In general, the result of the uniaxial tensile test is a curve similar to the engineering stress-strain curve, where the stress and the elongation is measured in real time. However, the elongation needs to be recalculated into strain by the gauge length. This parameter depends on the type and size of cross-sectional area of the test specimen. Subsequently, the engineering stress-strain curve can be designed.

It is worth mentioning that stress value at each point of the engineering stress-strain curve is determined in the undeformed initial state and this is the main reason, why it is not correct to use the engineering stress-strain curve as a material property for computer simulations based on finite elements method.

Because of the fact that the shape of the test specimen in the finite element analysis can deformed, it is necessary to transform the engineering stress-strain curve into a true stress-strain curve, in which the current deformed state of specimen is in correlation with the history of previously states. Without the conversion of the stress-strain curve from the engineering type to the true one, a substantial miscalculation can occur not only in the ultimate strength point of the curve, but in the whole plastic region of the curve.

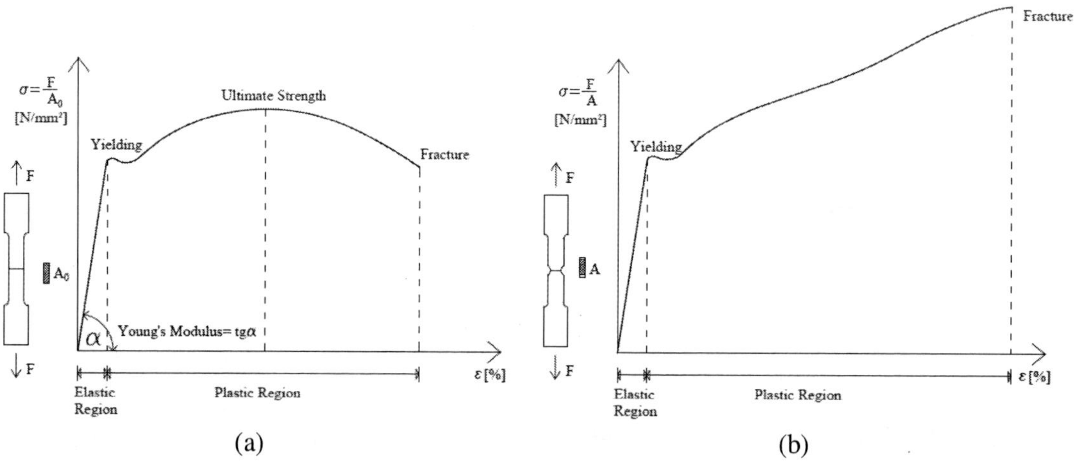

Figure 1. Stress-strain curve: engineering stress-strain curve (a), true stress-strain curve (b).

As already mentioned, engineering stress-strain curve can be converted to a true stress-strain curve using logarithmic equations. These can be applied in most cases of problems, because they cover a large region of the engineering stress-strain curves – from beginning of the test until the moment when the ultimate strength of the specimen is reached. Afterwards it is obligatory to use another method to gain the remaining part of the true stress-strain curve, where the necking of the test specimen is localized. The true strain is characterized by the division of the increment of the strain and the actual length (1). Because of the uniformity of the strain along the test specimen, the true stress can be calculated as a division of the axial force and the actual cross-sectional area of the test specimen (2)[1].

$$\varepsilon_t = \int_{L_0}^{L} \frac{1}{L} dl = \ln(1+\varepsilon_e) \tag{1}$$

$$\sigma_t = \frac{P}{A} = \frac{PL}{A_0 L_0} = \sigma_e \frac{L}{L_0} = \sigma_e(\varepsilon_e + 1) \tag{2}$$

One of the possibilities is to use an iterative method in finite element software, e.g. ANSYS, where the test specimen can be modelled. The main idea of the process is to continuously determine the necking part of the true stress-strain curve until the fracture point of the curve. The iterative procedure is as follows: increase the displacement of the test specimen and exceed the strain of the last known point; afterwards, adjustment of the slope of the true stress-strain curve to obtain the appropriate stress value at the engineering stress-strain curve at the corresponding strain value. The adjustment of the slope should be repeated until the value of the stress with the appropriate precision has been reached. Then, the displacement should be increased until the strain at the fracture point is obtained. Finally, the true stress-strain curve is completed and can be used in finite element analysis [1].

3. Model

In order to generate the true stress-strain curve, a rectangular cross-sectional test specimen was prepared of steel X60 (according to API 5L). The geometry of the specimen is specified in figure 2. In the hatched areas, the model was constrained in z-direction and the left hatched area was constrained in x-direction. The displacement was defined on the right hatched area.

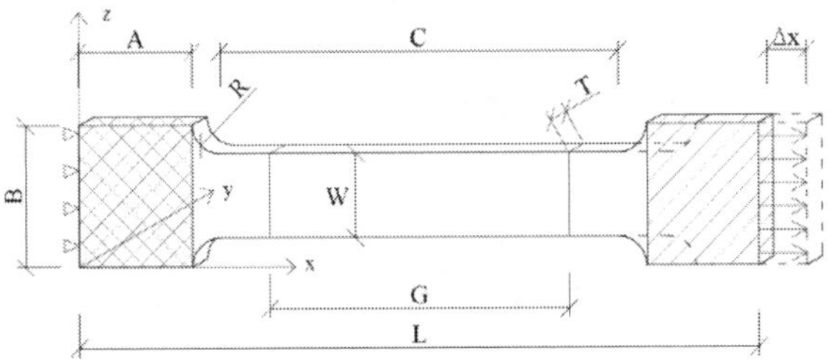

Figure 2. Geometry of the test specimen.

Table 1. Dimensions of the test specimen.

	A [mm]	B [mm]	C [mm]	L [mm]	G [mm]	W [mm]	R [mm]	T [mm]
TS	50.0	61.0	175.0	300.0	131.3	36.0	12.5	15.0

Subsequently, it was inserted in the tensile test machine in order to obtain the dependency curve of the change in elongation depending on the applied load. Using the gauge length [2], the elongation of the test specimen was recalculated to the strain and due to local inaccuracy, the shape of the curve was corrected.

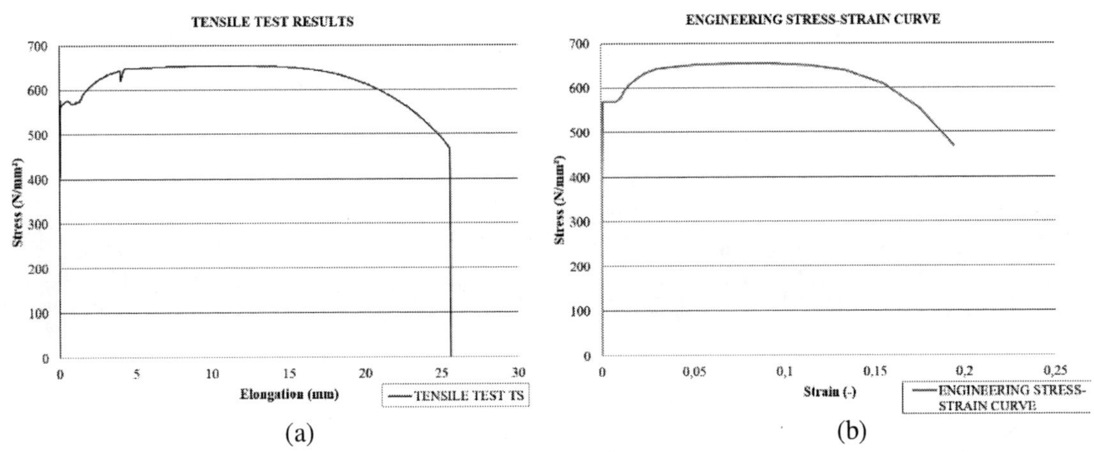

Figure 3. Original tensile test results (a), corrected and recalculated results into engineering stress-strain curve (b).

The work has also focused on the impact of the size of the finite elements and of the type of the hexahedral elements. Lower size of the elements can increase the accuracy but can significantly raise the duration of the analysis. Also, the element order is an important parameter that can affect the output data of the simulation. Linear elements (without the mid-side nodes) usually provide accurate solution in reasonable time and it is recommended to use this type of elements in nonlinear problems. Usage of the quadratic elements (with the mid-side nodes) considerably extended the duration of the simulations and in comparison with the equivalent mesh consisting of linear elements, they provide less accurate solution in more time.

The test specimen was modelled in ANSYS software in accordance with the abovementioned geometry as a 3D model. Four models of the test specimen were prepared: three with the linear elements order and one with the quadratic elements order. Also, the mesh size varied in the analysis – in the first group of the elements, three different size of finite elements were used – 1 millimetre, 1,5 millimetre and 2 millimetres. For comparison 1.5-millimetre elements were used for the model with quadratic elements order.

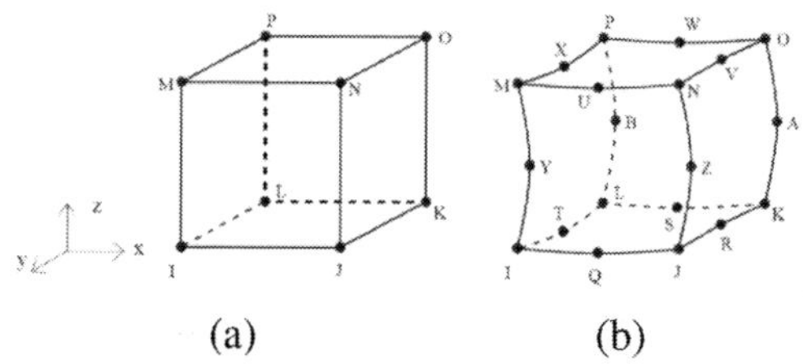

Figure 4. Linear element (a), quadratic element (b).

Because of the requirement to achieve not only the ultimate strength point, but also the fracture point, it was needed to run the analysis as displacement controlled. Additionally, the Newton-Raphson algorithm does not accept a decreasing slope of the stress-strain curve.

Firstly, the usage of the true stress-strain curve as the correct material input was needed to be proved. The engineering stress-strain curve prepared from the results of the tensile test was inserted as the material property in the first analysis, while the true stress-strain curve of the steel recalculated from the engineering stress-strain curve by the logarithmic equations was used in the second analysis. In both simulations, the intention was to continuously increase the value of the displacement until the ultimate strength point is reached and to monitor the development of the plastic strain.

From the simulation results it can be stated, that by using the true stress-strain parameters it is possible to achieve the ultimate strength point with high precision (figure 7). On the other hand, using the engineering stress-strain parameters, the ultimate strength point cannot be reached. Furthermore, the strain is not uniform along the gauge length, the necking of the test specimen starts to appear (figure 5), and the value of the plastic strain is more than 6 times larger than in the simulation with the true stress-strain parameters (figure 7).

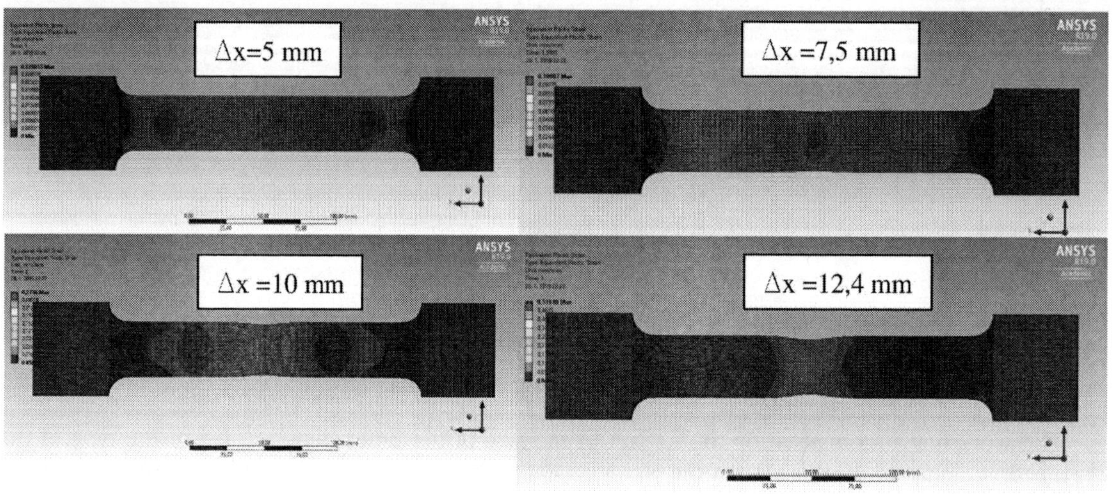

Figure 5. Time development of the plastic strain using engineering stress-strain material properties.

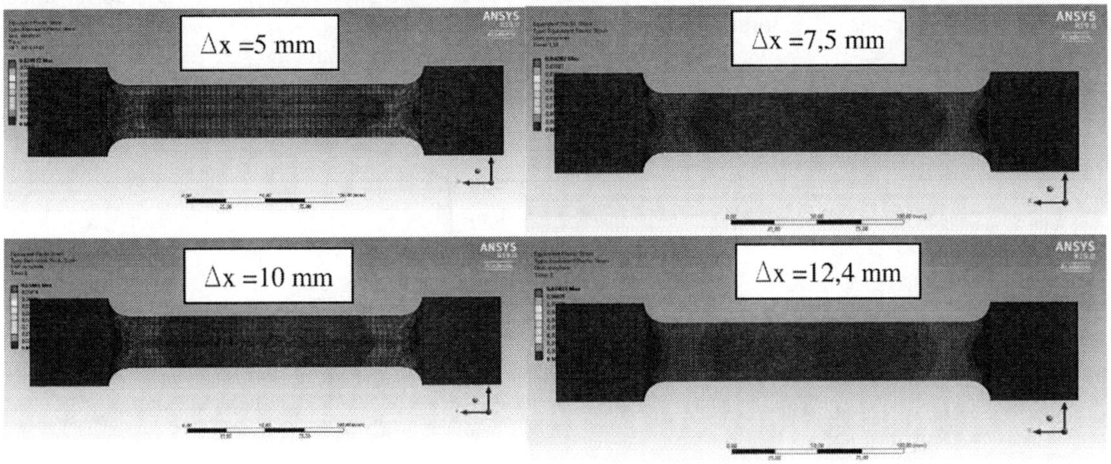

Figure 6. Time development of the plastic strain using true stress-strain material properties.

Secondly, the influence of the size function of the elements on the generation of the necking region of the true stress-strain curve was observed. It is important to say that for every different size of the element new true stress-strain curve needs to be calculated. The increasing number of nodes extends the time of solving the model; it can be reached by two processes – decreasing the size of the elements or the usage of higher order elements with the mid-side nodes. Four different test models were analysed with the abovementioned mesh variations and the variation of the elements order. The true stress-strain curves developed by the analysis are similar, but it can be claimed that with the decreasing element size, the plastic strain and the true stress raise. Also, the slope of the curves increased with the finer elements mesh (figure 8).

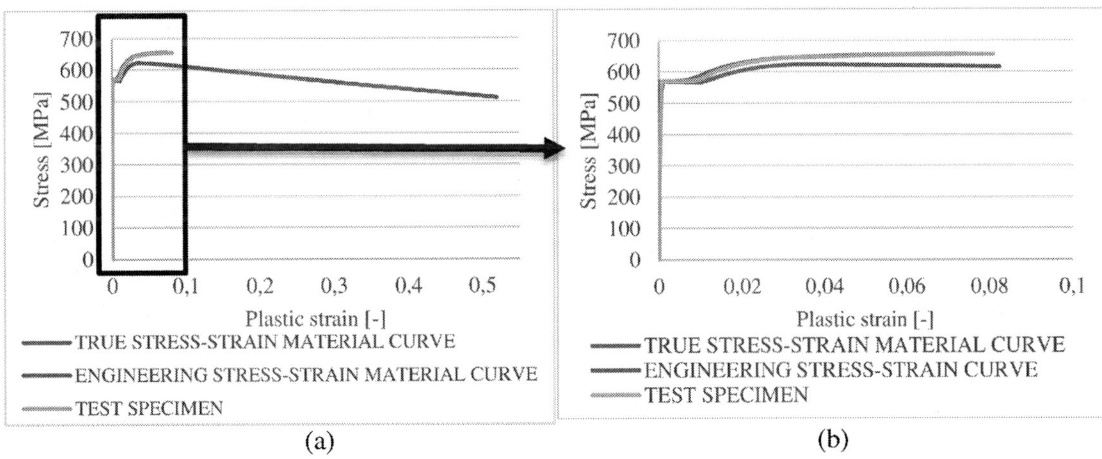

Figure 7. Comparison of material inputs in solution (a), detail of comparison (b).

Despite the fact that the quadratic elements order is not recommended for the nonlinear simulations, they were used in the last simulation. With the highest number of the nodes, the duration of this analysis was the longest among the whole research and the development of the true stress-strain curve was the most time consuming. The value of the true stress and the plastic strain at the fracture point is higher with the quadratic elements order than with the linear elements order with the same size of the elements. However, the usage of the higher density of linear elements makes the solution faster and the curves are

IOP Conf. Series: Materials Science and Engineering **566** (2019) 012025 doi:10.1088/1757-899X/566/1/012025

almost similar, there is no reason why not to follow the recommendation to use rather smaller linear elements order, than larger quadratic elements order.

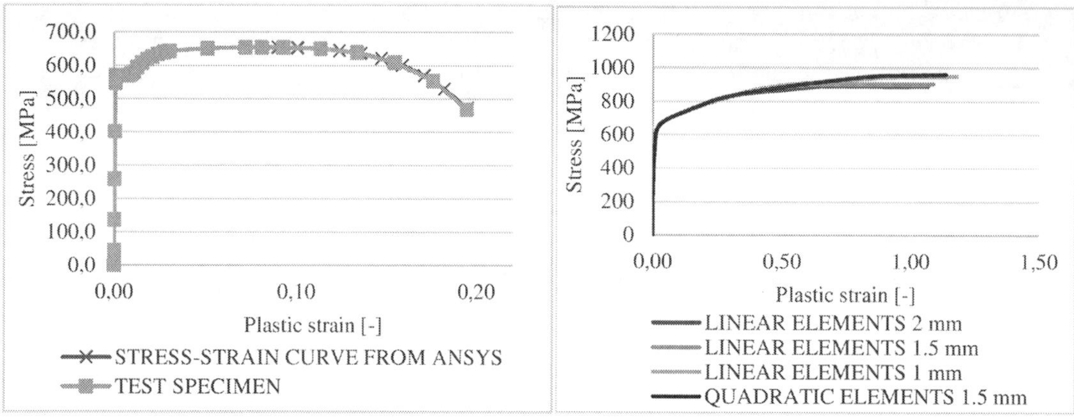

(a) (b)

Figure 8. Comparison of stress-strain curves obtained by the tensile test and the iterative method in ANSYS software (a), comparison of true stress-strain curves by the size of the elements (b).

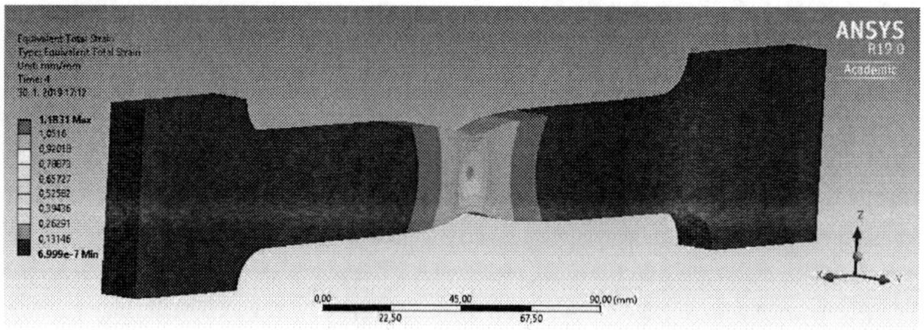

Figure 9. Strain of the specimen after reaching the fracture point.

4. Conclusion

The main aim of this research was to prove the usage of the true stress-strain curve as the correct material property for nonlinear simulations of steel components. After the analyses and their comparison, usage of the engineering stress-strain curve leads to strongly incorrect results. Also, the Newton-Raphson method for solution cannot cooperate with the decreasing slope of the stress-strain so the engineering stress-strain curve needs to be converted into the true stress-strain curve by analytic (logarithmic equations) and iterative methods (FE software).

Size function of the elements and elements order are the principle parameters that influence the precision of the solution. Increasing the mesh density of the elements also increases the duration of the solution, as well as the use of quadratic elements instead of the linear ones. For practical reasons, for every size function of the elements, a unique true stress-strain curve exists, and it needs to be generated before every simulation.

5. References

[1] Gannon L 2011 *Mesh Dependence of True Stress-Strain Curves in Finite Element Analysis of Steel Structures* (DRDC Atlantic TM 2001-283; Defence R&D Canada – Atlantic) pp 1-20

[2] International Organization for Standardization 2016 *Metallic materials – Tensile testing – Part 1: Method of test at room temperature* (ISO Standard No. 6892-1) retrieved from https://www.iso.org/standard/61856.html

[3] Ehlers S 2010 Strain and stress relation until fracture for finite element simulations of a thin circular plate *Thin-Walled Structures* **48** pp 1-10

Acknowledgements
This paper has been supported by the Scientific Grant Agency of the Ministry of Education, science, research and sport of the Slovak Republic and the Slovak Academy of Sciences – grant VEGA 1/0773/18.

IOP Conf. Series: Materials Science and Engineering **566** (2019) 012026 doi:10.1088/1757-899X/566/1/012026

Developing of the rainfall intensity-duration-frequency curves for the Kosice region using multiple computational models

A Repel[1]

[1] Institute of Environmental Engineering, Department of Environmental Engineering, Faculty of Civil Engineering Technical University of Kosice, Vysokoskolska 4, Kosice 04200, Slovakia

Abstract. There are several formulated methods for determining of calculative stormwater flow rates in the sewage system. Three basic characteristics can be attributed to rainfall, i.e. duration, intensity and frequency. The flow rate in storm drainage network is mainly dependent on the intensity of the rain. The main parameter in the IDF curves is rainfall intensity, which is a measure of rain falling in a given time. Extremeness of a rainfall event depends on the duration of rainfall. For this reason, Intensity–Duration–Frequency (IDF) curves are extensively used in water resources engineering. This paper is focused on developing new IDF curves using three models: Dub model, Blaszczyk model and Reinhold model in the Kosice region. From the calculated values of intensities, it is clear that the relations used in Slovakia and Germany have brought very similar results of intensities, but intensity calculated according to Blaszczyk model, used in Poland equals to quarter amount.

1. Introduction

Rainfall, and its duration, intensity and frequency are very important in urban hydrology, for example in designing of sewerage or drainage system is cities. Extremeness of a rainfall event depends on the duration of rainfall. For this reason, Intensity–Duration–Frequency (IDF) curves are extensively used in water resources engineering for planning and designing [1]. IDF curves provide estimates of return levels for the continuum of durations and return periods. The problem of creating IDF curves is that they need more than a hundred years of data collection, while the series that are available are mostly much shorter. Estimating the 100-year return level, for example, relies then on extrapolating using some statistical model. Uncertainty is inherent to this estimation because no model is perfect. This is particularly true for extreme value estimation – such as the 100-year return period – because it is based on few data, so a subsequent variability is induced by sampling. Risk evaluation should account for this uncertainty to avoid over-optimistic results [1]. Since current infrastructure dealing with flooding and precipitation (e.g. dams or dikes) are based on IDF curves, ignoring uncertainty may result in sharp underestimation of flood risk and failure risk of critical infrastructures [2].

The main parameter in the IDF curves is rainfall intensity, which is a measure of rain falling in a given time. The rainfall intensity is usually expressed by the height of the rainfall layer per minute (mm/min) or as the specific rainfall intensity in liters per second on one hectare (l/s·ha). The intensity of the precipitation, its duration and the area it has affected have fundamental importance in determining the outflow from the area and therefore there are very important in hydrology [4]. Between rainfall intensity expressed in mm/min (i) and specific rainfall intensity in l/s·ha (i') applies equation $i = 0.006 \cdot i'$ [5]. In general, the greater the intensity of rainfall is the shorter length of time it continues. In other words, shorter the rain is, the greater is its intensity. For a particular location, it is possible to draw a series of curves showing the probabilities of various intensities of rainfall occurring at that place in given periods. These curves are called IDF curves, which were mentioned above [6]. Proper selection of storm sewers geometry requires the determining of the duration of calculative stormwater flow, during which the established precipitation intensity will bring to a certain stormwater runoff from the catchment to the sewage system. There is a number of formulated methods for determining of calculative stormwater flow rates in sewage systems. Precipitation belongs to random events, and virtually every one of them has a different course in time and space. However, when comparing the rains observed over o long period of time, it can be noticed a certain relationship between them [7].

2. Material and methods

There are several methods of calculation, on the base of which the establishing of calculative stormwater flow in sewage system is carried out, taking into account a number of parameters characterizing mainly the catchment and accepted level of reliability of the designed sewage system, associated with the frequency of gravitational operating, action under pressure of the occurrence of flooding from network [7]. The following part of the paper presents different models for calculation specific rainfall intensity. This study is focused on developing IDF curves using three different empirical models, which are used in Slovakia, in Germany and in Poland. First empirical model is Dub model, which is using in Slovakia, second is Blaszczyk model used in Poland and second is Reinhold model, which is using in Germany.

2.1. Rainfall intensity according to Dub

In Slovakia, a relationship based on professor Dub's knowledge, and based on rain data processed by Šamaj and Valovič, is used to calculate the intensity. This relationship was developed by Urcikán and Horváth (1979) [5] and is described by equation (1).

$$q = K \times (t^a + B)^{-1} \tag{1}$$

where: q – specific rainfall intensity (l/s·ha),
K, B, a – local parameters for 68 ombrographic stations in Slovakia,
t – duration of precipitation (min).

2.2. Blaszczyk model

In Poland for determining of calculative stormwater flow rate when dimensioning of drainage and combined sewage systems the formula of Blaszczyk model is commonly used for more than 50 years. Blaszczyk model was developed in 1954 and coming out from 67 years of observations. On the territory of Poland for the dimensioning of drainage infrastructure Blaszczyk model is mostly used. It is described by the equation (2) [7].

$$q = \frac{6.631 \times \sqrt[3]{H^2 \times C}}{t_{dm}^{2/3}} \tag{2}$$

where: q – specific rainfall intensity (l/s·ha),
H – average annual amount of precipitation from multi-year period (mm),
C – frequency of calculative rainfall occurrence for storm drainage system dimensioning (years),
t_{dm} – calculative time of rain (min).

The use of this model is also possible in Slovakia because of similar latitudes. The average annual precipitation amount from the multi-year period for the territory of Slovakia has a value in the range of H = from 500 mm to 2000 mm in high altitudes in High Tatras [8].

IOP Conf. Series: Materials Science and Engineering **566** (2019) 012026 doi:10.1088/1757-899X/566/1/012026

2.3. Reinhold model

In Germany, the calculative rainfall intensity for the purposes of drainage systems dimensioning in urban areas the Reinhold model is often used. This model was published in 1940 and it is presented by the relationship described by equation (3) [7].

$$q = q_{15,1} \times \frac{38}{t+9} \times \left(\frac{1}{\sqrt[4]{p}} - 0.3684\right) \tag{3}$$

where: q – specific rainfall intensity (l/s·ha),
$q_{15,1}$ – reference rainfall intensity of duration t = 15 minutes and frequency C = one year (l/s·ha),
p – the probability of precipitation occurrence (-),
t – duration of precipitation (min).

The value of $q_{15,1}$ can be read out from the Atlas of rains developed in Germany and depends on the geographical location of the canalized catchment. This reference value varies from 90 to 170 l/s·ha.

3. Results and discussion

Each of these three empirical models (Dub, Blaszczyk and Reinhold) were used to calculating of rainfall intensity in the city of Kosice for periodicity C = 5 years (one time per five years) or probability p = 0.2, which is the same. Although not every model was developed for the area of Slovakia (only one of them is specially used in Slovakia), their use is possible as they have been developed under similar morphological conditions. Each model was used to calculating rainfall intensity for rain with probability 0.2 and with durations from 5 minutes to 120 minutes. The results of calculated rainfall intensities for rains with durations from 5 to 120 minutes by three models are in Table 1.

Table 1. Rainfall intensities for rain with probability 0.2 in city of Kosice.

	5 min	10 min	15 min	30 min	60 min	80 min	100 min	120 min
DUB	367.33	257.86	199.87	120.75	68.45	53.39	43.86	37.27
BLASZCZYK	298.98	188.35	143.74	90.55	57.04	47.09	40.58	35.93
REINHOLD	342.59	252.44	199.85	122.98	69.51	53.89	44.00	37.18

The highest values of rainfall intensities are when Dub model is used for empirical calculation. It means bigger dimensions of pipes when designing rainwater sewage systems, then when using Reinhold or Blaszczyk model. Values of rainfall intensities calculated by Dub and Reinhold are very similar. The lowest rainfall intensities are, when Blaszczyk model is used. From the calculated rainfall intensities, IDF curves were developed (Figure 1). IDF curves are graphical representation of table 1, where rainfall intensity is on vertical axis, duration of the rain is on horizontal axis, and computational methods are distinguished by different colors. These IDF curves shows rainfall event with periodicity 5 years, which is the same that probability 0.2.

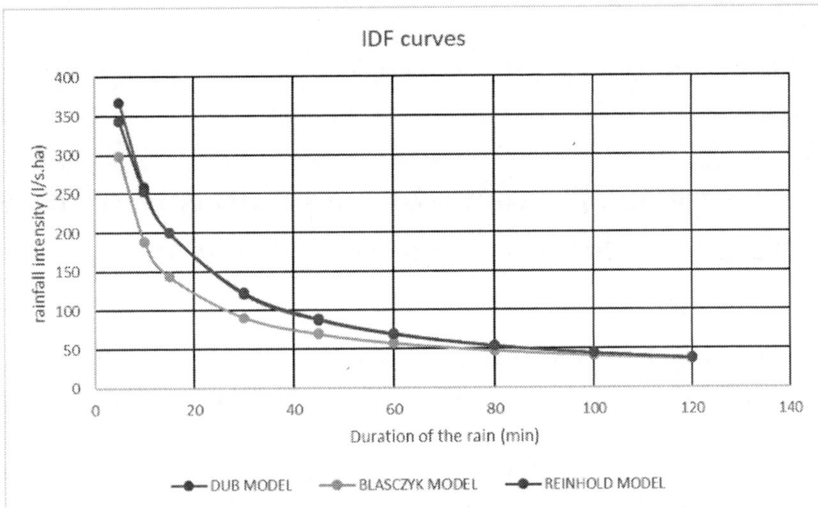

Figure 1. IDF curves for Kosice, developed by Dub, Blaszczyk and Reinhold model.

4. Conclusion

One of the most important parameter used to calculate design flow in designing urban sewerage and drainage systems is rainfall intensity, which is the main parameter of IDF curves. In this paper, rainfall intensities were calculated using three empirical models, and after that IDF curves for Kosice were developed. From the calculated values of intensities, it is clear that the relations used in Slovakia (Dub model) and Germany (Reinhold model) have brought very similar results of intensities, but intensity calculated according to Blaszczyk model, used in Poland equals to quarter amount. The highest values of rainfall intensities are when Dub model is used.

5. References

[1] Rantz S E 1971 Precipitation depth-duration-frequency relations for the San Francisco Bay region, California *US Geological Survey, Prof. Paper* 237-241

[2] Coles S and Pericchi L 2003 Anticipating catastrophes through extreme value modelling *Journal of the Royal Statistical Society: Series C (Applied Statistics)* ***52*(4)** 405-416

[3] Mélèse V and Blanchet J and Molinié G 2018 Uncertainty estimation of Intensity–Duration–Frequency relationships: A regional analysis *Journal of Hydrology* ***558*** 579-591.

[4] Dub O and Němec J 1969 *Hydrologie – technický pruvodce 34* SNTL – Nakladatelství technické literatury Praha ISBN 04-711-69

[5] Urcikán P and Rusnák D 2011 *Stokovanie a čistenie odpadových vôd – Stokovanie I* Slovenská technická Univerzita v Bratislave ISBN 978-80-227-3435-6

[6] Wilson E M 1969 *Engineering Hydrology* Macmillan-London p 1-49

[7] Dziopak J and Starzec M 2015 *Sewage systems – Basics and design* Oficyna Wydawnicza Politechniki Rzeszowskiej, Rzeszow ISBN 978-83-7934-032-3.

[8] Slovak hydrometeorological institute *Klimatické pomery Slovenskej republiky* (in Slovak) Online[12.2.2019]: http://www.shmu.sk/sk/?page=1064

Acknowledgments

This work has been supported by the Slovak Research and Development Agency by supporting the project SL-PL-18-0033.

IOP Conf. Series: Materials Science and Engineering **566** (2019) 012027 doi:10.1088/1757-899X/566/1/012027

Fracture response of cement-based composite with spherical glass aggregate exposed to high temperatures

I Rozsypalová[1], I Kumpová[1,2], T Majda[1], Z Keršner[1] and P Štancl[3]

[1] Brno University of Technology, Faculty of Civil Engineering, Veveří 331/95, 602 00 Brno, Czech Republic

[2] Czech Academy of Sciences, Institute of Theoretical and Applied Mechanics, Centre Telč, Batelovská 485, 486, 588 56 Telč, Czech Republic

[3] Statika 3 Structure s.r.o., Designing Company, Nám. ČSA 39, 676 02 Moravské Budějovice, Czech Republic

Abstract. This paper concerns an investigation into the effects of high temperatures acting on specimens made from specially designed fine-grained cement-based composite with spherical glass aggregate. The experimental programme was carried out on four sets of beam specimens with the dimensions 20 × 40 × 200 mm. The specimens were thermally loaded to pre-set temperatures of 100, 200 and 400 °C, after which these temperatures were maintained for 60 minutes (one of the sets contained three reference specimens which were placed in a laboratory at a temperature of 20 °C and were not exposed to heat). After the temperature loading had been performed, the specimens were left to cool down to ambient temperature. After that, fracture tests were used to determine the degree of damage caused to the specimens by temperature loading. Three-point bending tests were conducted on the previously loaded beam specimens, which were notched to provide them with an initial stress concentrator using a diamond blade saw before testing. The depth of the initial edge notch on the bottom side of the specimen was approximately 1/3 of the total specimen depth. The span length was 180 mm. Fracture response was evaluated via load versus deflection diagrams. The monitored parameters were modulus of elasticity, effective fracture toughness and specific fracture energy. Informative compressive strength values were also determined. Selected specimens were investigated using X-ray micro-tomography (micro-CT).

1. Introduction

Cement-based composites rank amongst those building materials which are frequently used for a wide range of applications [1]. Structures and construction components utilizing this composite are commonly designed assuming normal service temperatures [2]. However, the exposure of cement-based composite to high temperatures causes a wide range of physical and chemical processes to occur within it. These result in changes in the structure of the composite and thus affect its mechanical properties [3–7]. In this paper, the effects of high temperatures acting on specimens made from specially designed fine-grained cement-based composite with spherical glass aggregate are investigated. The monitored parameters were compressive strength, modulus of elasticity, effective fracture toughness, and specific fracture energy. A characterization of the degradation of this cement-based composite when exposed to high temperatures which was gained from data obtained via the ultrasonic pulse method was presented in [8]. In this paper, attention is mainly paid to mechanical fracture properties [9].

Selected specimens were investigated using X-ray micro-tomography (micro-CT) to supplement the results with visual information about the inner structure of the newly designed material before and after thermal loading.

2. Materials and methods

2.1. Specimens

The experimental programme was carried out on four sets of prismatic specimens with the nominal dimensions 20 × 40 × 200 mm. A total of 12 samples were tested. These specimens were formed from fine-grained cement composite. Sodium-potassium glass spheres with a diameter of 2 ± 0.2 mm were used as an aggregate in fresh concrete mixtures. Portland cement (type CEM I 42.5 R – from the Mokrá cement plant) was used as a binder. The mix proportion by weight used in this research was 3:1:0.35 aggregate:cement:water.

Specimens (age 28 days) were loaded using a Classic type 5013 laboratory furnace (with 3 sides of heating coils) with a temperature increase of 5 °C/min up to a nominal temperature of 100, 200 or 400 °C. The temperature was then maintained at that level for 60 minutes. One reference set was left without exposure to high temperatures.

2.2. Fracture test

After the temperature loading had been performed, the specimens were left to cool down to ambient temperature. After that, fracture tests were used to determine the degree of damage caused to the specimens by the temperature loading. Three-point bending tests were conducted on the previously loaded beam specimens, which were notched by a diamond blade saw before testing to provide them with an initial stress concentrator – see Figure 1. The depth of the initial edge notch on the bottom side of the specimen was approximately 1/3 of total specimen depth. The span length was 180 mm. Fracture response diagrams – load versus deflection (*L–d*) and load versus crack mouth opening displacement – were measured. The obtained *L–d* diagrams were corrected and evaluated for this paper via the Effective Crack Model [9] (modulus of elasticity, effective fracture toughness) and Work-of-Fracture Method [10] (specific fracture energy). Informative compressive strength values were determined from fragments of the beams after the fracture tests had been completed [11, 12].

Figure 1. Fracture test configuration in three-point bending: Overall view of the testing machine (left); detail of selected specimen with central edge notch.

2.3. X-Ray micro-tomography

The X-Ray investigation was carried out using the patented TORATOM (Twinned ORthogonal Adjustable Tomograph) device at the Institute of the Theoretical and Applied Mechanics, Centre Telč. The TORATOM system depicted in Figure 2 is a patented (EP2835631) multipurpose X-ray imaging device consisting of two independent pairs of X-ray tube-detectors in an orthogonal arrangement, with a shared rotational stage. All the components are independently adjustable via a multi-axial computer-controlled positioning system, and it is possible to adapt the position of each component according to

IOP Conf. Series: Materials Science and Engineering **566** (2019) 012027 doi:10.1088/1757-899X/566/1/012027

the required magnification or resolution and imaging method. The system also allows the observed object to be moved as required, as well as the scanning of objects larger than the detector area.

For qualitative assessment of the degree of damage suffered by the tested fine-grained cement-based composite during exposure to high temperatures, selected specimens were investigated before and after the thermal load using micro-CT. A microfocus X-ray tube (XWT-240-TCHR, X-ray WorX, Germany) operated at a voltage of 160 kV, target current of 281 μA and a power of 45 W was used for tomographic scanning. A flat panel (Dexela 1512NDT, Perkin Elmer, USA) with an active area of 145.4 × 114.9 mm, a pixel matrix of 1944 × 1536, and a resolution of 74.8 μm per pixel was used as a detector. The geometric parameters of the tomographic assembly were set in order to obtain the best possible resolution with respect to the object size and the detector area. The source-detector distance was fixed at 600.009 mm and the source-object distance was set to 199.971 mm, resulting in a geometric magnification of the object of about 3.0×, leading in the 3D reconstruction to a spatial point edge (voxel) size of 24.933 μm. CT scans were acquired focusing on the notched part of the specimens. Every CT scan had 600 projections with a 400 ms acquisition time.

Figure 2. Modular tomographic system TORATOM: 1 – detector holder with large area flat panel detector (145.4 × 114.9 mm, 74.8 μm pitch); 2 – detector holder with large area flat panel detector (409.6 × 409.6 mm, 200 μm pitch); 3 – nanofocus X-Ray tube (up to 160 kV voltage); 4 – microfocus X-Ray tube (up to 240 kV voltage); 5 – calibration filter holder; 6 – high precision rotary stage; 7 – investigated specimen; 8 – active damped anti-vibration platform with precise CNC positioning system.

3. Results and discussion

The test results can be found in Tables 1–4, which show the values obtained from individual measurements, their arithmetic mean values, standard deviations and coefficients of variation (calculated via functions in Microsoft Excel). For the mean values, relative values are given in parentheses – 100 % always corresponds to the arithmetic mean of the parameter found for specimens without thermal loading (20 °C). Table 5 lists the correlation coefficients (calculated via function in Microsoft Excel) between temperature and mean values for all mechanical fracture parameters.

The compressive strength values (Table 1) gradually decrease with increasing temperature to 40 %. It is possible to notice a very significant drop – at approximately 80–95 % – in the elasticity modulus value (Table 2) for all temperature loaded specimens compared to the reference specimens. A similar situation can also be seen in the case of effective fracture toughness (Table 3) – there is a drop at 90 %

– as well as in the case of specific fracture energy (Table 4) – where there is a decrease at about 80 %. Note that both of these parameter results were obtained with very high variability.

Volumetric images obtained from the CT measurements confirmed that while the structure of the specimen loaded up to 100 °C remains compact, at 200 °C the beginning of disintegration within the interfacial transition zone is observable. At 400 °C disintegration is well propagated, leading to loss of material cohesion. Expansion and interconnection of the pores occurs. Examples of these phases of damage are depicted in Figure 3. The findings are in very good agreement with the values gained for the monitored fracture parameters.

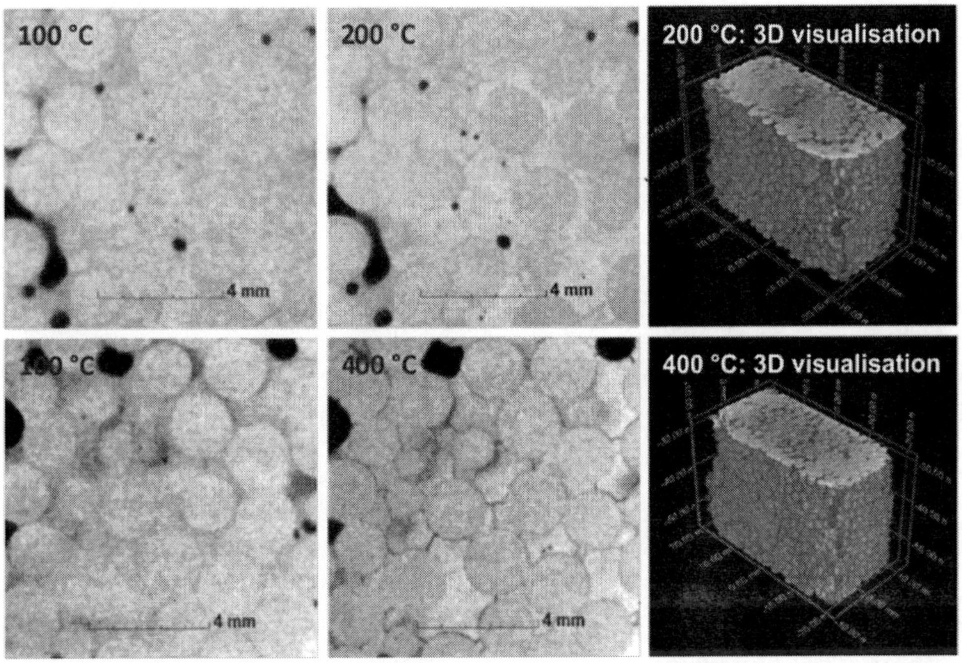

Figure 3. An example of tomographic results. Left – 2D slice visualization of part of a specimen loaded with a temperature of 100 °C; middle – corresponding 2D slice visualisation of part of a specimen loaded with temperatures of 200 °C and 400 °C; right – visualisation of a 3D representation of the volume of interest of a specimen loaded with temperatures of 200 °C and 400 °C.

Table 1. Compressive strength vs. temperature: individual values and basic statistics (relative values).

Maximum Temperature [°C]	Specimen Identification	Value [MPa]	Mean Value [MPa]	Standard Deviation [MPa]	Coefficient of Variation [%]
20	–	–	11.68 (100 %)	0.07	0.6
	29_20	11.73			
	30_20	11.63			
100	4_100	8.21	7.56 (65 %)	1.33	17.6
	25_100	6.03			
	27_100	8.44			
200	6_200	6.91	6.44 (55 %)	1.58	24.5
	7_200	4.68			
	8_200	7.73			
400	10_400	4.42	4.64 (40 %)	0.50	10.7
	11_400	4.29			
	12_400	5.21			

Table 2. Modulus of elasticity vs. temperature: individual values and basic statistics (relative values).

Maximum Temperature [°C]	Specimen Identification	Value [GPa]	Mean Value [GPa]	Standard Deviation [GPa]	Coefficient of Variation [%]
20	–	–	23.26 (100 %)	0.53	2.3
	29_20	23.26			
	30_20	22.88			
100	3_100	4.33	4.30 (18 %)	0.60	14.0
	4_100	3.69			
	25_100	4.89			
200	6_200	1.13	1.38 (6 %)	0.35	25.3
	–	–			
	8_200	1.63			
400	10_400	–	4.41 (19 %)	0.08	1.7
	11_400	4.46			
	12_400	4.35			

Table 3. Effective fracture toughness vs. temperature: individual values and basic statistics (relative values).

Maximum Temperature [°C]	Specimen Identification	Value [MPa·m$^{1/2}$]	Mean Value [MPa·m$^{1/2}$]	Standard Dev. [MPa·m$^{1/2}$]	Coefficient of Variation [%]
20	–	–	0.446 (100 %)	–	–
	29_20	–			
	30_20	0.446			
100	3_100	0.085	0.110 (25 %)	0.0301	27.3
	4_100	0.102			
	25_100	0.144			
200	6_200	0.041	0.038 (9 %)	0.0040	10.4
	–	–			
	8_200	0.036			
400	10_400	–	0.090 (20 %)	0.0444	49.5
	11_400	0.058			
	12_400	0.121			

Table 4. Specific fracture energy vs. temperature: individual values and basic statistics (relative values).

Maximum Temperature [°C]	Specimen Identification	Value [J·m^{-2}]	Mean Value [J·m^{-2}]	Standard Deviation [J·m^{-2}]	Coefficient of Variation [%]
20	–	–	10.33 (100 %)	–	–
	29_20	–			
	30_20	10.33			
100	3_100	3.84	5.57 (54 %)	1.79	32.2
	4_100	5.43			
	25_100	7.42			
200	6_200	3.61	2.30 (22 %)	1.86	80.8
	–	–			
	8_200	0.99			
400	10_400	–	6.46 (63 %)	3.79	58.6
	11_400	3.78			
	12_400	9.14			

Table 5. Correlation coefficients [–] for temperature and mean values of mechanical fracture parameters.

	T	f_c	E	K_{Ic}	G_F
Maximum Temperature T	1.00				
Compressive Strength f_c	–0.90	1.00		symmetry	
Modulus of Elasticity E	–0.62	0.90	1.00		
Effective Fracture Toughness K_{Ic}	–0.65	0.91	1.00	1.00	
Specific Fracture Energy G_F	–0.39	0.70	0.91	0.91	1.00

4. Conclusion

This work was focused on the evaluation of the degree of damage suffered by fine-grained cement-based composite specimens when exposed to high temperatures. After the exposure of the specimens to elevated temperatures, their mechanical fracture parameters were significantly reduced and their internal structure damaged. The relationship between fracture response and loading temperature for this special cement-based composite with spherical glass aggregate was confirmed: in relation to the increasing temperature, the compressive strength values showed a strong negative correlation, while the modulus of modulus of elasticity and effective fracture toughness values showed a moderate negative correlation and specific fracture energy values showed a weak negative correlation.

5. References

[1] Aïtcin P C and Mindess S 2011 *Sustainability of Concrete* (New York: Spon Press)
[2] Neville A M 2011 *Properties of Concrete* (Harlow: Pearson Education Limited)
[3] Bažant Z P and Kaplan M F 1996 *Concrete at High Temperatures: Material Properties and Mathematical Models* (London: Longman Group Limited)
[4] EN 1991-1-2:2002/AC:2009 *Eurocode 1: Actions on structures – Part 1-2: General actions – Actions on structures exposed to fire*
[5] Wang Y Burgess I Wald F and Gillie M 2012 *Performance Based Fire Engineering of Structures* (Hoboken: CRC Press)
[6] Hager I 2013 Behaviour of cement concrete at high temperature *Bull. Pol. Acad. Sci.-Te.* **61** 1–10
[7] Lim S 2015 *Effects of elevated temperature exposure on cement based composite materials* (Urbana)
[8] Rozsypalová I Vyhlídal M Dvořák R Majda T Topolář L Pazdera L Šimonová H and Keršner Z 2018 Characterization of cement based composite exposed to high temperatures via ultrasonic pulse method *Acta Polytechnica CTU Proc.* **15** 99–103
[9] Karihaloo B L 1995 *Fracture Mechanics and Structural Concrete* (New York: Longman Scientific & Technical)
[10] Rilem Determination of the fracture energy of mortar and concrete by means of three-point bending test on notched beams *Materials and Structures* vol 18 pp 285–290
[11] ČSN EN 196-1: 2016 *Methods of testing cement – Part 1: Determination of strength*
[12] BS 1881: Part 119: 1983 *Testing concrete – Part 119 Method for determination of compressive strength using portions of beams broken in flexure (equivalent cube method)*

Acknowledgment

This paper was produced under the "National Sustainability Programme I" project "AdMaS UP – Advanced Materials, Structures and Technologies" (No. LO1408) supported by the Ministry of Education, Youth and Sports of the Czech Republic and Brno University of Technology. The partial financial support of the Czech Science Foundation under project No. 19-01982S is gratefully acknowledged.

New aspects of shore stabilization

M Slezingr[1,2], P A Fernandez[1], V Ondrejka Harbulakova[3], P Pelikan[1]

[1] Mendel University in Brno, Department of Landscape Formation and Protection, Faculty of Forestry and Wood Technology, Zemedelska 3, 613 00 Brno, Czech Republic

[2] University of Technology Brno, Faculty of Civil Engineering, Zizkova 17, Brno 60200, Czech Republic

[3] Technical University of Kosice, Institute of Environmental Engineering, Vysokoskolska 4, 04200 Kosice, Slovak Republic

Abstract. The basis for a design of stabilization measures for banks of watercourses and reservoirs is an assessment of the bank from the perspective of the material it consists of, the existing slope inclination, and the current elements of natural or artificial stabilization. Another important step is to assess the potential cause of the bank damage, the probability of its occurrence, and the significance of possible damage with respect to the bank usage. If the basic assessment of the current state of the riparian area concludes it is really possible that erosion damage to the bank may appear and develop, one of the types of bank reinforcements has to be designed.

1. Introduction

Shoreline hardening practices such as bulkheading have several drawbacks. The structure may only assimilate 20 percent of the wave's energy, which means the majority of the energy is forced back out toward the lake and bottom, thus creating scour at the bottom of the structure, or creating accelerated erosion across from or adjacent to the structure. The chemicals used in pressure treated lumber are harmful to the aquatic life of our lakes. Riprap (rocks/stone), under most circumstances, adequately stabilizes shorelines and does not share all the negative characteristics of bulkheading, however, it does not provide optimal wildlife and water quality benefits as vegetated shorelines. It is important to always install filter fabric under riprap to prevent soil from washing out from underneath the rocks. Vegetative and bioengineering solutions (biologs) stabilize the shoreline, provide aquatic wildlife habitat, and create beautiful landscapes of varied textures, colors, and flowers. Vegetated shorelines also filter lawn runoff and uptake nutrients prior to the runoff entering the lakes. This nutrient uptake will help to reduce algal growth evidenced in the lakes each summer [1].

The impulse for a more intensive research interest in this issue was the request of the commercial sector to develop, test and implement an appropriate stabilization technique for banks of flooded gravel pits and sand pits, where mining has ceased but the highly unstable banks continue landsliding. These sites are very often located within the interests of the environmental authorities, which – rightly – put a great emphasis on an appropriate inclusion of the anthropogenically influenced sites in the landscape [1, 2]. The designing a levee should take into account various technical aspects, as well as interference in the country and the impact on flora and fauna. Solution should be developed in alternatives; the resulting design should be the most suitable option in terms of reliability, of economy, efficiency and environmental acceptability. Technical stabilization elements are not accepted, and the recommended elements of biotechnical and biological bank protection do not work, or show a very low efficiency at such sites [3]. The new approach to bank stabilization was thus oriented differently than most of the existing ones [4]. As said above, a slope of a beach type, reinforced with biological elements can resist damage to some extent. With the current two-meter high erosion wall at the experimental site, the slope of a beach type, i.e. approximately 1:10 to 15, would mean sloping in a length of 20 to 30 m. This is not possible in many cases. However, it is possible to design the so-called "half-dish profile" of the bank slope, which is based on three years of detailed monitoring at experimental sites of banks of sand pits near Hulin (Czech Republic) and others. The sloping in the form of three continual slope inclinations proved to be the best practice: 1:10, 1:5 and 1:2.

1.1 Importance of the stabilizing effect of vegetation

Bankside trees and shrubs are one of the building blocks of territorial systems of ecological stability (TSES). It is part of an ecologically balanced landscape, a form of spread green vegetation growing outside integrated forest complexes. It is created by tree species and herbs growing along streams. In relation to stream regulation, linear building along water streams etc., a lack of riparian and accompanying stands started to manifest negatively. We can say that only once it decreases, will we start to realize its indispensability in our landscape [5, 6, 7].

Figure 1. Abrasive damage to the shore (Hulin dam Czech republic). [foto M.Šlezingr 2015]

2. Material and methods

Apart from a construction of retaining walls and other purely technical elements of stabilization, the basis of a reinforcement is usually bank sloping, most often from 1:1.5 to 1:3. This is followed by a stabilization of the footslope forming the bank, or a direct stabilization of the bank at the site of the most likely damage. If it is possible to use technical elements of stabilization to the necessary extent (stone, concrete, steel elements or their combinations), a long term bank stability is not a problem. However, the situation is different if environment protection authorities are involved in the decision making process, i.e., if the stabilization is to be designed outside towns, in territories with some kind of protection, but also in areas that are inaccessible by machinery, etc. Even then, the basic schema is kept – a suitable sloping of the bank (see above). However, based on quite strict requirements of the authorities, the following stabilization is designed as only biological or biotechnical with the vegetation element dominating [4, 8].

However, a bank reinforced with the above mentioned procedures cannot resist long-term waving (meeting conditions 1 and 2). There is no available tree species in the Czech Republic whose root

IOP Conf. Series: Materials Science and Engineering **566** (2019) 012028 doi:10.1088/1757-899X/566/1/012028

system would be capable of withstanding the surface waving over the long term (demonstrated by the attempt to biologically stabilize parts of banks of Liptovska Mara Reservoir, Slovakia [9], etc).

More suitable than uniform slope inclination 1:1.5 or 1: 2 is gradual inclination 1: 1.5 (1:2) follows 1:5 and 1:10 shape “half-dish” (fig. 2)

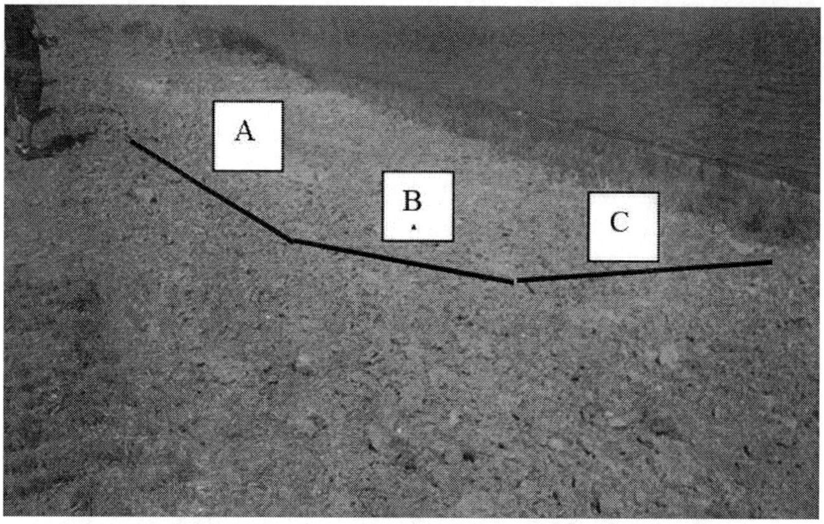

Figure 2. Slope adjustment: A-slope 1:2, B-slope 1:5, C-slope 1:10, detail of slope adjustment (Hulin dam, Czech republic. [foto M. Slezingr 2017]

Five experimental plots were established: two with biotechnical stabilization, two with biological stabilization and a reference one [10,11].

We present experimental areas with biological stabilization:

Experimental plot 1 – sloping shaped like the above-described "half-dish"; the part with the smallest inclination was stabilized using a belt of willow stem cuttings, *Salix viminalis*, 4–5 m wide, in a quincunx pattern of 0.4 x 0.4 m. Above this stabilization structure, it is assumed that appropriate grass mixtures will be sown and individual or groups of shrubs will be planted [12,13].

Experimental plot 2 – sloping shaped like the above-described "half-dish"; the part with the smallest inclination was stabilized gradually. First, it was planted with reeds (namely *Typha latifolia*) - supplied from local sources; then a belt of willow stands, 4–5 m wide, was planted - species *Salix viminalis*, a quincunx pattern of 0.4 x 0.4 m.

3. Results of Monitoring

As regards the stabilization procedures described within the description of experimental plots, the most appropriate and most effective technique proves to be the following: a belt approx. 1–2 m wide of reeds (namely *Typha latifolia*) followed by a stand of shrub willows in at least 5–7 rows in combination with a gravel layer. A stand of shrub willows of several (at least five) rows acts as a natural breakwater on the erosion platform - i.e., where erosion damage has already occurred and developed. This was concluded after evaluating the effectiveness of stabilization after 15 years. Combined with the distribution of slope inclination, it is a very good biological protection of the bank, preventing the formation of erosion wall. As has been found, the waves do not reach the third, i.e. the steepest, slope section significantly (resulting in deformation). The steepest section had an inclination of 1:2 at the experimental plots. Therefore, the monitoring (of 15 years at experimental plot Brno) has shown that the slope inclination designed for this section is irrelevant.

The planting of a willow belt manifested good effectiveness; however, the effectiveness was better in combination with reeds.

4. Conclusion

The following approach appeal to be the most appropriate:

First, an inclination of 1:10, 5–6 m long, planted entirely with shrub willows, the water edge planted with a belt of reeds; followed by an inclination of 1:5 (can be also 1: 4), seeded with a suitable grass mixture; the third inclination within the "half-dish" 1:2 (can be also variable from 1:1.5 to 1:2 with respect to possible maintenance, access to water, etc.). Further, a possible access to water provided by an artificial sloping (lightning-shaped) path, approximately 0.8 m wide was tested. This path did not become a preferred path for water during a torrential rain, the water was converted to the stand. It had no effect on breaking waves.

The proposed bank stabilization in the form of a half-dish in combination with appropriate planting can be regarded as very effective; additionally, it meets the requirements of environment protection authorities.

5. References

[1] Reston association 2018 *Shoreline stabilization guidelines* https://www.reston.org/portals/3/Parks-Recreation-Events/Nature/Publications/PDF/ShorelineStabilization.pdf

[2] Harbulakova V O Lechman M 2018 Flood control measures - dry basins - case study *Conference on Public Recreation and Landscape Protection - with Nature Hand in Hand (Krtiny Czech Republic)* pp 168-173

[3] Harbulakova V O, Harabinova S, Zelenakova M and Mydla D 2014 Modeling of levee slope stabilization in variants *Colloquium on Landscape Management* Brno: Mendel university in Brno pp 35-42

[4] Korytarova J, Slezingr M and Uhmannova H 2007 Determination of potential damage to representatives of real estate property in areas afflicted by flooding *Journal of Hydrology and Hydromechanics* **55** (4) 282-285

[5] Novak L et al 1986 *Vegetace v úpravách vodních toků* (Praha 1: STNL)

[6] Pelikan P and Slezingr M 2015 Parameters of wind driven waves on Nove Mlyny water reservoir *14th International Symposium on Water Management and Hydraulic Engineering (Brno Czech Republic)* pp 55-64

[7] Pelikan P, Slezingr M, Markova J et al. 2018 The Efficiency of a Simple Stabilization Structure in a Water Reservoir *Polish Journal Of Environmental Studies* **27** (2) 793-800

[8] Zelenakova M, Purcz P, Harbulakova V O et al. 2016 Determination of pollutant concentrations in the Krasny Brod River profile based on the Buckingham theorem *Desalination and water treatment* **57** (6) Special Issue: SI 2693-2701

[9] Slezingr M 2016 Lowering the water level at the dam reservoir Brno *In. Conference on Public Recreation and Landscape Protection - with Nature Hand in Hand (Krtiny: Czech Republic)* pp 102 – 104

[10] Gałaś S and Król E 2008 Indicators for environmental-spatial order assessment on the example of the Busko and Solec Spa *Communes Gospodarka Surowcami Mineralnymi - Mineral Resources Man Magement* Part: 2 **24** (2) 95-115

[11] Soldo B, Oreskovič M and Anioskin A 2010 An example of examination of the bank slope and water wavering contact *Journal of Landscape Management* No. 2 **1** 41 - 44

[12] Slezingr M 2007 Stabilisation of reservoir banks using an "armoured earth structure" *Journal of Hydrology and Hydromechanics* **55** (1) 64-69

[13] Junakova N and Balintova M 2014 Alkaline modified coal fly ash as an addition to concrete *Chemical Engineering Transactions* **39** 637-642

Acknowledgments

Supported by the Specific University Research Fund of the FFWT Mendel University in Brno, n. of Project: LDF-TP-2019002 Project FAST-S-18-5264. Thanks to The National Scholarship Programme of the Slovak republic there was opportunity to prepare this paper with the colleagues from University of Technology Brno.

IOP Conf. Series: Materials Science and Engineering **566** (2019) 012029 doi:10.1088/1757-899X/566/1/012029

Cost management of building constructions and works through knowledge-based database

J Smetanková[1], T Mandičák[1]

[1] Technical University of Košice, Faculty of Civil Engineering, Department of Construction Technology and Management, Vysokoškolská 4, 042 00 Košice, Slovakia

Abstract. Civil engineering represents an important economical segment. Higher pressure on raising effectiveness, productivity and automation is perceived in society. An important factor that influence raise of effectiveness and productivity is cooperation support between participants of building sector. Individual participants are connected between each other and they are allowed to gain, collect, share, rate and use available information's in required quality and required time. For this purpose, a lot of supporting tools are created and these dispose with automation features, such as knowledge system or knowledge database. One of the most important processes in building's life-cycle is cost setting through the whole life-cycle. Correct cost setting is an important indicator in deciding about size, shape, equipment and also on building realization itself. In building's life-cycle, besides economical aspect that is represented by costs connected with individual life-cycle phases, it's also important to consider other parameters that greatly influence building method and using or disposal of the building. This article, based on mentioned facts, brings a proposal of methods and structure to create knowledge database, that brings user a tool that dispose with information's needed for deciding about construction solution of future building and allows to manage costs for building operations and building works.

1. Introduction

Raising demands for prompt and effective realization of individual processes is displayed also in building sector. During preparation, realization and using of buildings, automatized processes are enforced more and more. One of the ways for raising effectiveness, productivity and automation is creation of knowledge systems. Knowledge systems are used for sharing and managing knowledge's. This system supports acquisition, organization and collection of knowledge's. Knowledge systems are a tool for knowledge management, that includes a number of strategies and experiences focused mainly on discovering competitive benefits and improving organization's efficiency [1].

Knowledge database is created on the ground of setting the specific goal or setting the need. After setting of the goal, it's important to acquire information's, accustom acquired data's, set the style of presentation and database management and constantly update the data's [2].

Availability and recency of data's positively influences the pricing process of building production. Into the process of calculation and building cost setting, there enters a high number of factors, for example wrong setting of starting costs, unexpectable technical problems, not specified changes, missing or wrong extensity definition and outer factors [3].

Building production pricing is an important process in building's life-cycle. Tools that supports building production pricing are on different level. There exists a lot of software applications that allows building production pricing. These applications works on a basis of allocating the price data's and information's to individual bill of quantities' items.

IOP Conf. Series: Materials Science and Engineering **566** (2019) 012029 doi:10.1088/1757-899X/566/1/012029

These limitations could be eliminated or reduced through knowledge system creation. In respect of current status in building production pricing, main goal of research was identified. The main goal of the research is method and structure design for creation of knowledge database that allows creation, identification, organization and sharing the information assets for support the building production pricing, and integration option of proposed solution into BIM environment. The goal was set in consideration of support of Slovak agency for research and development under the contract no. APVV-17-0549: "Research of knowledge-based and virtual technologies for intelligent designing and realization of building projects with emphasis on economic efficiency and sustainability. The main objective project is to develop the methodology for modelling the economic parameters in BIM environment. To propose the multidimensional structure of parametric interactions between intelligent construction element in the BIM model some on cost and economic parameters.

2. Cost estimating of construction production

Construction cost setting is one of the most important processes in building's life-cycle. Proper cost setting, construction price is important indicator in deciding about its 'size, shape, features and also realization itself. Cost is mostly set by price calculations. Success of each project depends on exact cost assessment.

Every building goes through life-cycle. This life-cycle is defined as a time lapse from first thought about building until its disposal. Based on this fact, life-cycle of a building is divided into four basic phases – pre-investment phase, investment phase, operational phase and disposal phase. In each of the life-cycle phases, cost demands are different [4].

Many factors enter the building cost calculation process. Most important reasons of cost raising are small starting cost assessment, unspecified changes, unpredictable technical problems, outside factors and missing or wrong range definition[3].

Result of calculation is the building estimate respectively building budget. Building budget is a form of setting the price in building works pricing. This budget represents priced bill of quantities, built-up according to technical documentation. Individual items are priced by relevant prices of construction items, itemized construction budget is created [5].

Building production pricing is noted for various specifications, which must be taken into account. Main factors are following: building production is custom-made and from bigger part the production is individual, long production cycle with many production processes, it's characterized by movable building production, character of used material, high amount of transport processes, building's place and character, imminent weather influence, performing activities on buyer's ground, investment severity and dependence on loans, transport, accommodation and catering pretence concurrence with building's occupation, realization time period, extent of equipment on project site, delivery character, reimbursement method, delivery and other conditions [6].

3. Knowledge database

Currently more emphasis is put on support the cooperation of institutions, processes, participants, etc. An important tool that supports cooperation is mutual connection of each part and an option to collect, share, rate and use available information's in required quality and required time. For this purpose, a lot of supporting tools are created, such as knowledge system respectively knowledge database. Knowledge database is so-called public library, or a database of relevant information's about particular subject[7].

Knowledge database is a database used for knowledge sharing and management. Database supports gathering, organization and acquisition of knowledge's. Knowledge database does not represent only a space for saving the data's, but through artificial intelligence tools, it supports intelligent determination. Determination is mediated by various knowledge presentation technologies, including frames and scripts. Two main types of knowledge database are known:

- **human readable knowledge bases** allow people to use knowledge's through saving information's, guides, information's about solving problems and saving the FAQ answers. This type of knowledge database can have interactive character, through which the database directly helps user with deciding. Database offers information's but the selection itself is performed by user.

- **machine readable knowledge bases** save knowledge's, but only in readable systems. Solutions are offered on the ground of automatically illative ideas and they are not so interactive because they rely on questions systems, that are included in sciential database basement, and thus the solution is constricted [7].

Knowledge database is an important tool of knowledge management. This management comprehends a number of strategies used in organization for creating, representing, analysing, distribution and accepting the experiences. It is focused mainly on competitive benefits and improving efficiency of organizations [1].

Knowledge database is created by knowledge engineer, that implement knowledge's into the database, based on real experiences from experts in this sphere. Database concept is created on the ground of "if-then" rules – if the condition is valid, then the conclusion is valid. During creation of the system, it's possible to use more rules. In this case, a rule-choosing conflict occurs, therefore is important to set the sequence of rules [8].

Knowledge databases are created in software applications for knowledge management. Software for knowledge management helps user to create, identify, organize and share informative assets. Software saves knowledge's, information's and manage policies, documents and databases thus all information's will be easily available, comprendious and continuously updated[9].

Knowledge databases can be created in many software's, such as: Sabio Knowledge Manage, Jira Service Desk, Dezide, Zoho Connect, Bitrix24, Freshdesk, Deskpro, Lighthouse Desktop, Enteprise Intelligence Platform and other [9].

4. Proposal of methodology and structure of the knowledge database

Knowledge database allows using, sharing and managing the knowledge's. Main purpose is to create a knowledge database of construction features like human readable knowledge bases, that allows users using the knowledge's included in knowledge system. Database will have an interactive character, that will help user in deciding with the system that offers options or information's and then it will offer user an option of particular pick. Database concept will be created on the ground of "if-then" rules - if the condition is valid, then the conclusion is valid. It's important to set the sequence of the rules, with which the conflict during rule choosing can be preceded.

Knowledge database structure will be chosen based on following aspects:

- allocation of an object into functional building parts,
- allowance of economical or financial aspect and time structure,
- allowance of object or feature durability
- allowance of environmental and sustainability aspect

4.1. Description of proposed structure of knowledge database

4.1.1. Allocation of an object into functional building parts. Building object is technically independent, spatially integrated part of a building, that subserves particular function. Functional building part is an integrated part of a building, that serves one or more particular, specified functions. During the building allocation into functional parts, it's important to take into account the building function, constructing method and project documentation structuring, and also in consideration of the connection to budgetary scale. Also important is to take into consideration the technological aspect. The technological aspect deals with the technology and parameters that will be used in each construction process. Technological structure of a construction process represents the object allocation into staging and fragmental construction processes. Based on these facts, an object will be divided into following functional construction parts:

- foundations construction and design of the substructure,
- hydroisolation,
- vertical load-bearing structures,
- partitions and partition walls,
- chimneys,

- ceiling constructions,
- stairways and communications,
- roof construction,
- roof drainage,
- isolations,
- surface modifications (interior, exterior),
- windows,
- doors (interior, exterior),
- floor,
- technical equipment of buildings (water supply, heating, electrical installation, ventilation, air and sanitary installations).

To every functional building all its types and kinds will be allocated. Every type and kind of a building is distinguished by different specific characteristics, therefore it's important to approach every type or kind individually and set particular specific characteristics, parameters and take into account f.e. production aspect, material aspect, technological aspect, financial aspect, option of burden diversity, object or feature durability, environmental aspect, etc. [4] [5].

4.1.2. Economical (financial) aspect and time allowance. Cost needed for the whole life-cycle of a building or a feature will be allocated to individual functional parts of knowledge database. Life-cycle of a building is divided into pre-investment and investment phase (acquisition costs), operating phase (costs for maintenance and operational costs) and disposal phase (disposal costs). Individual construction features will dispose with information's about costs needed in all life-cycle phases, thus costs connected with feature acquisition, maintenance costs, operational and disposal costs. Acquisition costs will be divided on the ground of calculation formula for direct material, direct wages, costs for machines and devices, other direct costs, production and administrative expenses, profit and risk. Besides cost parameters, each feature will include also time parameters, for example direct labour, technological breaks, etc.

4.1.3. Durability of buildings and features. Object ability to serve required functions until the reach of limiting state in maintenance system and reparation is expressed numerically by technical life, middle-technical life or a median use value[10].

Durability of an object is influenced directly from the start of an object predesign. Durability is influenced by quality of the materials, that undergo monitored check-ups and corresponds with required class, building feature realization method or object – technological methods, building standards and regular maintenance of a feature or an object.

Feature durability can be divided into two groups:

- technical durability – period from creation of building feature until its expiration. Technical durability is a period, when the mentioned feature serves its function, meet basic safety requirements and does not endanger building users. In this durability type, emphasis is put mainly on a material structure, that is divided into long-life durability features (bearing construction structures – basement, masonry, etc.) and short-life durability features (all other features - roofing, plumbing structures, windows and doors, floors, insulation, furnishings, etc.),
- economical durability – period from creation of building feature, object until the time when it's profitable to operate the building or a feature. After this time, the building become uneconomical, that means costs are higher than the profit. In this case, it's necessary to change the purpose of a feature, building or to dispose it and substitute it for new one. Economical durability is significantly shorter than technical durability [11].

IOP Conf. Series: Materials Science and Engineering **566** (2019) 012029 doi:10.1088/1757-899X/566/1/012029

4.1.4. Environmental aspects – feature sustainability. A care about the environment is an important activity about socio-economic significance. A healthy environment is the basis for the preservation of human existence, healthy development and influences the factors of the living standard of the population significantly [12][13].

Construction activity influences the quality of the environment greatly and requires enormous consumption of natural resources and energy. A huge amount of waste and pollution is generated. The negative consequences of construction work leaded to redundant pumping of natural sources, pollution, ecosystem disruptions, but also to negative socio-cultural impacts and changes in the environment [14].

Every building, every construction feature influences its surroundings. Influence of a building or construction feature on surrounding environment can have various displays, from this point of view it's possible to evaluate for example. spending resources, environmental pollution, inner environment quality but also functional, aesthetical and social context of a building or a feature. Based on mentioned facts, environmental effects can be integrated into following categories:

- PEE - total amount of primary energy,
- GWP- Global Warming Potential,
- ODP – Ozone Depletion Potential,
- AP – Acidification Potential,
- POCP –potential for photooxidant formation,
- NP – potential for eutrophication of soil and water [15].

An important factor that influences knowledge database's structure are environmental aspects. The term "sustainability" certainly belong between environmental aspects. The definition of "sustainability" is the study of how natural systems function, remain diverse and produce everything it needs for the ecology to remain in balance. Human civilization takes resources to sustain our modern way of life [16][17].

Sustainability consists of three basic pillars economic development, social development and environmental protection:

- to mitigate the effect of climate change, pollution and other environmental factors that can harm and do harm people's health, livelihoods and lives,
- increase health of the land, air and sea,
- sustainable economic growth while promoting jobs and stronger economies and other[18].

Currently, there exists several certification systems and standards, that helps to set the level of concordance of the buildings with sustainable construction principles. The impulse for their creation was the need to set, how are buildings beneficial and the need to evaluate the influence of creation and operation of the buildings on environment. Every building has different ratio of consumption places allocation, therefore evaluating systems were created for the purpose of calculation individual influences on an environment and possible mutual comparison. In Central Europe, four main certification systems are used BREEAM (1990, United Kingdom), LEED (1993, United States of America), DGNB – Deutsche Gesselschaft für Nachbhaltiges Bauen (2007, Germany), SBToolCz – Sustanable Building Tool CZ (2010, Czech Republic)[19].

The priority is to take into account and incorporate mentioned environmental aspects and certification systems into proposed knowledge system.

4.2. *Possibility integrating the proposed knowledge database into building information modelling.* Created knowledge database offers user a deciding tool that implements acquired knowledge's. User can choose correspondent construction feature type from functional, technological, economic and environmental point of view, or a sustainability view. Every feature will carry graphic and non-graphic information's. Graphic information's will be visualized through 3D model. Non-graphic information's of construction feature will represent information's about material, costs (wages, material costs, machine and device costs, other direct costs and operational costs) and information's about feature durability and its' environmental influence or sustainability. So-defined feature will carry similar information's and data's as are features created in BIM environment. Based on this fact, there will be an ambition to integrate proposed solution into BIM environment.

5. Conclusion

Raising requirements for prompt and effective realization of individual processes are displayed also in building sector. Automatized processes are used more and more in preparation, realization and using of the buildings. One of the ways for raising effectivity, productivity and automation is creation of knowledge systems. This article brings a proposal for method and structure of knowledge database, that will allow cost management for building constructions and works and helps with deciding about construction solution for future building. The result of compliance mentioned method and structure of knowledge database will be an evaluation tool that allows user to choose construction feature with consideration of functional, technological, economic and environmental aspect. In the database, user can visualize chosen construction feature, its' graphic and non-graphic parameters. Information's will be provided to user through 3D model that dispose with information's about material, costs, feature durability and environmental influence or sustainability.

6. References

[1] Technopedia *Knowledge base* [Online] Available at https://www.techopedia.com/definition/2511/knowledge-base-klog [accessed 25. Feb. 2019]

[2] Atlassian It Unplugged *What's a knowledge base and why you need it* [Online] Available at https://www.atlassian.com/it-unplugged/knowledge-management/what-is-a-knowledge-base [accessed 25. Feb. 2019]

[3] Venkataraman R R and Pinto K J 2018 *Cost and Value management in projects. In: John Wiley & Sons, Inc.* New Jersey USA ISBN 9780470069134).

[4] Tichá A, Aigel P, Hromádka V, Korytárová J, Marková L and Nový M 2008 *Optimalizace technickoekonomických charakteristík životního cyklu stavebního díla* Volume 1 Brno *Akademické nakladateľstvo* CERM s.r.o. 2008,ISBN 978-80-7204-599-0

[5] Tichá A, Tichý J and Vysloužil R 2008 *Rozpočtování a kalkulace ve výstavbě* Volume 2 Brno *Akademické nakladateľstvo* CERM s.r.o. ISBN 978-80-7204-587-7

[6] Čavojský J 2012 *Oceňovanie stavebných prác – Klasifikácie, kalkulovanie, oceňovanie a rozpočtovanie* ISBN 978-80-970678-1-6

[7] Search CRM *What is knowledge base* [Online] Available at https://searchcrm.techtarget.com/definition/knowledge-base [accessed 28. Feb. 2019]

[8] Ivanko Ľ 2017 *Návrh a optimalizácia kybernetického systému použitím expertných systémov* [Online] Available at http://www.wray.sk/other/dw/dw.pdf [accessed 15. Mar. 2019]

[9] Capterra *Knowledge Management Software FAQs* [Online] Available at https://www.capterra.com/knowledge-management-software/faq/ [accessed 15. Mar. 2019]

[10] Tichá A, Marková L and Puchýř B *1993 Ceny ve stavebnictví,* Volume 1 Brno

[11] ASB *Výpočet technického stavu a ceny vyjadrujúcej opotrebnie stavby* [Online] Available at https://www.asb.sk/stavebnictvo/vypocet-technickeho-stavu-aceny-vyjadrujucej-opotrebenie-stavby [accessed 15. Mar. 2019]

[12] Bezděk J and Svozil P 1987 *Stavební činnost a životní prostředí* Volume 1. SNTL – Ochrana životního prostředí 167 p.

[13] Jokl M and Kočí J 1986 *Výstavba jako faktor tvorby životního prostředí* Volume 1 SNTL – Ochrana životního prostředí 187 p

[14] Zaťková M 2012 *Stavebná činnosť a životnné prostredie POSTERUS - Portál pre odborné publikovanie* ISN1338-0087 [Online] Available at: https://www.posterus.sk/?p=13466&output=pdf [accessed 15. Mar. 2019]

[15] Nagy E 2004 *Environmental Assessment of Building Materials, Constructions and Buildings.* Volume 38, No. 4, p 182 – 187 [Online] Available at: http://147.213.211.222/sites/default/files/2004_4_182_187_nagy.pdf [accessed 15. Mar. 2019]

[16] United States Environmental Protection Agency *Sustainability* [Online]. Available at https://www.epa.gov/sustainability [accessed 15. Mar. 2019]

[17] Mason M *What Is Sustainability and Why Is It Important?* [Online] Available at https://www.environmentalscience.org/sustainability [accessed 25. Mar. 2019]

[18] Un Department of Public Information *General Assembly's Open Working Group proposes development goals* United Nations Conference on Sustainable Development 2014 [Online]. Available at https://sustainabledevelopment.un.org/content/documents/4538pressowg13.pdf [accessed 25. Mar. 2019]

[19] Aachinfo *Hodnotenie udržateľnosti budov – metodika CESBA, časť 6: Certifikačné nástroje* [Online] Available at https://www.archinfo.sk/diela/teoreticka-praca/hodnotenie-udrzatelnosti-budov-metodika-cesba-cast-6-certifikacne-nastroje.html [accessed 25. Mar. 2019]

Acknowledgments
The paper presents a partial research results of project APVV-17-0549 "Research of knowledge-based and virtual technologies for intelligent designing and realization of building projects with emphasis on economic efficiency and sustainability" and project VEGA 1/0828/17 "Research and application of knowledge-based systems for modelling cost and economic parameters in Building Information Modelling".

IOP Conf. Series: Materials Science and Engineering **566** (2019) 012030 doi:10.1088/1757-899X/566/1/012030

Evaluation of the antimicrobial efficiency of slag based composites

M Smolakova[1] and H Vojtkova[2]

[1] Technical university of Kosice, Faculty of Civil Engineering, Department of Environmental Engineering, Slovakia
[2] VSB-Technical University of Ostrava, Faculty of Mining and Geology, Institute of Environmental Engineering, Czech Republic

Abstract. Degradation of cement mortar composites occurs due to a variety of causes such as corrosion (sulphate or chloride attack) cracking or shrinkage. Chemical agents often play a major role in corrosion processes, however recent research has revealed the importance of biological agents. Microbiological corrosion is caused by the presence and activities of microorganisms, including algae, bacteria, yeasts and fungi. These microorganisms produce substances like biogenic organic and mineral acids that can disrupt the matrix of mortar composites. The most effective method for improving the resistance of cement based composites is the addition of pozzolans like blast furnace slag. Slag is often used in preparation of cement composites to improve the acid resistance, impermeability and crack resistance of mortars. This article presents the investigation of antimicrobial efficiency of blast furnace slag regarding the selected species of bacteria (*Pseudomonas aeruginosa, Micrococcus luteus*), yeasts (*Candida albicans, Saccharomyces cerevisiae*) and algae (*Chlorella vulgaris*). Antimicrobial properties were studied on mortars with different share of slag (65% - 95%). Antimicrobial efficiency for the tested species of microorganisms was found moderate to low.

1. Introduction

Building materials are progressively and inevitably subjected to biological colonization [1,2]. For some years now, it has been well known that microbiological contamination and growth on building materials may lead to the severe issues, including the structural degradation of materials (biodeterioration) [3,4]. More especially, the sewers become more and more bioreceptive due to their exposition to waste water [2,5].

Prolonged exposure of any solid surface to water results in a series of physical, chemical, and biological events at the surface that form a community of microorganisms referred to as biofilm. The process starts with initial settlement of free floating microorganisms on the surfaces followed by the accumulation of an organic material layer, generally the extra-cellular polymeric substances (EPS) excreted by the attached microorganisms [6].

Several biological organisms can be involved in the biocolonization of sewer network, such as bacteria, cyanobacteria, algae, fungi and yeast. Among these microorganisms, cyanobacteria and algae are predominant in the biofilm formed on the substrates, and they are the first colonizers after bacteria [2,7]. The intensity and kinetics of biological colonization of building surfaces depends on the environment and substrate [2,5,8,9]. Physical (roughness, porosity) and chemical (mineralogical/chemical composition, surface pH) characteristics of the substrate influence the biological colonization [2,5,10].

In attempts to prevent or limit microbial development on building materials, various solutions have been studied like using of biocide or using pozzolanic materials like blast furnace slag, fly ash or silica fume, to improve the quality of these materials [4,11].

Recently, there has been a growing trend for the use of supplementary cementitious materials, whether natural, waste, or by-products, in the production of composite cements because of ecological, economical, and diversified product quality reasons. Blast furnace slag, a by-product of the transformation of iron ore into pig-iron in a blast furnace, is one of these materials whose use in cement manufacture goes to as far back as 1880 [12]. Since then its use has expanded because it has various advantages over other cementitious materials. Firstly, blast furnace slag has a relatively constant chemical composition compared to fly ash, silica fume etc. Besides, it has advantages like low heat of hydration, high sulfate and acid resistance, better workability, higher ultimate strength, etc. [1,13].

It was shown that cement mortars with the addition of blast furnace slag exhibits a greater resistance than ordinary Portland cement to inorganic or organic acids produced by microorganisms [2,14]. These properties are beneficial for specialised objects associated with water and waste water management such as hydroelectric dams, sewer networks, large bridges and harbours [13].

The aim of the present study is to investigate the potential antimicrobial efficiency of slag based cement mortars regarding the selected species of bacteria (*Pseudomonas aeruginosa* and *Micrococcus luteus*), yeasts (*Candida albicans* and *Saccharomyces cerevisiae*) and algae (*Chlorella vulgaris*).

2. Material and methods

2.1. Mortar composites

Five different cement composite mixtures were investigated in the bacterial experiment. The R represented the reference sample with 100 wt.% of Portland cement CEM I 52.5 R, with the content of main components as follows: 57.15 wt.% of CaO, 18.11 wt.% of SiO_2, 4.02 wt.% of Al_2O_3, 2.69 wt.% of Fe_2O_3, 1.49 wt.% of SO_3, 1.37 wt.% of MgO, 1.12 wt.% of K_2O, 0.33 wt.% of P_2O_5, 0.18 wt.% of TiO_2, 0.06 wt.% of Cl.

The samples with different share of blast furnace slag were BFS1 (65 wt.% of slag), BFS2 (75 wt.% of slag), BFS3 (85 wt.% of slag) and BFS4 (95 wt.% of slag), respectively. The content of main components of incorporated blast furnace slag was as follows: 39.55 wt.% of CaO, 38.95 wt.% of SiO_2, 10.11 wt.% of MgO, 8.33 wt.% of Al_2O_3, 0,74 wt.% of MgO, 0.57 wt.% of SO_3, 0.54 wt.% of Fe_2O_3, 0.48 wt.% of K_2O, 0.37 wt.% TiO_2, 0.04 wt.% of P_2O_5, 0.02 wt.% of Cl.

2.2. Microorganisms

Microscopic bacteria *Pseudomonas aeruginosa* and *Micrococcus luteus*, yeasts *Candida albicans* and *Saccharomyces cerevisiae* and algae *Chlorella vulgaris* were obtained as new isolates from the environment of metal-contaminated sludge ponds after ore extraction [15,16,17].

2.3. Methods

The composite samples were crushed and powdered and then sterilized in sterilizer Stericell (BMT Medical technology, Czech Republic) at 180°C during 1 hour (figure 1). Soya Casein Digest (HiMedia Laboratories, Mumbai, India) represented the growth medium for bacteria, Algae Culture Agar (HiMedia Laboratories, Mumbai, India) was used as a growth medium for algae and for growth medium for yeast the Sabouraud Dextrose Broth (HiMedia Laboratories, Mumbai, India) was used. All of the growth mediums were used for inoculation and cultivation of each microorganism and the composition of each medium is shown in table 1 .

Table 1. The composition of used growth mediums.

Medium	Composition (in 1 L H_2O)	Final pH of medium
Soyabean Casein Digest Medium	17g tryptone, 3g soya peptone, 5g NaCl, 2.5g dextrose (glucose) and 2.5g K_2HPO_4	7.3 ± 0.2

Sabouraud Dextrose Broth	20g dextrose (glucose) and 10g special peptone	5.6 ± 0.2
Algae Culture Agar	1g $NaNO_3$, 0.25g K_2HPO_4, 0.513g $MgSO_4$, 0.05g NH_4Cl, 0.058g $CaCl_2$, 0.003g $FeCl_3$ and 15g agar	7.0 ± 0.2

Sterilization of growth medium was realized in laboratory autoclave at 121°C for 15 minutes. In total, five Erlenmeyer flasks with growth medium were prepared, each for every powdered composite sample. Immediately after the sterilization of growth medium the powdered composite samples were added and thoroughly mixed to the broths and this mixture was then placed in the Petri dishes (with an average Ø 90.0 mm). The antimicrobial activity was determined by dilution methods in agar growth media so that the resulting concentration of tested composite sample in growth media was 10%, by mass.

On the surface of solidified growth media bacteria, yeasts and algae were evenly dispersed (figure 2). Each microorganism was placed on separate Petri dish with growth medium. The duration of experiments was one week for all studied microorganisms.

Figure 1. Sterilization of mortar samples in Stericell.

Figure 2. Inoculation of tested microorganisms on growth medium with composite samples.

The scale of evaluation of the microorganism growth on building materials is according to [18] by where the degree of the growth is expressed as follows:

- No growth - No growth of microorganisms.
- Poor growth - The growth of microorganisms is negligible (the colonies of microorganisms are dispersed).
- Fair growth - The growth of microorganisms is gradual (numerous small colonies of microorganisms, which cover 25% of the sample surface).
- Good growth - The growth of microorganisms is intensive (colonies of microorganisms cover up to 50% of the sample surface).
- Excellent growth - The growth of microorganisms is very intensive (colonies of microorganisms cover up to 75% of the sample surface).
- Full growth - The sample surface is fully covered by fungi (colonies of microorganisms cover 100% of the sample surface).

3. Results and discussion

The antimicrobial efficiency of composites samples with blast furnace slag at inhibition of growth of selected microorganisms are given in table 2. The highest antimicrobial efficiency regarding the degree of growth of all tested microorganisms was observed for the reference sample R with no addition of slag and similar results was detected for sample with 65% of blast furnace slag.

Table 2. The antimicrobial efficiency of composite samples.

Microorganism	Sample				
	R	BFS1	BFS2	BFS3	BFS4

Pseudomonas aeruginosa	No growth	No growth	Full growth	Fair growth	Full growth
Micrococcus luteus	No growth	No growth	Good growth	Good growth	Good growth
Candida albicans	No growth	Poor growth	Full growth	Fair growth	Fair growth
Saccharomyces cerevisiae	No growth	No growth	Good growth	Poor growth	Fair growth
Chlorella vulgaris	No growth	No growth	Excellent growth	Poor growth	Fair growth

Figure 3 shows the effect of *Pseudomonas aeruginosa* on growth media with addition of cement composite with BFS. Samples R, BFS1 and BFS3 were found to have no to minimal amount of dispersed colonies of *P. aeruginosa* and no antimicrobial activity was found for samples BFS2 with 75 wt.% of slag and sample BFS4 with 95 wt.% of slag.

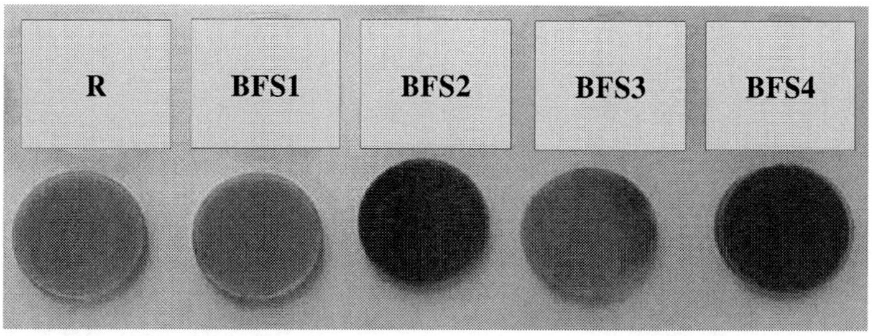

Figure 3. Antimicrobial efficiency of composite samples regarding to strain *Pseudomonas aeruginosa*.

Antimicrobial efficiency of mortar samples regarding the strain *Micrococcus luteus* is shown on figure 4. The highest antimicrobial efficiency was found for the samples with ordinary Portland cement R and sample BFS1 with 65 wt.% of slag. For the other 3 samples good growth of microorganisms was detected.

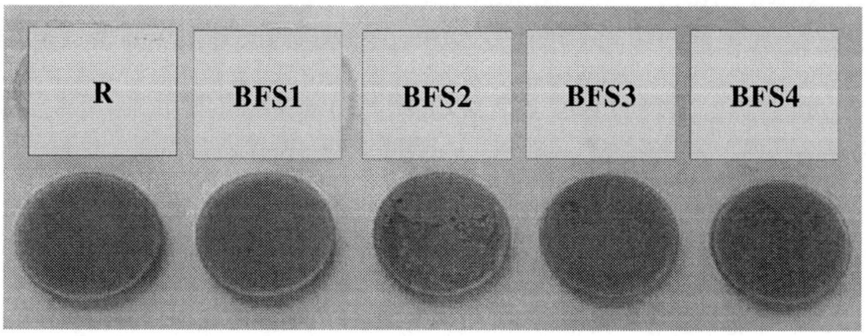

Figure 4. Antimicrobial efficiency of composite samples regarding to strain *Micrococcus luteus*.

The results of determining the antimicrobial properties of mortars against strain of yeast *Candida albicans* are given in figure 5. The most resistant sample was the reference sample R. The growth of colonies of *C. albicans* was detected on all of the samples with blast furnace slag, with the most effective antimicrobial properties for the sample BFS1 with 65 wt.% of slag.

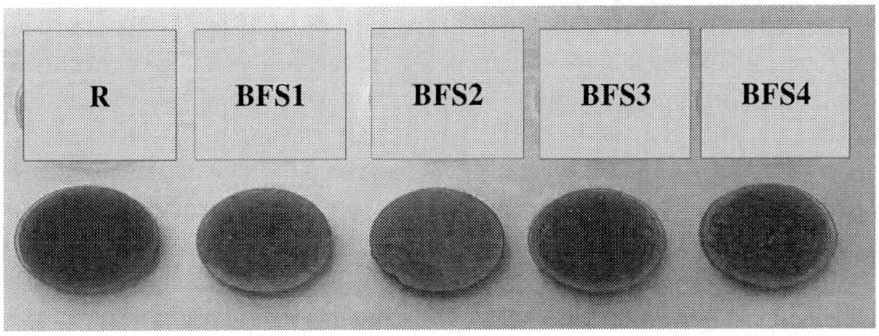

Figure 5. Antimicrobial efficiency of composite samples regarding to strain *Candida albicans*.

Best results regarding to the inhibition of growth of microorganism was evaluated for the strain of yeast *Saccharomyces cerevisiae* (figure 6). The growth was determined to be ranging from no growth to fair growth of colonies of yeasts, except for the sample BFS2 with 75 wt.% of slag, where the inhibition was only around 50%.

Figure 6. Antimicrobial efficiency of composite samples regarding to strain *Saccharomyces cerevisiae*.

Figure 7 shows the effects of algae *Chlorella vulgaris* on composites samples, the highest antimicrobial properties was obtained for reference sample (R) and sample with 65 wt.% of slag (BFS1), respectively.

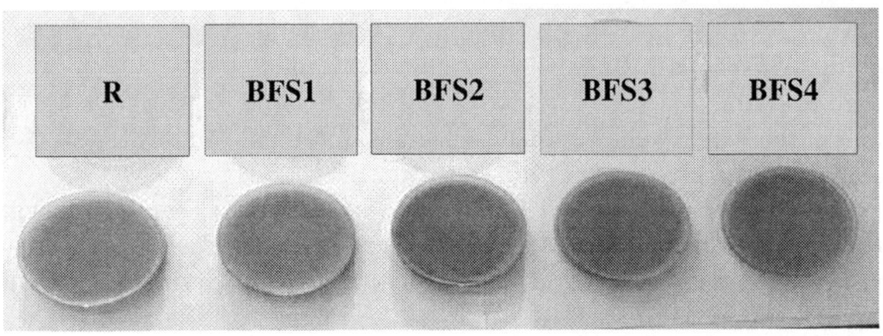

Figure 7. Antimicrobial efficiency of composite samples regarding to strain *Chlorella vulgaris*.

Overall the best results for determining the antimicrobial efficiency had the reference sample R with ordinary Portland cement. For the composites with blast furnace slag, the most resistant was found to be the sample BFS1 with 65 wt.% of slag. Results obtained for the sample BFS2 with 75 wt.% of blast furnace slag showed little inhibition efficiency. This outcome is more likely linked with possible biological fouling of sample during milling process and from this point of view additional experiments regarding this sample are needed.

4. Conclusion

The aim of the paper was the evaluation of possible antimicrobial efficiency of cement composites with different share of blast furnace slag. The inhibition properties were studied on five different strains of microorganisms (*Pseudomonas aeruginosa, Micrococcus luteus*, *Candida albicans, Saccharomyces cerevisiae* and *Chlorella vulgaris*). According to the results, the addition of blast furnace slag is suitable for achieving microorganism-free environment mostly regarding yeast *Saccharomyces cerevisiae* and also *Candida albicans*. The highest inhibition efficiency was found for the reference sample with ordinary Portland cement. For the samples made with slag, the sample with 65 wt.% was found to be most effective for all studied strains of microorganisms. Since in the experiment a low concentration of studied composites in the growth media was used, the mechanism of the antimicrobial effects of blast furnace slag should be the subject of further study.

5. References

[1] Harbulakova V O, Estokova A and Luptakova A 2018 An Investigation of the Bacterial Influence of Acidithiobacillus Thiooxidans on Concrete Composites *E3S Web Conf.* **45** 1-6

[2] Govin A, Grosseau P, Lors C and Damidot D 2018 Impact of a thin coating layer of calcium aluminate and sulfoaluminate cements on algal biofouling *Proc. Int. Fin. Conf. of RILEM TC 253-MCI: Microorganisms-Cementitious Materials Interactions (Toulouse)* vol 2 (Paris: RILEM Publications SARL) p 319

[3] Bertron A 2014 Understanding interaction between cementitious materials and microorganisms: a key to sustainable and safe concrete structures in various contexts *Mater. Struct.* **47** 1787-1806

[4] Verdier T, Bertron A and Roques CH 2018 Antibacterial properties of a photocatalytic paint-like coating for the protection of building materials against microbial proliferation *Proc. Int. Fin. Conf. of RILEM TC 253-MCI: Microorganisms-Cementitious Materials Interactions (Toulouse)* vol 2 (Paris: RILEM Publications SARL) p 355

[5] Estokova A, Ondrejka Harbulakova V, Luptakova A, Prascakova and Stevulova N 2011 Sulphur oxidizing bacteria as the causative factor of biocorrosion of concrete *Chem. Eng. Trans.* **24** 1-6

[6] Alum A, Rashid A, Mobasher B and Abbaszadegan M 2008 Cement-based biocide coatings for controlling algal growth in water distribution canals *Cem. Concr. Comp.* **30** (9) 839-847

[7] Tomaselli L, Lamenti G, Bosco M and Tiano P 2000 Biodiversity of photosynthetic microorganisms dwelling on stone monuments *Int. Biodeterior. Biodegrad.* **46** 251-258

[8] Crispim C A, Gaylarde P M and Gaylarde C C 2003 Algal and cyanobacterial biofilms on calcareous historic buildings *Curr. Microbiol.* **46** 79-82

[9] John D M 1988 Algal growth on buildings: A general review and methods of treatment *Biodeterior. Abstr.* **2** 81-102

[10] Estokova A, Harbulakova V O, Luptakova A, Stevulova N, Palascakova L and Repka M 2013 Analysis of the selected characteristics changes of cement composites exposed to the sulphate environment *Chemical Engineering Transaction* **32** 1-6

[11] Harbulakova V O, Estokova A and Luptakova A 2016 June) Study of dependencies between concrete deterioration parameters of fly ash-based specimens *Proc. Int. Conf. on Dependability and Complex Systems (Wroclaw)* (Cham: Springer International Publishing AG) pp 229-238

[12] Higgins D D 1995 Ground granulated blast furnace slag *World Chem.* **6** 51-52

[13] Öner M, Erdoğdu K and Günlü A 2003 Effect of components fineness on strength of blast furnace slag cement *Cem. Concr. Res.* **33** (4) 463-469

[14] Herrison J, Van Hullebusch E D, Moletta-Denat M, Taquet P and Chaussadent T 2013 Toward an accelerated biodeterioration test to understand the behavior of Portland and calcium aluminate cementitious materials in sewer networks *Int. Biodeterior. Biodegrad.* **84** 236-243

[15] Simonovicova A, Ferianc P, Vojtkova H, Pangallo D, Hanajik P, Krakova L, Feketova Z, Cernansky S, Okenicova L, Zemberyova M, Bujdos M and Pauditsova E 2017 Alkaline technosol contaminated by former mining activity and its culturable autochthonous microbiota *Chemosphere* **171** 89-96

[16] Simonovicova A, Petkova K, Jurkovic L, Ferianc P, Vojtkova H, Remenar M, Krakova L, Pangallo D, Hiller E and Cernansky S 2016 Autochthonous microbiota in arsenic-bearing

technosols from Zemianske Kostolany (Slovakia) and its potential for bioleaching and biovolatilization of arsenic *Water Air Soil Poll.* **227** (9) 336

[17] Smolakova M, Estokova A and Vaclavik V 2018 Antifungal efficiency of slag based cement composites *Proc. Inter. Multidisciplinary Scientific GeoConference: SGEM: Surveying Geology & mining Ecology Management (Bulgaria)* vol 18 (Sofia: STEF92 Technology) 27-34

[18] CSN 72 4310: 1977 *The testing of mould proofness of building product and materials*

Acknowledgements
This paper has been prepared within the Grant No. 2/0142/19 of the Slovak Grant Agency for Science.

IOP Conf. Series: Materials Science and Engineering **566** (2019) 012031 doi:10.1088/1757-899X/566/1/012031

The Influence of corrosion on the stresses in members of open deck railway bridges

R Štecák[1]

[1] Slovak University of Technology in Bratislava, Faculty of Civil Engineering, Department of Steel and Timber Structures, Slovakia

Abstract. Nowadays, many old riveted railway bridges, mainly their members of the open deck, are corroded. The paper demonstrates the influence of corrosion on a stringer beam and on a cross-girder beam of a particular steel truss railway bridge in Slovakia. This is an important step in the assessment of the loading capacity and the residual lifetime of this bridge. A simple beam FEM model was refined by modelling the investigated members as plate elements. In such a way the real boundary conditions of the members were modelled most realistically by keeping the model as simple as possible at the same time. The stresses in the members were first calculated with the original cross-sectional dimensions and the results were compared with the real measured corroded cross-sections. A future corrosion rate was estimated and calculation of this case was also performed. The influence of stringer to cross-beam rigidity, the size of 2D FEM elements, and the steel material model (linear, bilinear with 1% hardening) were investigated too.

1. Introduction

Many bridges in Slovakia are in poor conditions. Their complete diagnostics is therefore required. Instead of designing new bridges, it is useful to verify the load-carrying capacity and the residual life of the old ones. The exact procedures are difficult. The software model of the bridge has to agree with reality to a high degree so that realistic internal forces and stresses in the members can be assessed.

A finite element method (FEM) beam model is frequently used for an analysis. Such a beam model can assess only internal forces and moments in the members. The stresses can then be calculated by means of an engineering theory. Stresses in the elements of the cross-section are different. That is the reason, why a beam model is often insufficient. A FEM model using plate or solid elements can solve this problem. With a plate FEM model it is possible to detect stresses in every structural detail.

2. Finite element method

With FEM it is possible to divide members into many elements. The mesh density can be defined by the user (Figure 1). The mesh of elements can be shaped in various ways. Triangles, rectangles, squares, trapezoids etc. It is necessary to use the right shape of elements for the mesh. Usually a rectangular (square) shape is the most appropriate. In the case of corrosion hole (or gap) modelling, triangles or trapezoids are needed. Generally, the program can define the type of mesh automatically, but the user is advised to control it manually.

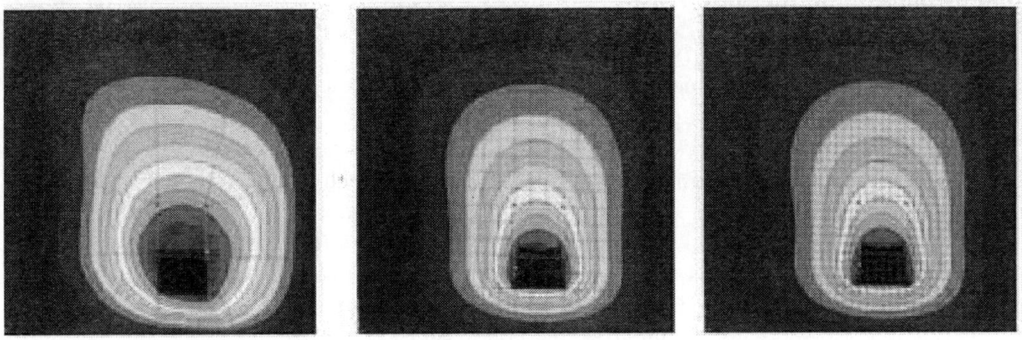

Figure 1. Effect of mesh density on results.

The mesh density of elements is also very important in obtaining proper data. Insufficient division can lead to incorrect results. This is caused by low approximation of elements. The more dense the mesh is, the more precise the results will be.

It is very demanding to analyse bridge structures with FEM plate model. For each plate element a proper mesh density has to be defined. Creation of such a model can take several weeks. In practice there is usually not enough time to make this type of calculation. Common engineers could find these calculations very demanding.

3. Modelling and the results

Calculations are much easier with a beam model. Instead of many hours of calculation time, it only takes several minutes. This model is appropriate for calculation of internal forces, moments and deformations. It is not possible to find out stresses in details. The best solution is a combination of both types of models. A beam model combined with a plate FEM model. The critical investigated member can be modelled as a plate member and incorporated into a beam model. However, the connection of these two members is the most problematic issue. With improper connection the results could become inaccurate or even wrong. In this case, it is very important to choose the right mesh density.

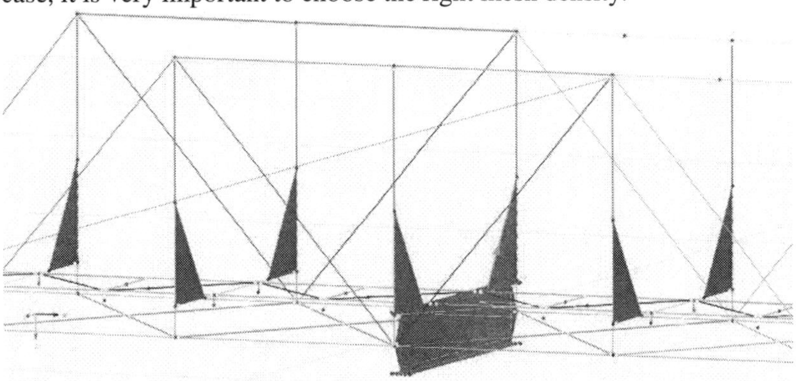

Figure 2. Combined model.

A riveted railway steel bridge in Eastern Slovakia was assessed. It is a 129-year old bridge. The bridge was reconstructed in the 60´s. Corrosion is one of the biggest problem of all old bridges, this particular one was no exception. The current state of the members is quite different in comparison with their original condition. A large part of the steel plates were corroded. A thorough check of all members was necessary. The loss of material in some places was almost half of the thickness of the plates. All elements of the bridge were measured and compared with the drawing documentation. Actual condition of all elements was classified. The first class was in perfect condition without corrosion, the second one with a corrosion of approximately 2 mm of the horizontal surfaces of the plates and the third class with corrosion of about 4 mm. There were not many elements in the third class, but these elements, without strengthening, will develop corrosion gaps in the near future. Complete repair or exchange of this elements will be needed. A complete review was therefore necessary to be done, in order to be aware of the effects of such unpleasant features to the construction.

Program Dlubal RFEM was used for all calculations. This program could calculate a beam model and a 3D FEM model. A space beam model of the structure was made first to properly model the structure, because the bridge is a skew one. The loads were then modelled. Effect of permanent loads, traffic loads (standard load model UIC 71, braking forces, nosing force, etc.) and wind were taken into account. Then load combinations were made with the rules of Eurocode. Later the cross-beam and the stringer were separately analysed in more detail with FEM plate models. Original dimensions were used first, then with corrosion class 2. Due to actual condition of some stringers also a model with gaps in the web was modelled (Figure 3).

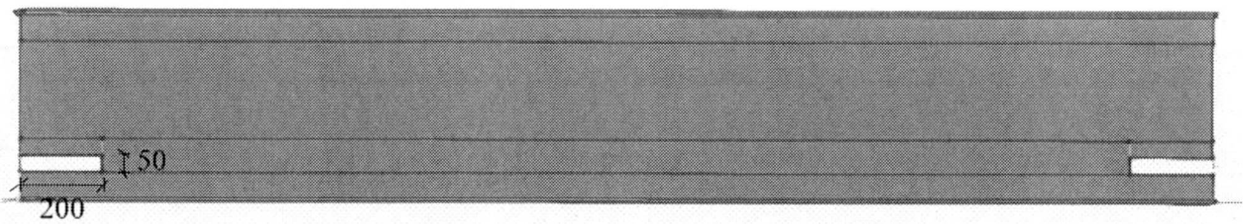

Figure 3. Stringer with gaps at the supports.

The FEM model was created by plate surfaces. The cross-sections are riveted. Connections between each surface were made using rigid surfaces, because a line of rivets is rigid enough (up to a very high level of corrosion). The rigid surface was located at the line where the actual rivets are placed. Connection of web with the angles was simplified. A plate with a thickness of both angles and web was used. Due to a better transfer of forces a rigid surface was placed also into the middle of the cross-section above the member web (Figure 4). Steel S235 with a yield stress 235 MPa was used for calculation.

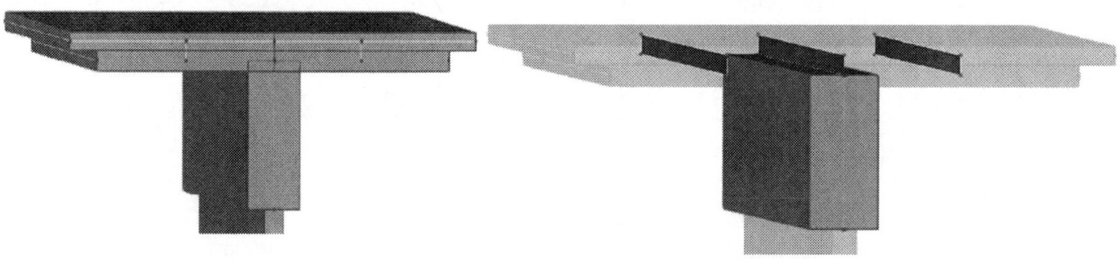

Figure 4. Connection between components.

It was necessary to properly model the boundary conditions and the loading of the separately investigated member. The internal forces and moments from the beam model were used. These forces and moments were used to calculate the imposed line support displacements and rotations. In combination with the loading we arrived to the same internal force and moment distribution as in the beam model.

The loading of the stringer was modelled by a surface load to the top of the member. The cross-beam has a different distribution of bending moment. This was slightly varied to make it easier for modelling. This is shown in Figure 5.

The load was divided between the elements. One third of the load was acting on the flange. The rest of the load was acting on the web. Calculation was without any problem with this division.

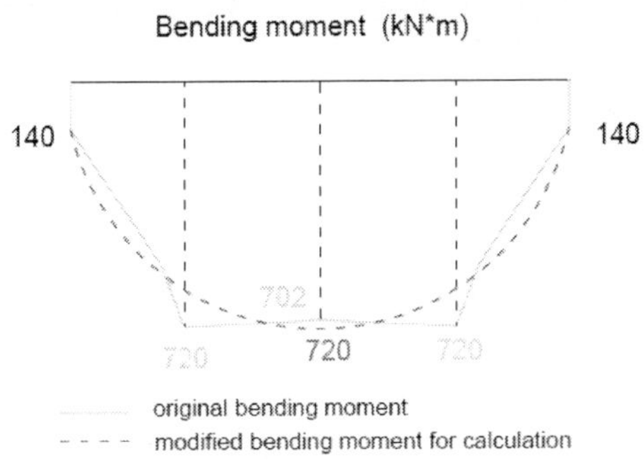

Figure 5. Bending moment of cross-beam.

The members had pinned supports. In cross-beam there was a nodal support inserted in the place of the stringers. These supports were free in the direction of axis y and z. They were prevented to move in the direction of axis x, that means out of the web of the cross-beam. Two models were created. The first one, the original, the second one with corrosion. The member was classified as class 2.

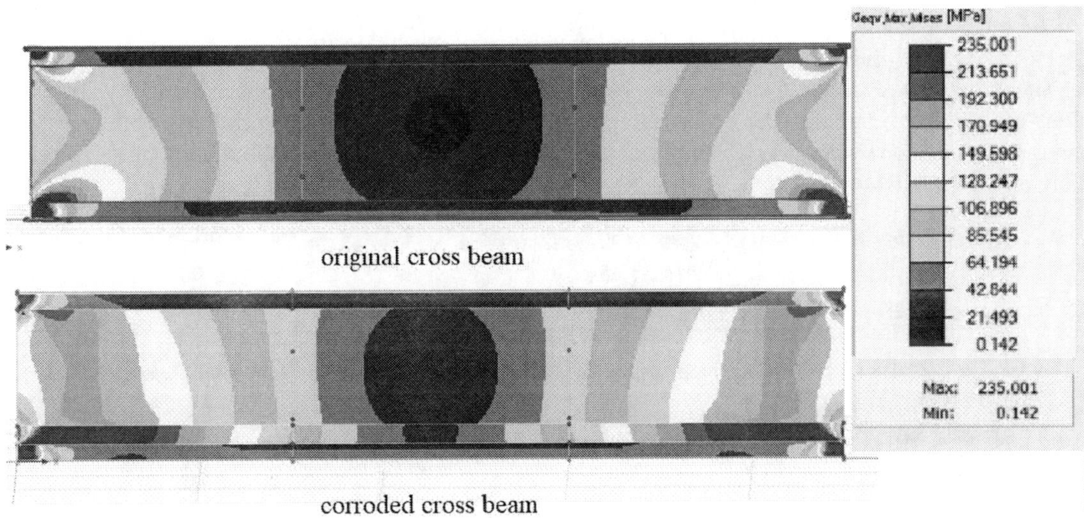

Figure 6. Stresses in cross-beam.

The results show that the effect of corrosion is significant. Stresses in the bottom of web reached 235 MPa and plastification occurs. Stresses at the supports got higher too. The original state beam performed in an acceptable way, but the actual condition of the member can be dangerous. Some members of the bridge were even in a worse condition. Different results have been achieved by different mesh density.

Final mesh density was 10 mm. Diagram 1 is provided for the comparison. It shows time dependence of mesh density.

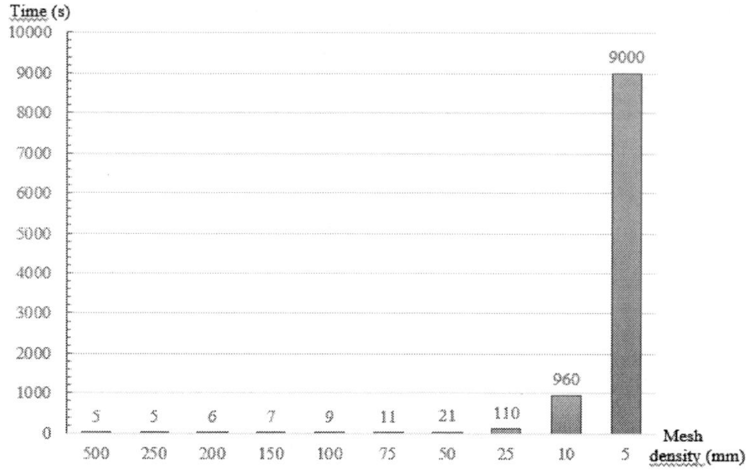

Figure 7. Time dependence of mesh density.

The stringer was created the same way. Nodal supports were used, however, without any stiffening along the length of element. There were three models. The original one, the corroded one in class 2 and the corroded one with gaps. Results were the same as for the cross-beam, but gaps caused a much more change in stresses.

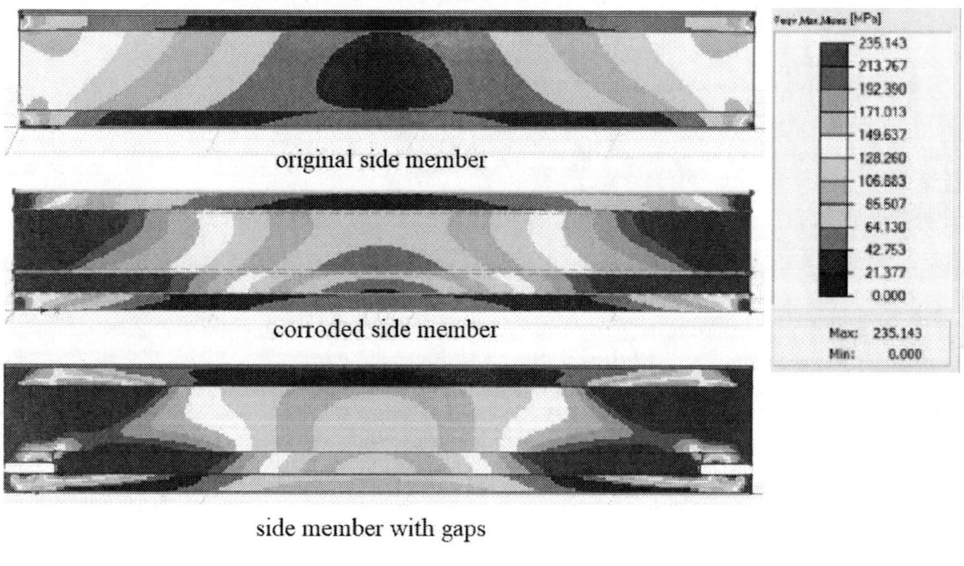

Figure 8. Stresses in stringer.

As it is shown in Figure 7, element with gaps has high stresses not only at the supports, but across the entire cross-section as well. This was only an experimental model, but the current state of some stringers approach already this condition.

The combined model was created on the basis of the beam model. The cross-beam plate model was embedded into the beam model. Connection between both model types was made by means of a rigid area. Triangle shape of area was used to provide connection between members. This type of connection ensured the right transfer of forces. A problematic detail was the connection between cross-beam and stringer. Local peaks of stresses appeared there. Deformations of the plate model were similar as of the beam model.

The results depend on many factors. First one is a problematic connection between the beam and plate model. As shown in Figure 8, local stresses exceed the limit in these places. With a mesh density of about 10 mm we obtained lower stresses than with the one with 20 mm density. Problem was calculation time.

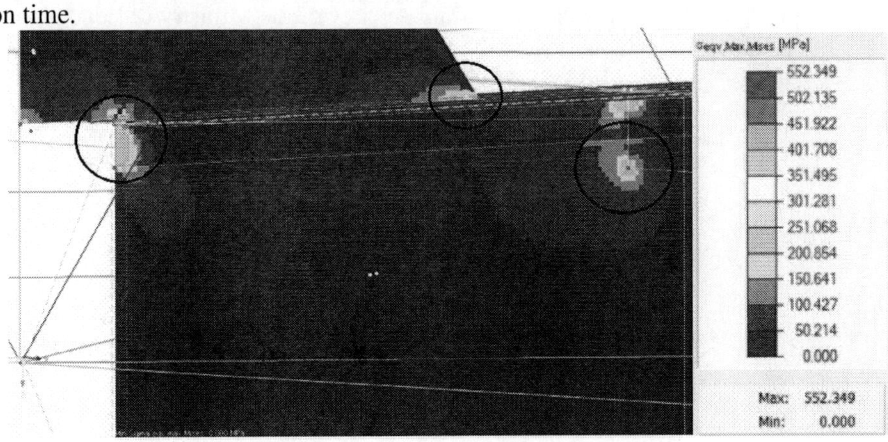

Figure 9. Local areas of stress peaks.

In addition, calculation with linear and bilinear diagram of steel was compared. It is considered with hardening of steel in bilinear diagram. For this time hardening about value of 1 % was used. It has no effect to the results. Stresses in cross-beam were the same.

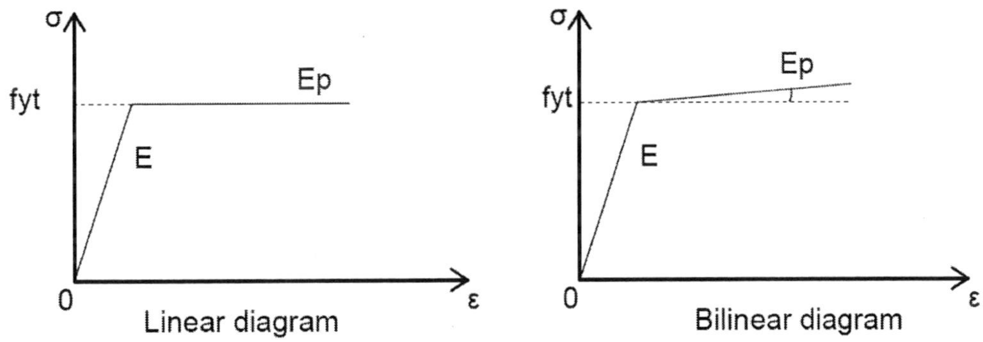

Figure 10. Diagrams of steel [3].

4. Conclusion

Assessment and design of steel bridges and other more complicated structures is not easy. The best way is to use a combined model, but it is not always possible. The biggest problem is the necessity of very sophisticated information technology. For the beginning the best way for exact computing is to make a beam model. With this model it is possible to detect internal forces, moments and deformations. Then to select members for the plate FEM model. With the internal forces and moments from the previous model it is possible to assess the stresses. The best way to get the right stresses is to input internal forces by imposed deformations into the member.

5. References

[1] EN 1993-1-1. 2005 Eurocode 3: Design of steel structures- Part 1-1: General rules and rules for building
[2] EN 1993-2. 2006 Eurocode 3: Design of steel structures- Part 2: Steel Bridges, October
[3] RFEM 5 2016 Program Description (Prague: Dlubal Software s.r.o.)

IOP Conf. Series: Materials Science and Engineering **566** (2019) 012031 doi:10.1088/1757-899X/566/1/012031

[4] Kocharík J Imrich V 2007 Metóda konečných prvkov v teórii a príkladoch Department of production technologies Košice

Acknowledgement
This paper has been supported by the Scientific Grant Agency of the Ministry of Education, science, research and sport of the Slovak Republic and the Slovak Academy of Sciences – grant VEGA 1/0773/18.

IOP Conf. Series: Materials Science and Engineering **566** (2019) 012032 doi:10.1088/1757-899X/566/1/012032

Proposal of utilized to the lift for evacuation of persons

M Szénay[1]

[1] Technical university of Kosice, Faculty of Civil Engineering, Institute of Architectural Engineering, Slovak Republic

Abstract. In 2010 Averill defined in his work five challenges for the solution of evacuation of persons in buildings to 2020. One of the challenges is to implement helpfull technologies during evacuations from buildings – lifts. The aim of the work was to find the ideals of the design of the necessary type and number of lifts. The numerical calculation method was used using simulation software. A model of a fifteen-storey building with multiple versions of the lifts was created. The results showed what type and number of lifts brought the desired shortening of the evacuation time. Conclusion is the recommendation of an ideal lift design.

1. Introduction

Potential of passenger lifts is not sufficiently used as it could be. Passenger lifts are not allowed for evacuation of people in the world and in the Slovak Republic except for lifts directly intended for evacuation. Lifts directly designed for evacuation are fire and evacuation lifts that also operate during a fire or power failure. However, the legal regulation of the Slovak Republic [1] commands the design of these lifts only in exceptional cases. In most cases related to buildings, people are forced to use staircases for evacuation. For example, in a building with more than two stories high with up to ten person of limited movability, evacuation lifts may not be proposed [2]. In this way many buildings (not only in Slovak Republic) represent a series of challenges for safe evacuation, mainly based on 2 reasons - limitations of movement from the point of view of physical handicaps of people [3], [4], or fatigue during movement downstairs [5]. Within the framework of the legal regulation of the Slovak Republic [6], three movement properties of people are considered. The first group includes persons able to move alone. Each person, whose movement is not limited from the health point of view, can be categorized to this group. The 'persons of limited movability' belong to the second group. Persons, whose evacuation is more demanding than in the case of other persons, belong to this group. Old people, children, disabled persons of limited movability are an example of this group. The third, last group includes 'immobile persons'. Persons, whose evacuation is possible only with help of other persons, are categorized to this group. For example, mentally affected persons, in-patients at clinics with serious health states, infants, toddlers and the like belong to this group. According to the Statistical Office [7], almost 20% of inhabitants are persons in the old pension age (62 years and more), and 12% of inhabitants are persons of the school age (4 to 15 years) in Slovakia in 2017. Altogether at least one third of inhabitants of the Slovak Republic can be considered persons of limited movability in the case of the proposal of escape routes. Moreover, since 2002 the legal regulation states that all public building in the Slovak Republic must be accessible for persons of limited movability and orientation. For that reason, lifts are inevitable for evacuation of persons of limited movability.

In 2010, Averil [8] defined several necessary steps to cope with large challenges presented by the solution of evacuation in buildings. The quantification of missing data dealing with evacuation of persons using lifts is one of these steps. Up to the present day, several papers and studies on evacuation of persons using lifts were carried out based on simulated calculations or in-situ experiments. Study by

IOP Conf. Series: Materials Science and Engineering **566** (2019) 012032 doi:10.1088/1757-899X/566/1/012032

Ding Y. et al [10] can served as another example in this field, where simulation software was used to ascertain, beside others, which ratio of persons using the lift and staircase has the most favorable effect on the course of evacuation. Therefore, they modeled in their work 28-storey building with the staircase and 2 lifts. Their work also included different age groups of people. They said that the simulation results show that the optimal percentages of the occupants evacuated by the lifts, when achieving the shortest evacuation time, is almost not related to the number of evacuated persons and floors. Further, when focusing on age groups, it was ascertained that if older persons used the lift, the staircase will not be significantly overloaded. At the same time, if the lift is used by children, the rate of its use will be improved. Therefore, the selection of portion of persons by age categories, who should be evacuated by the lift can reduce congestions on staircase, and effectively speed up the evacuation process. For example, in the next study using the simulation tool, Andrée et al. [11] focused on the high-rise building, selection of exits and waiting time for evacuation lifts. Their objective was to examine, inter alia, effect of the lighting system on the selection of exits and to quantify waiting time for lifts. The study resulted into the fact that the well and simply designed system of marking escape routes will affect the selection of an exit. This system can also be used for increasing the portion of people, who will decide for the evacuation lift as the first choice of escape. Further, their results have demonstrated that people were willing to wait for the lift up to 5 minutes, but if they decided to wait, their waiting is usually extended to 20 minutes and more.

Results of previous examinations have demonstrated potential of lifts for evacuation of persons. However, in all aforementioned papers they call attention to the need of quantification of other movement parameters and description of the entire process using a suitable algorithm that could be used for simple, but also complex calculation models. Therefore, the work was focused on the design of an ideal solution for the use of the lift for evacuation. The calculation method was used with building Exodus simulation software in which the example of the building was modelled. By various variations of input quantities, combinations that represent a model of evacuation lift are subsequently obtained. As a result, the ideal combinations were subsequently selected to reduce the total evacuation time.

2. Method

For ease of reference, some terms have been simplified as follows:

- t_u – total / final evacuation time
- $t_{u,I}$ – total / final evacuation time without the use of the lift
- $t_{u,II}$ – total / final evacuation time with using the lift
- $t_{u,V}$ – the time of evacuation of the last person who used the lift
- $t_{u,S}$ – the time of evacuation of the last person who used the stairway

Model of building

In simulation calculations, a building with a occupancy of no less than 4 $m^2 \cdot p^{-1}$ was used. An example of this may be an office building with open offices (Figure 1). In this case, 110 people were placed on each floor. The number of floors was varied in the range from one to fifteen. The ground floor of the model does not contain any person. The building always contained one staircase with the same capacity parameters (1 m width) and one, two or three lifts with varying maximum capacity and speed.

The assumption of the use of the lift was chosen so that the same number of persons used the lift on each floor. The basic alternative is the following number of E_{PXX} people who use the elevator:

- 20% of the floor – 22 persons
- 40% of the floor – 44 persons
- 60% of the floor – 66 persons

The percentage of persons on the E_{PXX} floor was based on a logical proposal. It is thus possible to argue that there will never be situations where 80% and 100% of persons throughout the building will use the lift. For this reason, these alternatives were not used. On the contrary, EP20 proposals were based on the likely occurrence of persons of limited movability in the building. According to the Slovak law [1, 2], one escape route can be used only if less than 10% of persons with movement restrictions are on it.

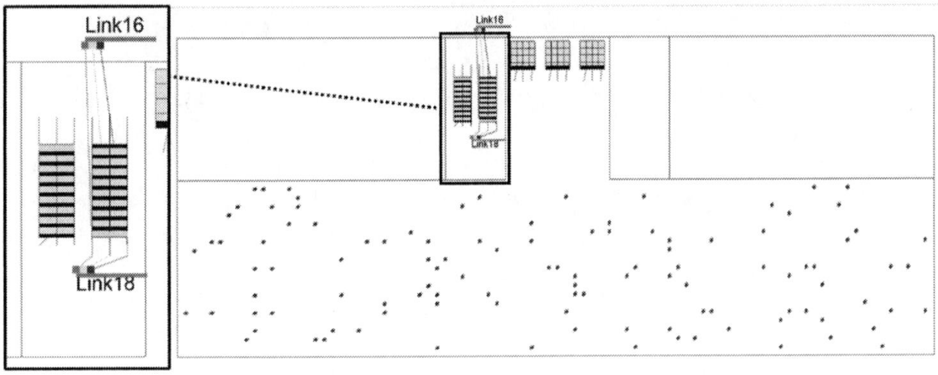

Figure 1. Schematic of the geometry of the floor of the model building.

Lift(s) were in the core of the floor. The choice of elevator parameters was developed and adjusted according to market availability. The maximum capacity of the C_X cabin was 6 persons (C_6 small lift), 13 people (C_{13} medium lift) and 21 people (C_{21} high capacity lift). The maximum speed of the lift S_X was 0.5 m·s^{-1} ($S_{0.5}$ low-speed lift), 1.5 m·s^{-1} ($S_{1.5}$ medium-speed lift) and 2.5 m·s^{-1} ($S_{2.5}$ high speed lift). Situation of the number of lifts is shown in Figure 1. The model was thus divided into different combinations according to the following key:

- Lift capacity "C"
- Number of lifts "N"
- Lift speed "S"
- Percentage of people on floor "E_P"

3. Results and discussions

3.1. Comparison of the number of lifts

For a lift C_6 and $S_{0.5}$ (Figure 2 A), it can be seen, that with the highest number of floors, neither combination produced a shortening of the resulting time. The difference in the evacuation times (at the highest floor) between two and three lifts was up to 20% and between one and two lifts up to 50%. Two and three pieces of such lifts bring shorter time t_u only to 4-storey building. If lift speed is increased to 1.5 m·s^{-1} (Figure 2 B) it can be seen that at fifteen floors the final time t_u is shortened with the use of three lifts. In this case, the difference in the evacuation times (at the most storey) between two and three lifts was 15% and between one and two lifts 40%. With two lifts, the time is shortened up to the thirteenth floor. One lift does not produce any time reduction. At lift's speed 2.5 m·s^{-1} (Figure 2 B), it can be seen, that two and three lifts will always reduce the time t_u. On the other hand, even at this speed one lift does not bring any shortening of time t_u. The difference in the evacuation times (at the highest floor) was between two and three lifts up to 10% and between one and two lifts 40%. When setting E_{P40}, the total time was not shortened nor in one combination at the highest floor. For this reason, these shares of individuals were not included in the results and were not considered in other parts of the work.

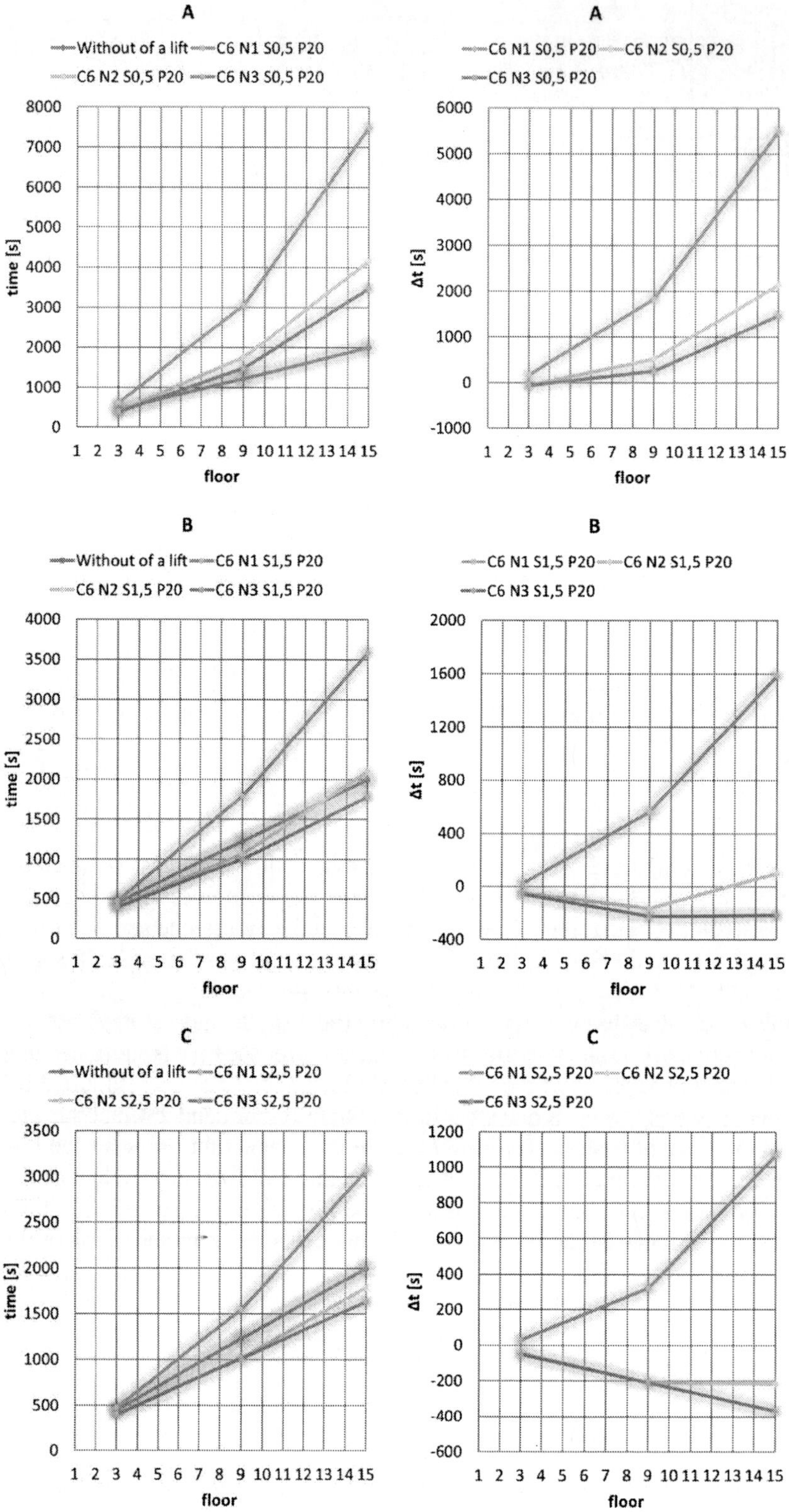

Figure 2. Evacuation times t_u and Δt for combinations with the C_6 lift.

At the lift C_{13} and $S_{0.5}$ (Figure 3 A), it can be seen, that at fifteen floors the final time t_u is shortened here only by using three lifts. The difference in evacuation times (at the highest floor) was between two and three lifts up to 12% and between one and two lifts up to 40%. Combination with two lifts brought shortening of the building time to twelve storeys and one elevator only for a building up to four floors. When changing the speed to 1.5 m·s^{-1} (Figure 3 B), the resulting time is shortened even when one lift is used. The difference in evacuation times (at the highest floor) was between two and three lifts up to 2% and between one and two lifts to 17%. If the lift speed increased to 2.5 m·s^{-1} (Figure 3 C), the difference in times (at the highest floor) was between two and three lifts up to 2% and between one and two lifts up to 3%. With E_{P40}, the total time t_u was shortened to the highest number of floors in combinations of two and three lifts. On the other hand, when setting the E_{P60}, the total time was shortened by using only three lifts. However, the total times t_u at E_{P60} were never shorter than with E_{P40}. On the other hand, the total time t_u of E_{P40} was shorter than E_{P20}, with a difference of 7%, which can be considered approximately the same time. Moreover, the results of the combination with E_{P40} from the eighth floor approximate the result with E_{P20}. For this reason, these shares were not included in the results and were not considered in other parts of the work.

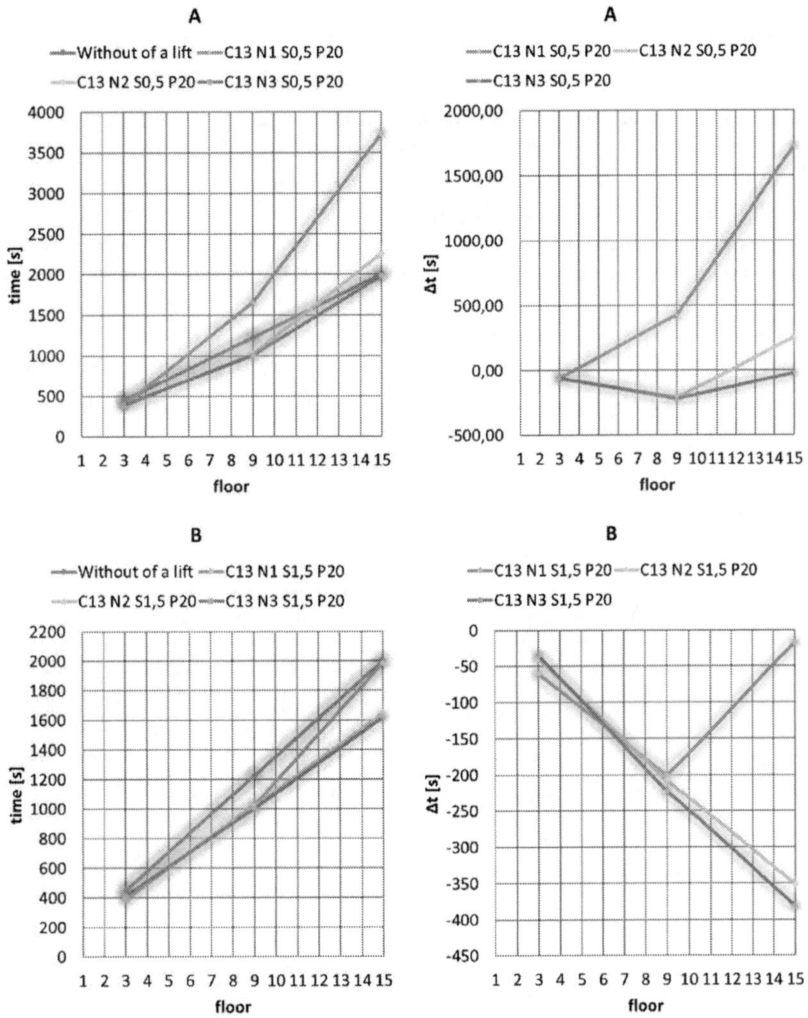

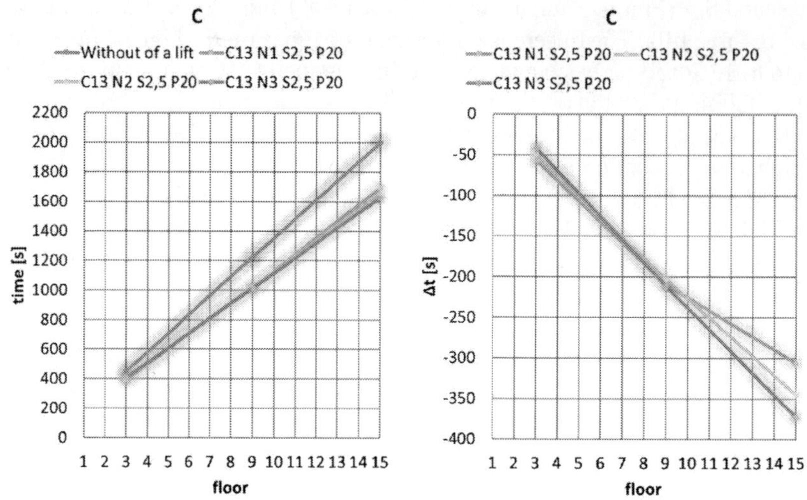

Figure 3. Evacuation times t_u and Δt for combinations with the C_{13} lift.

For the lift C_{21} and $S_{0.5}$ (Figure 4 A) it can be observed that at fifteen floors total time t_u is shortened using two or three lifts. The difference in evacuation times (at the highest floor) was between two and three lifts up to 1% and between one and two lifts up to 34%. Combination with one lift brought shortening of the total time t_u to only tenth floor. When changing speeds to 1.5 $m \cdot s^{-1}$ and 2.5 $m \cdot s^{-1}$ (Figures 4 B and C), the total time t_u is shortened even when using one lift. The difference in evacuation times (at the highest floor) was 2% between two and three lifts. Total times t_u using one and two elevators were almost same (with a difference of 0.5%). With E_{P40} setting, the total time t_u was shortened to the highest number of floors in combinations of two and three lifts. In both cases, the total times at E_{P40} were always shorter than with E_{P20}. In these cases, it can be argued that with a large lift capacity it is important to increase the ratio of the number of persons using the lift to 40%. However, the design of 2 to 3 large capacity lifts in 15-storey buildings is virtually impossible in real life. For this reason, we will not consider these alternatives in the next part of the work.

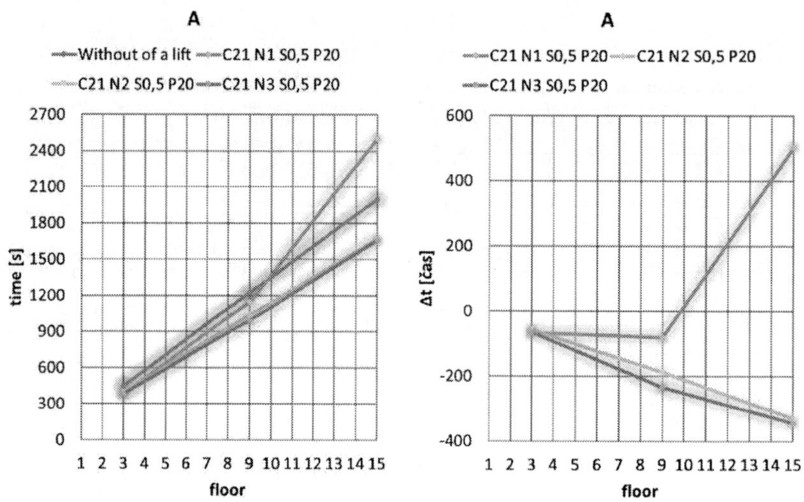

IOP Conf. Series: Materials Science and Engineering **566** (2019) 012032 doi:10.1088/1757-899X/566/1/012032

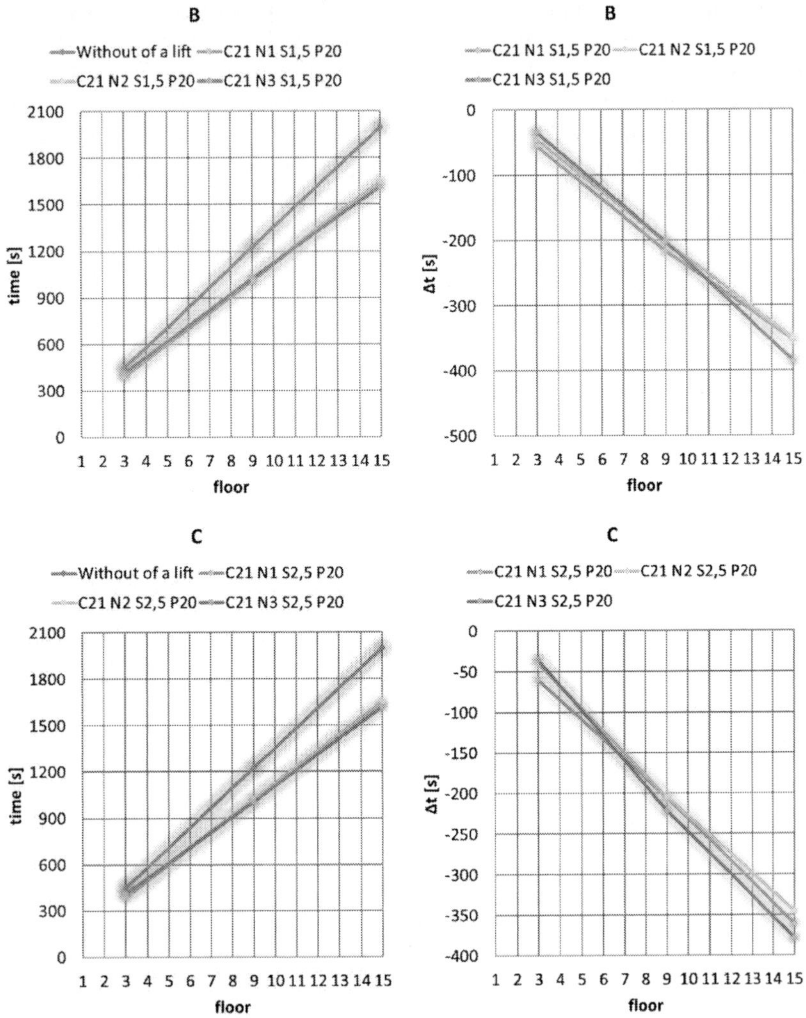

Figure 4. Evacuation times t_u and Δt for combinations with the C_{21} lift.

3.2. Evaluation of the ideal design

The time ratio $t_{u,V}$ and $t_{u,S}$ is a comparison of time values. If the ratio greater than 1.0 means that time $t_{u,V}$ is longer than $t_{u,S}$. The same goes for the opposite. The most ideal use of the elevator takes place when the last person in the lift evacuates together with the last person from the staircase at the same time. If one person decides to change the path of evacuation, the ratio will no longer be 1.0. In the next procedure, it was necessary to choose the ratio of the ratio when the lift is ideally utilized. The range from 0.9 to 1.1 (90% to 110%) was selected. If the ratio less than 0.9 is elevated, the lift design is oversizing and if the ratio is greater than 1.1, the design of the lift is under dimensioned (Figure 5).

$t_{u,V}/t_{u,S} < 0,9$ 90% or less	$0,9 \le t_{u,V}/t_{u,S} < 1$ 90% to 100%	1 (100%)	$1 < t_{u,V}/t_{u,S} \le 1,1$ 100% to 110%	$1 < t_{u,V}/t_{u,S}$ 110% or more
$t_{u,V} < t_{u,S}$	$t_{u,V} < t_{u,S}$	$t_{u,V} = t_{u,S}$	$t_{u,V} > t_{u,S}$	$t_{u,V} > t_{u,S}$
Oversizing proposal	Ideal proposal	The shortest evacuation time	Ideal proposal	Undersigned proposal

Figure 5. Schematic of the suitability of the lift design.

In Figure 6, combinations that represent the ideal use of the elevator or are close to it can be seen. It is also possible to see oversizing and underdimensional proposals. Areas highlighted in red and labelled "X" are combinations for which lifts (at first ride) were occupied on the highest floor less than 80% of their capacity. For example, 2 lifts with a capacity of 21 people can take up to 42 people on the top floor. However, for 22 persons on the floor (E_{P20}), these lifts are only used for 52% of their capacity. This is an unnecessary oversizing of the lift capacity design in terms of the number of people on the floor, despite the significant reduction in the evacuation time. Therefore, for these combinations, the ratio of times $t_{u,V}$ and $t_{u,S}$ was not calculated. Therefore, these combinations are ideal if you use 44 people on the floor using the lift (E_{P40}). Areas highlighted in blue and labelled "X" are combinations for which there has been no overall reduction in evacuation time, despite the fact that the lifts were occupied (at first ride) at the top floor over 80% of their capacity. This is a underdimensional of the lift design in terms of its total capacity for the number of persons on the floor. These combinations are ideal if the lift uses up to 11 people, representing about 10% of the floor. The remaining cells in the table are combinations for which elevators were occupied (at first ride) by more than 80% of their capacity on the top floor, and the overall evacuation time was shortened. In the next step, attention was paid to these combinations. This indicates an incorrect lift design.

Capacity	Speed	Floor	One lift	Two Lifts	Three lifts
		3	X	96%	72%
	0,5 $m\cdot s^{-1}$	9	X	X	X
		15	X	X	X
		3	X	59%	59%
6 p.	1,5 $m\cdot s^{-1}$	9	X	103%	89%
		15	X	X	110%
		3	X	62%	53%
	2,5 $m\cdot s^{-1}$	9	X	93%	72%
		15	X	109%	91%
		3	93%	65%	X
	0,5 $m\cdot s^{-1}$	9	X	93%	X
		15	X	X	X
		3	72%	44%	X
13 p.	1,5 $m\cdot s^{-1}$	9	101%	67%	X
		15	120%	77%	X
		3	67%	51%	X
	2,5 $m\cdot s^{-1}$	9	88%	58%	X
		15	102%	66%	X
		3	81%	X	X
	0,5 $m\cdot s^{-1}$	9	113%	X	X
		15	X	X	X
		3	58%	X	X
21 p.	1,5 $m\cdot s^{-1}$	9	75%	X	X
		15	87%	X	X
		3	60%	X	X
	2,5 $m\cdot s^{-1}$	9	66%	X	X
		15	78%	X	X

Figure 6. Ratio values $t_{u,V}$ and $t_{u,S}$ for individual combinations with EP20.

4. Conclusion

It can be said that the following combinations represent the ideal design of lift:

- For building up to three floors:

  - two lifts with a capacity of 6 persons and a speed of 0.5 $m \cdot s^{-1}$
  - one lift with a capacity of 13 persons and a speed of 0.5 $m \cdot s^{-1}$
- For building up to nine floors
  - two lifts with a capacity of 6 persons and a speed of 1.5 or 2.5 $m \cdot s^{-1}$
  - two lifts with a capacity of 13 persons and a speed of 0.5 $m \cdot s^{-1}$
  - one lift with a capacity of 13 persons and a speed of 1.5 $m \cdot s^{-1}$
- For building from nine to fifteen floors
  - two or three lifts with a capacity of 6 people and a speed of 2.5 $m \cdot s^{-1}$
  - one lift with a capacity of 13 persons and a speed of 2.5 $m \cdot s^{-1}$

The lift (speeds of 1.5 and 2.5 $m \cdot s^{-1}$) with a capacity of 21 people is always oversizing. However, when designing this lift at a speed of 0.5 $m \cdot s^{-1}$, the ideal lift design can be achieved by adjusting the number of people using the lift.

5. References

[1] Ministry of Interior of the Slovak Republic 2004 *94/2004 Decree of the Ministry of Interior of the Slovak Republic laying down technical requirements for fire safety in the construction and use of buildings*

[2] STN 92 0201-3 2000 *Fire protection of buildings, General regulations, Part 3, Escape routes an evacuation of occupants*

[3] Dunlop K E Shields T J Real fire emergency evacuation of disabled people *Proceedings of CIB W14 International Symposium and Workshops on Engineering Fire Safety in the Process of Design Equivalency* 13-16 September 1993 Newtownabbey Northern Ireland pp 157–164

[4] Shields T J Boyce K E McConnell N 2009 *The behavior and evacuation experience of WTC 9/11 evacuees with self-designated mobility impairments* vol 44 pp 881–893

[5] Averill J D Mileti D Peacock R D Kuligowski E Groner N Proulx G Reneke P Nelson H E 2018 *Final report of the federal building and fire safety investigation of the World Trade Center disaster* http://wtc.nist.gov/oct05NCSTAR1-7index.htm

[6] STN 92 0101 1997 *Structural fire protection, Vocabulary*

[7] Statistical Office of the Slovak Republic 2018 https://slovak.statistics.sk

[8] Averill J D *Five grand challenges in pedestrian and evacuation dynamics* 5th International Pedestrian and Evacuation Dynamics Conference 8-10 March 2010 Gaithersburg Maryland p 11

[9] Ding N Chen T Zhang H 2017 *Experimental study of elevator loading and unloading time during evacuation in high-rise buildings* vol 53 pp 29–42

[10] Ding Y Yang L Weng F Fu Z Rao P 2015 *Investigation of combined stairs elevators evacuation strategies for high rise buildings based on simulation* vol 53 pp 60–73

[11] Andrée K Nilsson D Eriksson J 2015 *Evacuation experiments in a virtual reality high-rise building: Exit choice and waiting time for evacuation elevators* vol 18 pp 554–567

Acknowledgments

This work was supported by grant project 2/0042/17 'Solar energy influences and integrated envelopes on the quality of the environment in buildings and cities.'

This paper was created thanks to the financial support from the EU Structural Funds, through the R&D Operational Program and project OPVaV-2008/2.2/01-SORO 'Architectural, engineering, technological and economic aspects of the design of energy efficient buildings', codenamed ITMS: 26220220050; which is financed by EC funds.

IOP Conf. Series: Materials Science and Engineering **566** (2019) 012033 doi:10.1088/1757-899X/566/1/012033

How and where to use technology of laser scanning and aerial photogrammetry for building purposes – case study

M Tkáč[1]

[1]Technical university of Kosice, Faculty of Civil Engineering, Department of Construction Technology and Management, Slovakia

Abstract. Laser scanning is a method of high – accuracy mapping or reality capture that uses laser beams to quickly capture complete detail of the entire building construction project in a short time. Laser scanners are a great tool for measuring existing buildings but not always because they have the same problem as visual surveys-if you can't see it, you can't scan it, i.e. the laser beam from the scanner can't measure a certain part of the surface of the structure if this part isn't in the scanner's field of view. In particular, this limitation applies to the surfaces of the roof structures and certain parts of the facades. In these cases, the resulting point cloud may contain no information from the building surface what may be a problem for the final results. A possible solution is a combination of two technologies, terrestrial laser scanning and aerial photogrammetry. Aerial photogrammetry (or airborne imagery) means taking of photographs from an unmanned aerial vehicle (i.e. "drones"). The aim of this paper is to describe the use of these two technologies, it means, when it is better to use these technologies separately and when it is better to make their integration. As a reference building the Faculty of Civil Engineering, located in Kosice, Slovakia was chosen

1. Introduction

The rich, complete data captured with a laser scanner or a drone can provide the required information on existing building conditions with the accuracy needed for construction planning or for creating new as-built documentation what is in many cases the main purpose for digital surveying [1]. The purpose of a 3D scanner and a drone in the civil engineering is usually to create a point cloud of geometric shapes from the surface of the building because measuring with using a laser beam or aerial images are non-contact measurement of surface of buildings, allows enhancing the design process, speeds up and reduces data collection errors, saves time and money [2]. The survey yields a digital data set, which essentially a dense 'point cloud', where each point is represented by a coordinate in 3D space. All of these points are placed into the same local coordinate system to make up a point cloud, which represents the area, building, or object being scanned in a 3D space [3]. With the appearance of digital cameras photogrammetry entered a new era. Photogrammetry is the science of making measurements from photographs. The input to photogrammetry is photographs, and the output is typically a map, a drawing, a measurement, or a 3D model of some real-world object (e.g. some building structure) or scene [4]. Despite the considerable progress of these technologies, there are still some limitations, which have an effect on the quality of the final 3D model in the form of a point cloud. Even though current laser scanners can produce large point clouds fast and reliably, on the other hand many building objects contain places, which the scanner cannot measure, e.g. roof structures. In this situation the final product can be a problem because the point cloud does not contain all spatial information. In contrary, the digital aerial photogrammetry can provide solution for this problem [5].

IOP Conf. Series: Materials Science and Engineering **566** (2019) 012033 doi:10.1088/1757-899X/566/1/012033

2. Digital as-built surveying by laser scanning and aerial photogrammetry

In the field of digital spatial surveying of building constructions by terrestrial laser scanning technology, there are often limitation factors with the measurement of the surfaces from roof structures because laser scanning technology is characterized as noncontact surface measurement technology. These limitation factors may be, e.g., type of roof (flat, sloping or steep roof), roof shape, position of the roof towards to the terrain, angle of the roof, height, layout plan of the building, etc. and these limitations often have a significant impact on the quality of the resulting point cloud, especially in situations where the roof surface is not possible to measure by a terrestrial laser scanner from the terrain. The aim of this chapter is to describe the results of the methodology of integration of terrestrial laser scanning and aerial photogrammetry. As a reference building for this methodology the Faculty of Civil Engineering, located in Kosice, Slovakia was chosen. Looking Fig. 2 it can be seen that the Faculty of Civil Engineering has a simple layout in the L shape with a built-up area of 3150 m^2, a height of 20 meters and a roof angle of 24°.

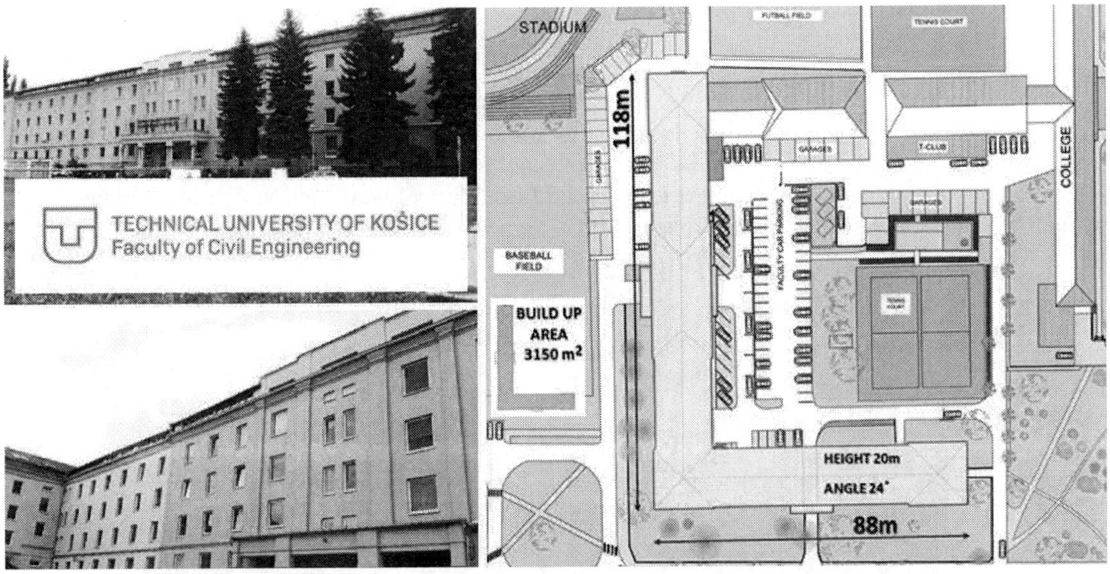

Figure 1. The situation of the Faculty of Civil Engineering, *source: author.*

The measurement of the Faculty of Civil Engineering was divided into two parts:

- **Terrestrial Laser Scanning - parts of the building selected for the laser scanner:** Exterior - facade of the building – completely,
- **Aerial photogrammetry - parts of the building selected for the drone:** Exterior - roof construction – completely.

For terrestrial laser scanning the laser scanner FARO Focus 3D X130 with a carbon telescope tripod, 12 reference spheres with a diameter of 145 mm and 6 reference spheres of 200 mm diameter was selected. The laser surveying took place in early October, i.e. the work had to be done by about between 6.00 a.m. - 6:30 p.m., because at that time it was almost dark in outside. The overall scanning time was from 6:45 a.m. to 6:30 p.m. (approximately 12 hours) and together was created 48 scanner positions. The dron DJI MAVICO PRO was used to create aerial images from the roof of the Faculty of Civil Engineering. Next was used the remote controller, DJI GO application and free online application the DroneDeploy. As a first step, it was necessary to set the drone flight plan. This plan can be either automatic or manual, i.e. aerial photographs can be created in automatic or manual mode during flight. The total time of data collection using the drone from the roof of the building was approximately 30 minutes (flight plan of the drone, aerial images in automatic mode, aerial images in manual mode). The overall of 302 aerial photographs were created and entire data collection process was created with one battery (*max flight time with one battery - 27 minutes - no wind*). The phase of data processing in this case consisted from three stages.

IOP Conf. Series: Materials Science and Engineering **566** (2019) 012033 doi:10.1088/1757-899X/566/1/012033

2.1. The 1st stage - creating two point clouds from two technologies

For the first point cloud from laser surveying was used the Faro SCENE software. The overal time of these processes was about 3 hours. For the second point cloud from aerial surveying the Agisoft Photoscan Professional software was used. The overall time for creating the point cloud from 302 aerial photographs was 31 hours. The reason was that the point cloud was created in the highest possible quality. Of course, lesser quality means less time but also the computer's parameters are decisive for the overall time (in this case: processor Intel(R) Core(TM) i7, memory 64 GB, graphic card NVIDIA GeForce GTX 1070). At the end of this stage, two point clouds were created in two different local coordinate systems.

2.2. The 2nd stage - integration of two local coordinate systems into one,

The point cloud coordinate system of aerial images was adapted to the point cloud coordinate system from laser scans based on a few equal points from both point clouds (e.g. window corners). In this way, 6 points were created in a cloud of aerial photographs. If the Agisoft software is used to process aerial photos, it is important to note that working with coordinates is only possible with professional edition, not with standard edition. Thanks to these processes, both point clouds have one coordinate system.

2.3. The 3rd stage - connection of two point clouds into one.

Now both point clouds are ready for the final connection, i.e. in the last step, it is necessary to connect the point cloud from the facade with the point cloud from the roof construction. For this connection the CloudCompare software was selected and the final connection was automatic.

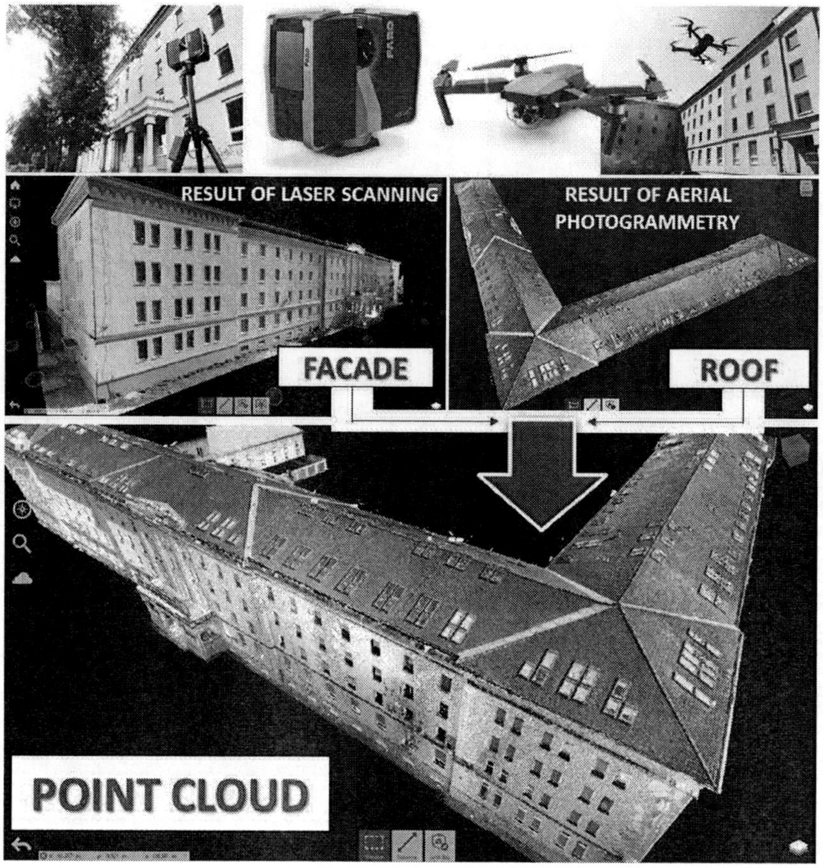

Figure 2. The result of integration of terrestrial laser scanning and aerial photogrammetry for the purpose of digital as-built surveying.

3. Digital as-built surveying by aerial photogrammetry

On the Faculty of Civil Engineering was also made the second digital as-built surveying but in this case only using a drone. The aim was to measure the complete building exterior including the roof and also the facade. In this case was overall created 1050 aerial images from the drone and overall data collection time was only 2 hours by contrast with the overall data collection time 12 hours by the laser scanner from previous case study. But these 12 hours was focused only on the facade of the building. If would by necessary to measure also the roof of the building with the laser scanner it would be necessary to create extra scanner positions at a distance of about 50-70 m from the building. In this case the overall laser scanning time would be increased maximum about 1 day; overall 2 days.

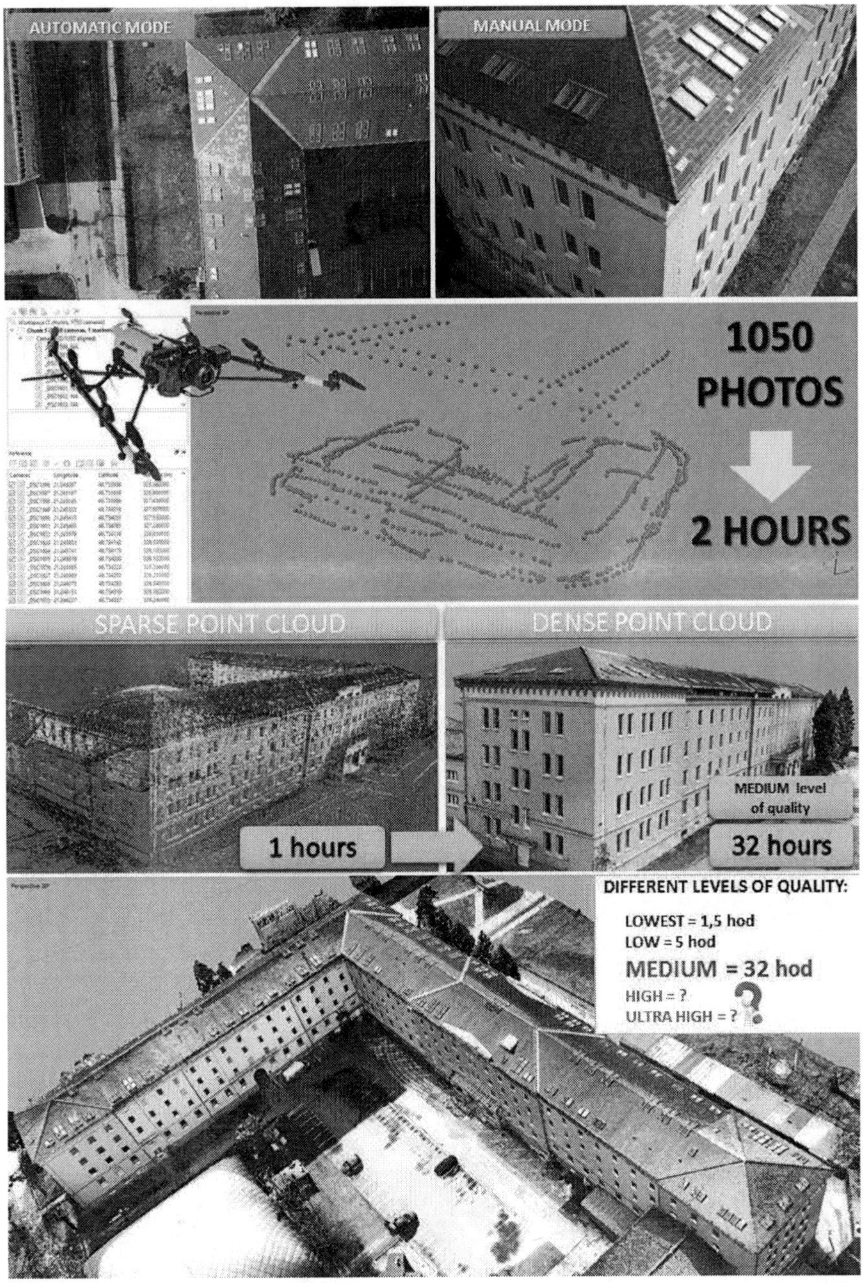

Figure 3. The result of aerial photogrammetry which was focused on the complete exterior of the Faculty of Civil Engineering.

For data processing of the aerial images from the drone was used the software Agisoft Photoscan but professional edition. The overall time of data processing was 33 hours. For the phase of data processing of aerial images from the drone, the quality of the dense point cloud is decisive for the overall time of data processing. In this case, only medium point cloud quality was selected and the overall time was 33 hours (1 day and 9 hours). The overall time for Ultra high quality of the point cloud? Expected time would by approximately 1 week or more. This implies that the phase of data collection by a drone is faster approximately about 90% by contrast with a laser scanner. But on the other hand the phase of data processing by a laser scanner is faster approximately about 60% (*medium quality of a dense point cloud*), 80-90% (*high quality of a dense point cloud*) by contrast with a drone.

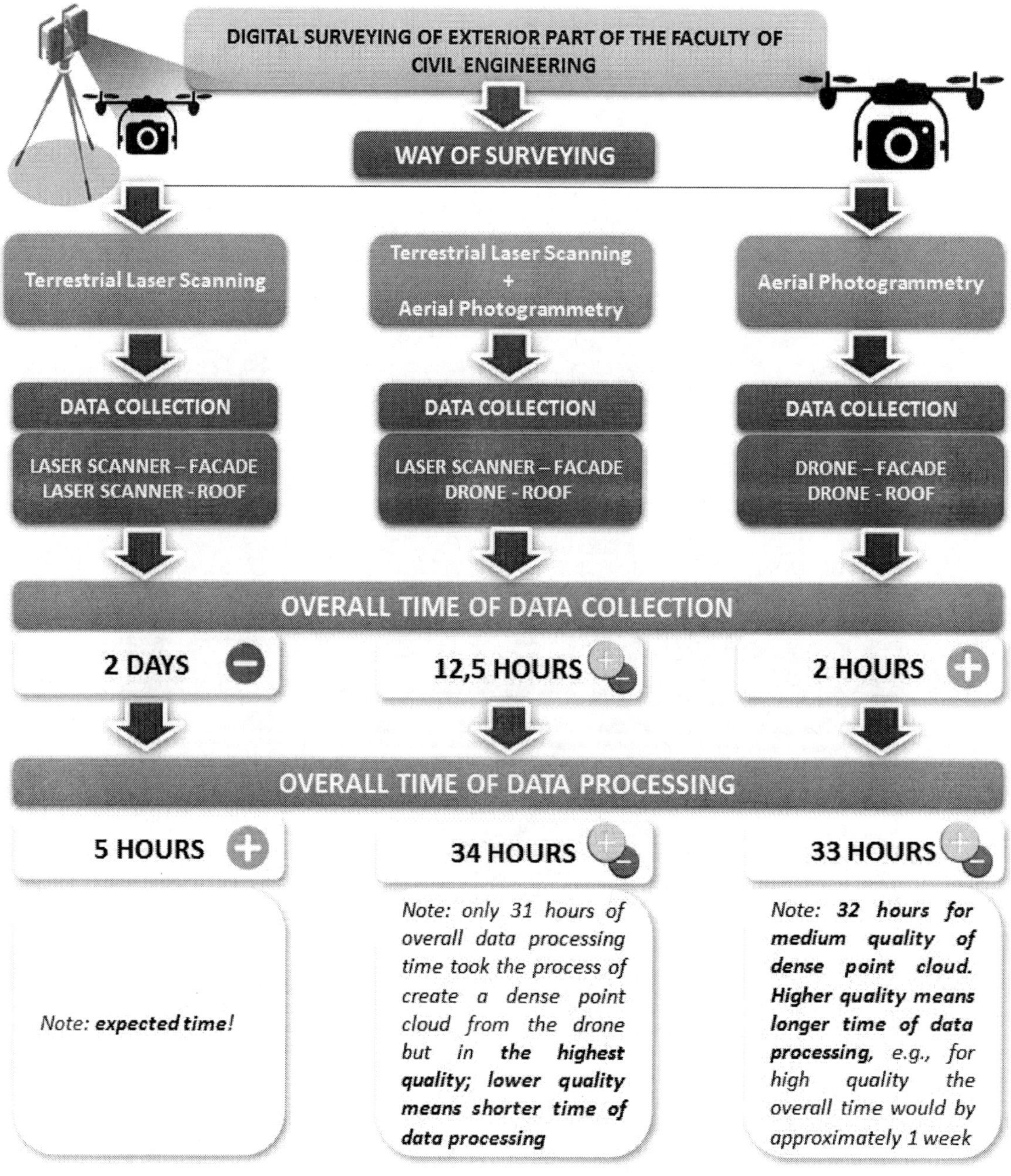

Figure 4. Digital surveying of exterior part of the Faculty of Civil Engineering – results

4. Conclusion

Technology of laser scanning and photogrammetry in general are definitely progressive technologies in the area of 3D digital spatial measuring in the construction industry but also in other sectors. The use of these two technologies is certainly huge and of course both technologies have their advantages and disadvantages. But there are situations where it is advantageous to use both technologies on one building. Of course in this case, each technology is used for another purpose or for another part of the building. Concretely for this article, terrestrial laser scanning and aerial photogrammetry technologies have been selected in order to bring in the form of a methodology a solution for the digital space measuring of those roof structures that cannot be measured by a laser scanner partially or completely. The solution of this methodology has been applied to the reference building at the Faculty of Civil Engineering of the Technical University in Kosice. This building could be measured in three ways. The first way is to measure this building by terrestrial laser scanning technology and aerial photogrammetry. The laser scanning was focused only on the facade of the building and measuring by a drone was focused only on the roof of the building. The combination of these two technologies for this building was very advantageous because with this way 100 % of exterior of the building was measured. Overall time for this process were 12.5 hours for data collection and 34 hours for data processing. The second way would be to measure the building using only a 3D scanner. In this case the overall scanning time would be 2 days and the overall data processing time approximately 5 hours but the shape, angle and height of the building represents non-ideal conditions for laser scanning, it means, that the final point cloud will be contain a lot of missing information from the roof space. The last way is to use for overall measuring only a drone. In this case the overall data collection time would be 2 hours and the overall data processing time 33 hours or more. This implies that the phase of data collection by a drone is faster by contrast with a laser scanner but on the other hand the phase of data processing by a laser scanner is faster by contrast with a drone.

5. References

[1] Hayes C and Richie E *When to use laser scanning in building construction: A guide for general contractors*, http://constructrealityxyz.com/test/ebook/LGS_AU_When%20to%20Use%20Laser%20Scanning.pdf, (last visited 10 February 2019)

[2] O'Day E *3D laser scanning: different type of scanners*, https://www.ideateinc.com/blog/2013/07/3d-laser-scanning-different-type-of, (last visited 10 February 2019)

[3] Aziz M A, Idris K M, Majid Z, Ariff M F M, Yusoff A R, Luh L C, Abbas M A and Chong A K 2016 *A study about terrestrial laser scanning for reconstruction of precast concrete to support qlassic assessment*, The International Archives of the Photogrammetry, Remote Sensing and Spatial Information Sciences, Vol. XLII-4/W1, pp.135–140, (last visited 10 February 2019)

[4] Corrigan F *Introduction to UAV photogrammetry and Lidar mapping basics*, https://www.dronezon.com/learn-about-drones-quadcopters/introduction-to-uav-photogrammetry-and-lidar-mapping-basics/, (last visited 10 February 2019)

[5] Haala N and Alshawabkeh Y 2006 *Combining laser scanning and photogrammetry - A hybrid approach for heritage documentation*, Conference on the 7th International Symposium on Virtual Reality, Archaeology and Intelligent Cultural Heritage, Nicosia, Cyprus, October 30 – November 04, pages 8, (last visited 10 February 2019)

Acknowledgments

The paper presents a partial research results of project APVV-17-0549 "Research of knowledge-based and virtual technologies for intelligent designing and realization of building projects with emphasis on economic efficiency and sustainability" and project VEGA 1/0828/17 "Research and application of knowledge-based systems for modelling cost and economic parameters in Building Information Modelling".

IOP Conf. Series: Materials Science and Engineering **566** (2019) 012034 doi:10.1088/1757-899X/566/1/012034

Analysis of current situation in field of LCA.

Ž Tomková[1] and S Vilčeková[2]

[1] Technical university of Košice, Faculty of Civil Engineering, Department of architectural engineering
[2] Technical university of Košice, Faculty of Civil Engineering, Department of environmental engineering

Abstract. With building industry and buildings large consumption of energy and environmental impact on planet, it is crucial to minimalize future consequences of our decisions. With help of LCA analysis, we can evaluate current situation and predict future impact of the design. This allows us to make better decisions in design stage of building process. There needs to be a lot of analysis to relevantly compare different types of buildings, in different environment and using different materials. This paper concentrates on current situation in this field of study, in global scale and with focus on Slovak republic.

1. Short history of the LCA method

We could say that first attempts in LCA problematics came in 1960s and early 1970s. These studies focused on energy consumption linked to manufacture of final product, especially packaging alternatives. Later on, the aspects examined expanded on material consumption, emissions and waste production. One of the first studies were financed by Coca Cola and their aim was to compare environmental impact of different types of soda packaging materials. Then in 70s, in Great Britain, Ian Bousted counted total consumed energy required for production of different packaging materials such as glass, plastic, steel, aluminium. He then consolidated the methodology and published a Handbook of Industrial Energy Analysis. The term REPA started to be recognized in USA in this context.

Between 1970 and 1990, the methodology of research did not have steady form. Standardization came gradually between 1990 and 2000. At the beginning, it was SETAC (Society of Environmental Toxicology and Chemistry), which, at a workshop, created a uniform methodology based on international agreement. Result of this effort was publishing the Guidelines for Life Cycle Assessment: A code of Practice in 1993. This codex became the base for creation of standard ISO 14 040 Environmental management: Life cycle assessment, principles and framework, from 1997. [1]

The interest about LCA has increased in new millennium. SATEC and UNEP (United Nations Environment Programme) published International Life Cycle partnership, known as Life Cycle Initiative. Aim of this initiative was to enhance support of LCA tools through enhancement of data and indicators and bring the whole analysis to wider awareness. Importance of this analytic approach grown also in European politics. In 2005 The European Platform on Life Cycle Assessment formed to support wider awareness in EU. In USA LCA was promoted by U.S. Environmental Protection Agency and simultaneously the environmental politics spread around the world. Unfortunately, methodologies started to differ. Besides LCA, LCC (Life Cycle Costing) and SLCA (Social Life Cycle Assessment) were formed. [2]

IOP Conf. Series: Materials Science and Engineering **566** (2019) 012034 doi:10.1088/1757-899X/566/1/012034

2. LCA (Life Cycle Assessment)

Structure of LCA analyses is guided by international standard ISO 14040. [3] For the processing, the commercial process and material flow databases are used. [3] The LCA is the main tool for environmental certification type III. [4] One of the main features and requirements for LCA are:

- Systematic and adequate dedication to aspects of production, from getting raw material to final waste disposal, so called "cradle to grave" system.
- Depth of detail and time frame can differ depending on definition of study aim and analysis scope.
- Scope, assumptions, description of data quality, methodology and output should be clear, transparent and well documented.
- Depending on LCA study used, it is necessary to have in mind even confidentiality of some information.
- This methodology should include scientific findings and improvements in modern technology.
- Specific requirements are included in analysis, so that it is possible to produce comparative claims that could be presented.
- LCA analysis could not be reduced to one total score or number, due to complexity of the whole methodology.

One unified method for LCA analysis doesn't exist. Standardization is just guidance for processing of this analysis, but organizations should have opportunity to flexibly implement LCA for practical use in specific application and be able to fulfil users' needs. (ISO 14040:19997(E))

3. Methodics of LCA

The frame structure consists of 4 major phases. Phases are often interdependent. Results of one inform about completion of other phases. (ISO 14040:19997(E)) [1] [5]

1. Definition of goal and scope – it defines, how big part of product life cycle, respectively building will be included in evaluation and what is evaluation's purpose. It describes evaluation criteria and selected time horizon.
2. Inventory analysis phase – includes collection of material and energy flows within product system and mainly its interactions with environment, used amount of raw material and produced emissions to environment. It describes all significant processes and sub-flows of energy and materials.
3. Impact assessment – in this phase results of all studied impact categories indicators are evaluated.
4. Interpretation phase – in this phase critical review and analysis of data sensitivity should be made and the result of the analysis should be presented.

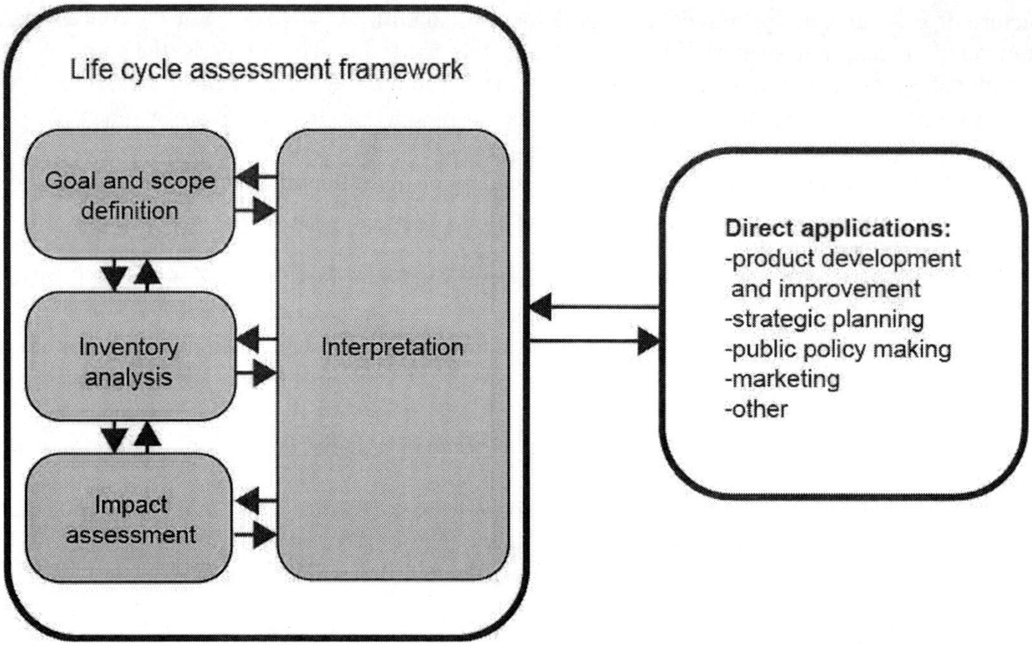

Figure 1. Phases of LCA according to STN EN ISO 14040.

4. LCA variations

From the point of view of the studied time frame we recognize these types of LCA:

From cradle-to-grave

Full life cycle assessment from extraction of raw material ("cradle"), through stage of use to demolition, or so-called "grave." In the case of buildings, it is from the extraction of raw material, through transport to the production, the building material production, transport to site, in-built, stage of use and possibly renovation stage to demolition.

From cradle-to-gate

This partial life cycle assessment begins with extraction of the raw material ("cradle") and ends at production gate. It means it only focuses on evaluation until the product leaves factory. In case of building industry we consider the gate as the point from were stage of use begins. Stage of use, reconstruction and demolition are not included. Cradle-to-gate assessment is sometimes base for environmental product declarations (EPD) called business-to-business EPD.

Cradle-to-cradle or Open Loop Production

It is specific case of assessment, when, in the cycle, we also consider the stage of recycling after the end of building's life. It is method used for minimizing impacts on environment by implementation of sustainable development, usage and demolition.

Gate-to-gate

In this type of analysis we focus just a part of the whole life cycle, just one chosen process. For example it focuses just on stage of use, or transport from factory to site. These gate-to-gate modules could be then assembled together and create wider assessment such as cradle-to-gate.

Well-to-wheel

It is another specific LCA used for analysing transport phases of life cycle. This type of analysis is often divided on stages called "well-to-station" or "well-to-tank" and "station-to-wheel", or "tank-to-wheel", or "plug-to-wheel". First stage, which includes entry materials or fuel productions, fuel supply or energy transport, is so called "upstream" phase. While the phase that deals with the actual operation of the vehicle is called "downstream". Well-to-wheel analyses are normally used to assess total consumption of energy or efficiency of energy conversion. For emission impact of navy ships, airplanes and car emissions (including carbon footprint as well as fuel consumption typical for any of these types of transport), similar assessment are used. [3]

Also LCA divides into two types due to other criteria. It is attributional and consequential LCA. Again, it is focused on a time frame but also on the amount and type of data input. Attributional type of LCA focuses on impact linked to production and usage of the product (building), service or process in certain point in time (usually a near past). This data then interprets mostly facts. Amount of data is than limited. Consequential type of LCA analysis wants to identify environmental consequences due to certain decisions or planned changes in the system according to study. It orients on future and predicts future impact. Amount of data is not limited, because multiple variants should be examined, which are compared to achieve optimal decisions.

5. Studied impacts

It is possible to evaluate multiple impacts, some of them are:

Global Warming Potential (GWP) – Greenhouse gasses are warming up earth due to absorption of energy. This results in slowed release of energy to outer space. GWP was developed in order to compare this ability between different gasses, because it is not equal. More exactly, it measures how much energy is absorbed by 1kg of gas emissions in certain time compared to 1kg of CO_2 emissions. Higher the GWP, higher the warming effect of certain gas on planet compared to CO_2 in certain time. Usual estimated time is a 100 year. Unit is kg CO_2 eq or CO_2 equivalent. [6]

Ozone Depletion Potential (ODP) – It is ability of a chemical to deplete ozone layer. Unit set similarly to GWP. It has comparing character. It determinates ability of a certain chemical to deplete ozone layer compared with referential compound CFC-11 ($CFCl_3$). Different compounds are active in different layers of atmosphere and with different impacts in time. Unit is kg CFC-11 eq. [6]

Acidification Potential (AP) – It refers to rate of pH decrease in rain water and fog, which causes damage to ecosystem leading to, for example, drainage of nutrients from soil and increased dissolving of metals to soil. It is measured in weight equivalent to SO_2. The highest impact is from burning fossil fuels, this causes release of SO_2 and nitrogen, which are dissolved in condensed water in atmosphere and fall down as rain. Unit is kg SO_2 eq. [6]

Eutrophication Potential (EP) – In general, eutrophication potential shows degree of nutrient enrichment in water or terrestrial environment, this leads to ecosystem damage from overreached enrichment and rate of quantity of phosphates equivalent. Unit is kg $PO4^{3-}$ eq. [6]

Photochemical Ozone Creation (POCP) – Nitrogen oxides contained in atmosphere (NOx, common polluting element) and organic dust particles (VOCs) could form ozone with presence of sunlight. Even if ozone in high atmosphere is crucial for protection against ultraviolet light (UV light), ozone in low levels of atmosphere causes different impacts, such as damage to plants and higher appearance of asthma and other respiratory difficulties. Potential of photochemical ozone creation (also known as summer smog) by compound emissions to atmosphere is measured in equivalent C_2H_4 (kg C_2H_4 eq). [6]

Abiotic Depletion Potential – Elements (ADP - elements) – It refers to depletion potential (or deficiency) natural resources (or elements), unused as source of energy, in earth, such as iron ore, aluminium, pressure metals and represents the final geological reserves and predicts amount of their depletion. It is measured in antimony equivalent. [6]

Abiotic Depletion Potential – Fossil fuels (ADP – fossil fuels) – It is similar to ADP for elements only ADP for fossil fuels counts rate of depletion potential for natural resources (elements) used for energy creation (fuels). Unit is MJ. [6]

Net use of fresh water (FW)) – It indicates overall net water consumption during whole building's life cycle. Unit is m^3 water.

Bre Eco Points – British rating system in environmental impact sector for buildings/products.

6. Tools and software

It was necessary to create databases of elements and materials to create LCA analysis and mainly to speed up and simplify whole process. This databases had to be updated according to production development, application of new and more effective ways of mining and extraction and many other factors. Databases should be locally divided, due to non-uniform production process methods and conditions in each country. For simplifying and effective LCA performing, the development of software tools for model processing. It is necessary to collect all information about used materials, their volumes, information about technical building equipment, operational information and other. In software those information are located and summarized. Databases then generate results. With this tools the whole process of analyse is simplified and more accessible. In short time from LCA development, many software tools were found. List of some of them is mentioned in table 1.

Table 1. Types of LCA programs.

Tool	*Link*	*Use*
Athena (Canada)	http://www.athenasmi.org/our-software-data/impact-estimator/	Building
Arquimedes (Spain)	http://arquimedes.cype.es/	Building
BEES (USA)	http://www.nist.gov/el/economics/BEESSoftware.cfm/	Building
Bilan Produit ADEME (France)	http://www.base-impacts.ademe.fr/bilan-produit	General
Carbon Footprint (UK)	https://www.carbonfootprint.com/	General
COCON (France)	http://eosphere.fr/COCON-comparaison-solutions-constructives-confort.html	Building
eToolLCD (Australia)	http://etoolglobal.com/	Building
Eco-bat (Switzerland)	http://www.eco-bat.ch/index.php?option=com_content&task=blogcategory&id=14&Itemid=30	Building
EcoCalculator (Canada)	http://www.athenasmi.org/tools/ecoCalculator/	Building
EcoEffect (Sweden)	http://www.ecoeffect.se/	Building
ECOSOFT (Austria)	http://www.ibo.at/en/ecosoft.htm	Building
EIME (France)	http://codde.fr/en/our-software/eime-en/eime-presentation	General
ELODIE (France)	http://www.elodie-cstb.f/default.aspx	Building
envest 2 (UK)	http://envestv2.bre.co.uk/	Building
EQUER (France)	http://www.izuba.fr/logiciel/equer	Building
GaBi (Germany)	http://www.gabi-software.com/	General
GaBi-Build-IT (Germany)	http://www.pe-international.com/sweden/services-solutions/green-building/building-lca/	Building
GreenCalc+ (The Netherlands)	http://www.greencalc.com/	Building
Klimagassregnskap (Norway)	http://www.klimagassregnskap.no/	Building
LEGEP (Germany)	http://www.legep-software.de/	Building
One Click LCA (Finland)	http://www.oneclicklca.com/green-building-software/	Building
OpenLCA (Germany)	http://www.openlca.org/	General
SimaPro (The Netherlands)	http://www.pre-sustainability.com/	General

IOP Conf. Series: Materials Science and Engineering **566** (2019) 012034 doi:10.1088/1757-899X/566/1/012034

SBS (Germany)	http://www.sbs-onlinetool.com/	Building

One from the two most common programs is SimaPro, with 25 year experience on market. But it is more oriented on product LCA. This program is licenced and contains a lot of databases. It uses standard ISO 14000 and generates his diagrams and other data according to it. Outputs are divided to three categories: emissions to atmosphere, emissions to water and emissions to soil. This emission dividing simplifies understanding of their impact effect. Demo licence is also available but only provides around 100 processes. Besides that it is still great tool for understanding and getting to know full functional programme.

Another tool is German GaBi. It is also more general but it contains specific building analyses. Is the most widely used tool on market and offers database of more than 8000 processes from different fields. Thinkstep Company revises data and methods used by this program every year. All entries are according to ISO 14040/44. Program is also licenced and offers 30 day trial of full version and demo version. [12]

Last but not least there is on-line tool eTool. It is open-use program founded in 2009. Reports from this program are accordance with ISO 14044 and EN 15978. It generates many impact factors. Besides already mentioned impacts there is also ability of recycling or radioactive potential. This is just part of many programs used for analyse processing.

7. Application example of LCA analysis from Czech Republic.

Very effective solutions were developed with rising demands on decreasing building energy performance. Materials with great thermo-technical features, new more effective systems of heating, wider use of renewable resources. Building energy performance decreased in stage of use phase, but this doesn't mean, that environmental impact is necessarily lower with the use of some of these systems/materials. Production of thinner, more effective heat-insulating materials could be more energy demanding and ecologically worse than benefits from using it. In this case LCA is required. Different scenario comparison and evaluation of complex impact will help in decision making process. Only then, is possible to objectively evaluate energy and ecological impact. Example for this study is life cycle assessment of panel residential building type T06B from Czech Republic 1985, made by ČVUT.Characteristics are in table 2.

Table 2. Panel building geometric characteristics. [7]

Characteristic	*unit*	
Built-up area	m^2	409
Usable floor area	m^2	1465
balconies area	m^2	47,7
Heated usable floor area	m^2	1065
Internal gross area	m^2	5120
Heated internal gross area	m^2	3480

Residential house has 14 apartments in total and around 45 inhabitants. The base is made of monolithic concrete slabs, on them prefabricated floor sills. Horizontal load bearing constructions are reinforced concrete panels with height of 2640-2660 mm and width of 140 mm. Vertical constructions are reinforced concrete ceiling panels 120 mm high and spanning 3460 mm. Staircase is double and prefabricated. Building envelope is made by large sandwich panels, inner wall is made of reinforced concrete 130 mm thick, then polystyrene 60 mm thick and finally outer wall made of reinforced concrete plate 70 mm thick. Roof is flat, single skin with waterproofing made of asphalt stripes. Windows are wooden and doubled. Non-load bearing walls are from plain concrete 80 mm thick. Floors are: 20 mm polystyrene, 35 mm spread concrete stepping layer with zero floor (this system is completed in loadbearing ceiling panel). Surface finishes are made by external washed-out teraco plaster, interior finish is by stucco plaster, cement milk (secondary rooms), lime pad and cement polished plaster. Building technical equipment: hot-water radiators, gas boiler, steel gas piping, PVC waste and water

IOP Conf. Series: Materials Science and Engineering **566** (2019) 012034 doi:10.1088/1757-899X/566/1/012034

piping. In early 90. final energy consumption decreased (entrance energy to building about 5%). Gables were insulated with contact insulation 60 mm thick and covered with plastic facade plates in 2000.

Energy performance decreasing scenarios

Scenario A - thermal-technical reconstruction by insulation of remaining walls with 80 mm polystyrene, thermal-technical reconstruction of roof and related structures, leaving the structure and insulating with 120 mm thick foam polystyrene, attic insulating by 50 mm thick foam polystyrene, change of windows – plastic windows Uw=1,4 W/(m^{2}K), boiler reconstruction (increasing efficiency of energy conversion to approx. 92%) and application of heating control by external temperature.

Scenario B - thermal-technical reconstruction by mineral fibre contact insulation 140 mm thick (current insulation will be removed), thermal-technical thick reconstruction of roof and related structures (removing of all layers and creating new structure with mineral fibre 200 mm thick, attic insulation with 50 mm thick mineral fibre), change of windows – wooden frames with panel polyurethane insulation layer and double glaze - Uw=1,0 W/(m^{2}K), natural ventilation, boiler reconstruction (condensing boiler - increasing efficiency of energy conversion to approx. 102%) and application of heating control by external temperature - thermostatic.

Scenario C - thermal-technical reconstruction with use of passive building technology knowledge (it doesn't necessarily means decreasing heat consumption under 15 kWh/(m^{2}a)). Estimated change of object system boundaries, because it is optimal to unite not heated areas of first floor with heated areas to create compact shape and eliminate heat bridges. Thermal-technical reconstruction by mineral fibre contact insulation 220 mm thick, insulation of all details linked to construction by 50mm thick insulation, plinths insulation with 120mm thick extruded polystyrene, thermal-technical reconstruction of roof and related structures (removal of all layers and creating new structure with mineral fibre 350 mm thick, attic insulation with 50 mm thick mineral fibre), change of windows – wooden frames with panel polyurethane insulation layer and triple glaze (with plating and filling with noble gas) - Uw=0,8 W/(m^{2}K), floor insulation on terrain (existing layer would be removed) by 100 mm mineral fibre plates, insulation of loggia panels and closing of loggia with pre-set glass wall, removing of current balconies and their replacement with completely separated construction, installation of mechanical ventilation system with recuperation, replacement of current boiler by biomass boiler, ½ heated water consumption covered from solar collectors and use of energy-saving appliances (estimated decrease of electric energy consumption by 50%).

Table 3. Panel building energy balance from realization to present. [7]

Parameter	*unit*	*realization*	*1992*	*2000*
Heat requirement for heating	kWh	178 800	171 650	152 770
Countable heat requirement for heating	kWh/(m^{2}a)	168	161	144
Final consumption of energy	kWh	324 980	308 730	285 130
Countable final consumption of energy	kWh/(m^{2}a)	305	290	268
Primal operation energy consumption	kWh	552 479	513 369	477 964
Countable primal operational energy consumption	kWh/(m^{2}a)	519	482	449

Table 4. Panel building energy balance of proposed scenarios. [7]

Parameter	*unit*	*Scenario A*	*Scenario B*	*Scenario C*
Heat requirement for heating	kWh	100 334	78 384	38 224
Countable heat requirement for heating	kWh/(m^{2}a)	94	74	36
Final consumption of energy	kWh	195 692	158 556	95 747
Countable final consumption of energy	kWh/(m^{2}a)	184	149	90

Primal operation energy consumption	kWh	343 814	288 111	68 923
Countable primal operational energy consumption	kWh/(m^{2}a)	323	271	65
Primal operation energy consumption (energy from renewable and non-renewable sources)	kWh	343 814	288 111	140 235
Countable primal operational energy consumption (energy from renewable and non-renewable sources)	kWh/(m^{2}a)	323	271	132

Table 5. Environmental impact quantifying of each scenarios – absolute values. [7]

Parameter	*unit*	*Original building*	*2000*	*2004*		
				Scenario A	*Scenario B*	*Scenario C*
Used material volume	m^3	1 499	1 513	1 623	1 726	1 939
Used material weight	kg	3 047 276	3 051 999	3 078 376	3 084 839	3 171 247
Inbuilt energy consumption	GJ	3 692	3 806	4 717	4 412	4 800
Inbuilt CO_2 emissions	kg	412 939	417 458	450 037	457 256	491 192
Inbuild SO_2 emissions	kg	1 700	1 733	1 946	1 896	2 051

When we look at table 5, we can see that inbuilt energy is higher in A scenario then in B scenario. It is caused by use of polystyrene as contact insulation. Production of this material is more energy consuming then production of mineral wool. There are visible differences between inbuilt emission CO_2 and SO_2. Despite the relatively similar volume changes in scenario A and B, even grater in case of B, the derogations are relatively small due to the choice of elements with smaller environmental impact during their life cycle.

Table 6. Cumulated values of primal energy from non-renewable sources during life cycle. [7]

Parameter	*unit*	*2000*	*2004*		
			Scenario A	*Scenario B*	*Scenario C*
Primal energy consumption from non-renewable sources	TJ	39,4	98,5	88,8	52,1
CO_2 emissions (operational + inbuilt)	t	1 931	4 520	4 163	2 440
SO_2 emissions (operational + inbuilt)	t	7,7	21,3	21,3	15,7

Big impact difference in selected time is clear from this table after realization of each scenario. Case like this shows importance of analysis as important tool in decisions making process. Such a calculation should be one of the most important factors in the decision-making process for the protection of our environment and the pursuit of sustainable development.

8. Conclusions

LCA analysis is relatively young research method but it should be more widespread to achieve sustainable development. In Slovakia, it is necessary to contribute more to the creation of databases and increase awareness about this method and to implement the LCA in the building design process, so that we can reach the more advanced countries were these projects are more used. Analysis and projects from Košice, Zvolen, Bratislava and SEVS (Central European University) are greatly contributing to popularization and spreading of LCA. Potential of this analysis is clearly visible in the above-mentioned Czech example.

9. References

[1] Dado M 2013 Analýza životného cyklu LCA *Eko-dizajn* ISSN 1338-6190

[2] Guinée J B Heijungs R et al. 2011 Life Cycle Assessment: Past, Present, and Future *Environmental Science & Technology* 45 (1) pp 90-96
[3] Gažová J 2012 Posudzovanie životného cyklu - LCA. *Odpady: odborný časopis pre podnikateľov, organizácie, obce, štátnu správu a občanov* 12(4), pp.32-33.
[4] Rusko M and Korauš A 2013 types I, II and III of ecolabels Journal of environmental protection, Safety, education and management N. 2 Vol.1 ISSN 1339-5270
[5] Stano Š 2011 Life cycle assessment – Hodnotenie životného cyklu 1.časť *Posterus* ISSN 1338-0087
[6] Tomková Ž 2019 Analýza súčasného stavu v LCA (Life cycle assessment) *11. Ročník seminár doktorandov vodbore:Pozemné stavby* ISBN978-80-553-3276-5
[7] Vonka M.: Sanace panelového domu s důrazem na snižování provozní energetické náročnosti – případová studie (Czech Technical University in Prague)

Acknowledgments
This paper was created with the support of grant project 1/0307/16 "Investigation of indoor environmental quality of educational institutions intent on health protection of children and young people" and 1/0674/18 "Theoretical and experimental analysis of architectural and structural shapes and fragments of building envelopes designed for demanding climatic conditions. " („*Výskum kvality vnútorného prostredia budov pre školstvo a vzdelávanie zameraný na ochranu zdravia detí a mládeže "a 1/0674/18 "Teoretická a experimentálna analýza architektonicko-konštrukčných tvarov a fragmentov obalových konštrukcií budov určených pre náročné klimatické podmienky.* ")

IOP Conf. Series: Materials Science and Engineering **566** (2019) 012035 doi:10.1088/1757-899X/566/1/012035

Heating demand – simulation and calculation

P Turcsanyi[1]

[1]Technical University of Kosice, Faculty of Civil Engineering, Institute of Architectural Engineering, Slovakia

Abstract. Energy efficient house is a term known very well to engineers and architects all over the world. In these times of minimizing the energy loads and negative emissions, energy efficient houses (ultra-low-energy, net-zero houses, positive houses) are one of the best ways to meet these targets. The main topic of this paper is a comparison between dynamic simulation and static calculation of heating demand of a family house designed for Slovak Republic climate. Two-storey family house was designed taking in account architectural, environmental and constructional requirements of today´s European directives focusing on energy performance and energy efficiency. Our goal was to find out the differences between heating demand calculated using national standards and heating demand using dynamic simulation software.

1. Introduction

Building energy performance assessment is crucial to ascertain the efficiency of energy use in buildings and is the basis to make any decision for enhancing energy efficiency. Globally, buildings consume nearly half of the total energy produced, and consequently responsible for a large share of CO2 emissions. The adoption of highly efficient energy production systems as well as high performance materials is being encouraged more and more in order to achieve the Nearly Zero Energy Buildings target. Energy efficient design (to create ultra-low-energy building, net-zero building or even energy-plus building) is a response to local climate conditions and global energy consumption. Thermal building simulation is a powerful tool to assess the energy performance of a building. Design Builder is currently the most comprehensive user-friendly Energy Plus interface. To sum it all, simulation software Design Builder is one of the most effective tools on the market, which can play very important part in the process of building design. When using Design Builder correctly, engineers can predict building behaviour throughout a reference year loaded to database.

1.1. Object information

Presented family houses are located in Kocice – Krasna, in the residential area of similar family houses. Two-storey family house represents a typical type of residential buildings built in this area. The floor layouts were designed for a family of two parents and three kids and the concept of a whole houses were made according to the latest energy efficient standards. Buildings are based on concrete footing foundation. Both envelopes of the houses were designed from materials with high thermal capacity insulated from outside with a thick layer of polystyrene (ETICS system) to minimize the heat loses during the heating period as well as to prevent the heat infiltration during the summer period.

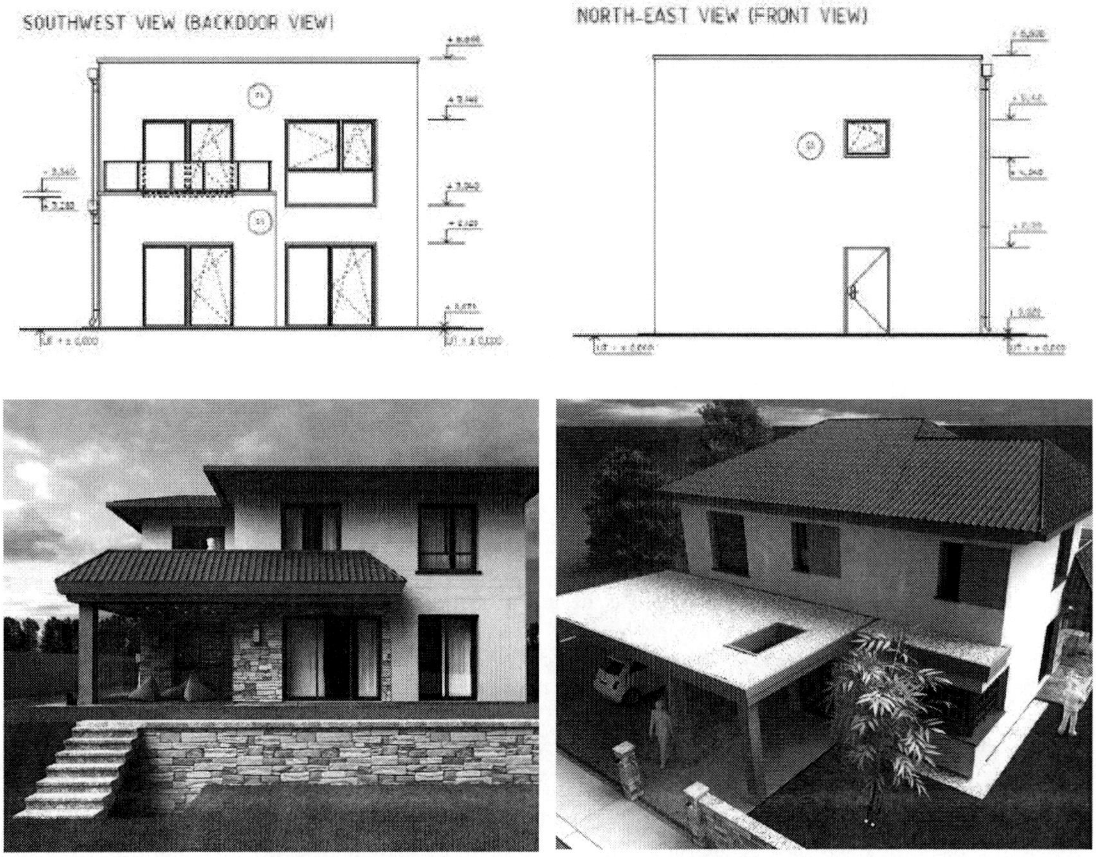

Figure 1. Family houses views (top A, bottom B).

Transparent parts of building envelopes were designed to meet the highest insulating standards.

Table 1. Family house "A" parameters.

A/V Ratio	0.43 1/m
Building volume	254,10 m3
Building envelope	149,80 m2
Total floor area	108,64 m2
Uwall	0.100 W/(m2.K)
Uroof	0.120 W/(m2.K)
Rfloor	6,25 (m2.K)/W
Uw – Schuco Alu Inside	0.60 – 0.76W/(m2.K)

IOP Conf. Series: Materials Science and Engineering **566** (2019) 012035 doi:10.1088/1757-899X/566/1/012035

Table 2. Family house "B" parameters.

A/V Ratio	0.78 1/m
Building volume	726.18 m^3
Building envelope	569.48 m^2
Total floor area	222.09 m^2
U_{wall}	0.148 W/(m^2.K)
U_{roof}	0.123 W/(m^2.K)
R_{floor}	4.38 (m^2.K)/W
U_w – Internorm HF 350	0.60 – 0.95W/(m^2.K)

1.2. Locality

Košice lies in the North Temperate Zone and has a borderline continental and marine climate with four distinct seasons. If defined as marine due to the winters just above −3°C (27°F), it would be one of the farthest inland areas with this climate type.

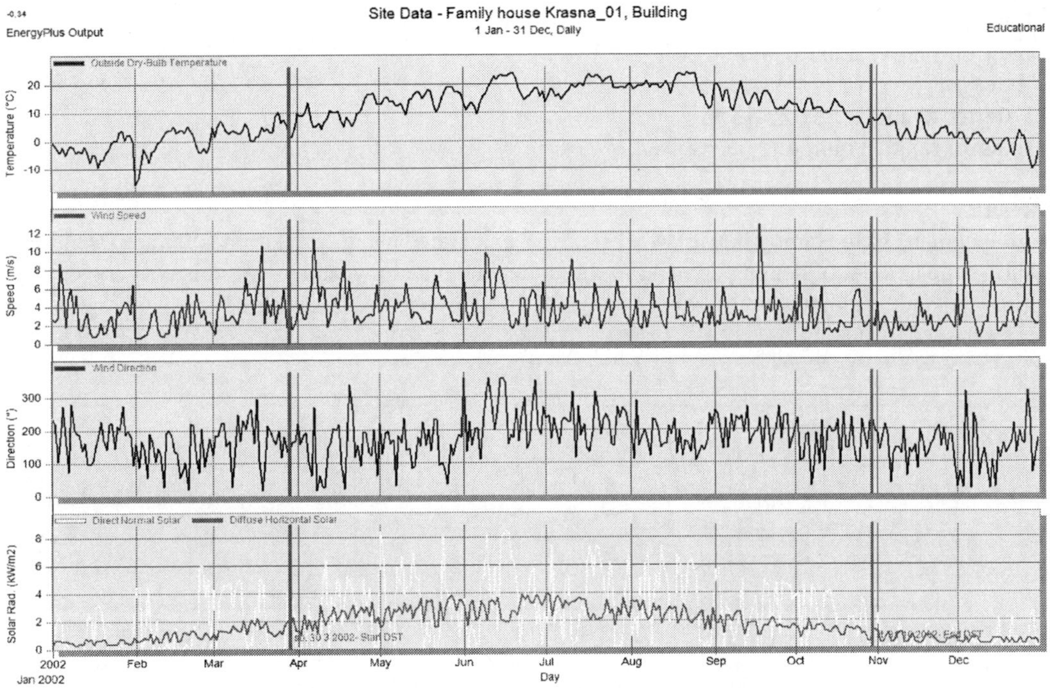

Figure 2. Weather data used in the simulation.

2. Methods

Both buildings simulations were set up with the DesignBuilder v4 software, in which building performance data is generated by the simulation engine EnergyPlus. EnergyPlus is a building simulation program used by United States Department of Energy and has been extensively tested by field measurements and empirical methods

.

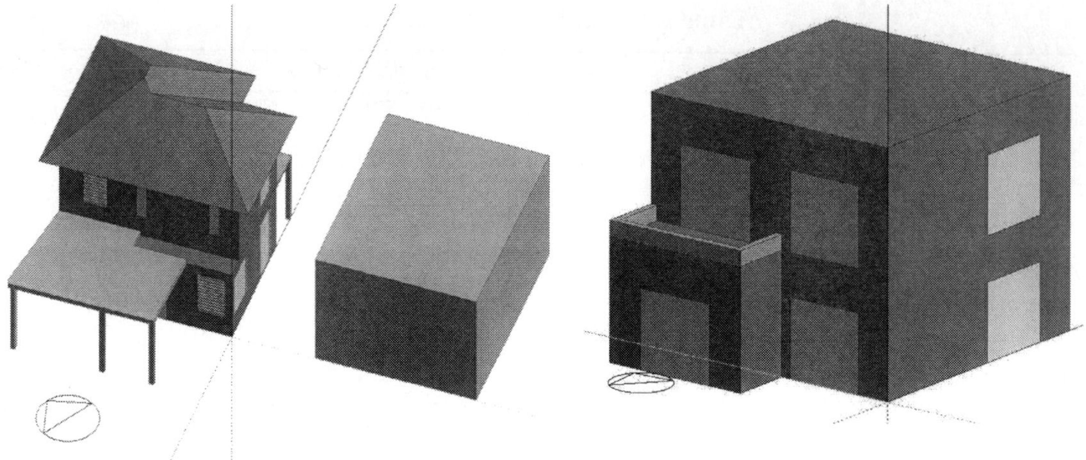

Figure 3. Left family house “B”, right family house “A”.

Second method, and the most commonly used one, is calculation of heating demand using national standards. When calculating heating demand of the family houses, these exterior and interior conditions were used (based on geographical situation):

Interior:
Interior temperature θ_i= 21°C
Interior relative humidity ϕi = 55%
Exterior:
Exterior temperature θ_e= -13°C
Exterior relative humidity ϕe = 84%

3. Results

Heating demand of the family house “A” (Figure 4), calculated using simulation software DesignBuilder v4 shows the amount of heating demand in direct dependence to weather – what is only logical. Yearly heating demand calculated on the square meter of flood area is 34,03 kWh/m^2.

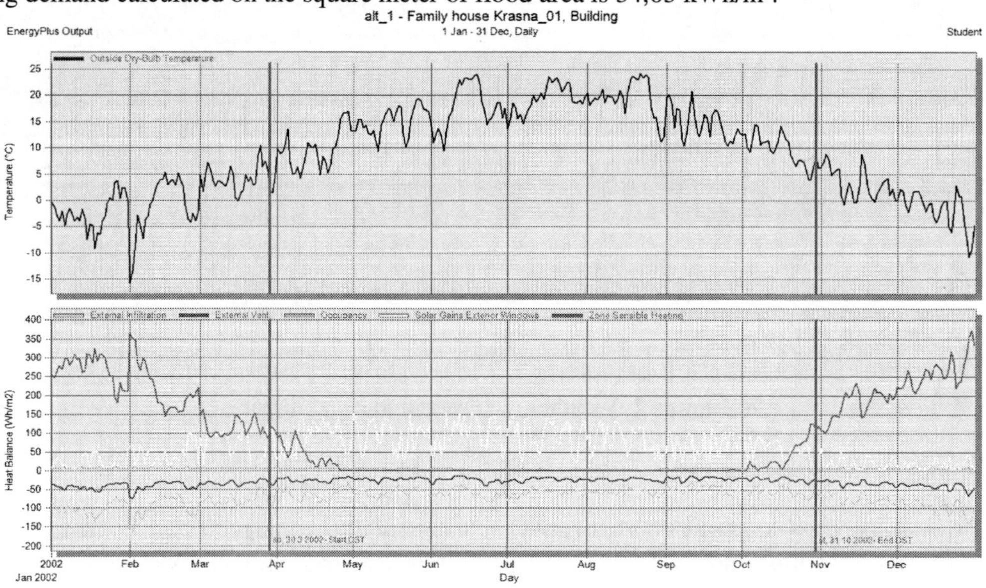

Figure 4. Heating demand /red line/ of the family house “A” – daily method.

Heating demand of the family house "B" (Figure 5), calculated using simulation software DesignBuilder v4 shows the amount of heating calculated on the square meter of flood area as 34,03 kWh/m^2.

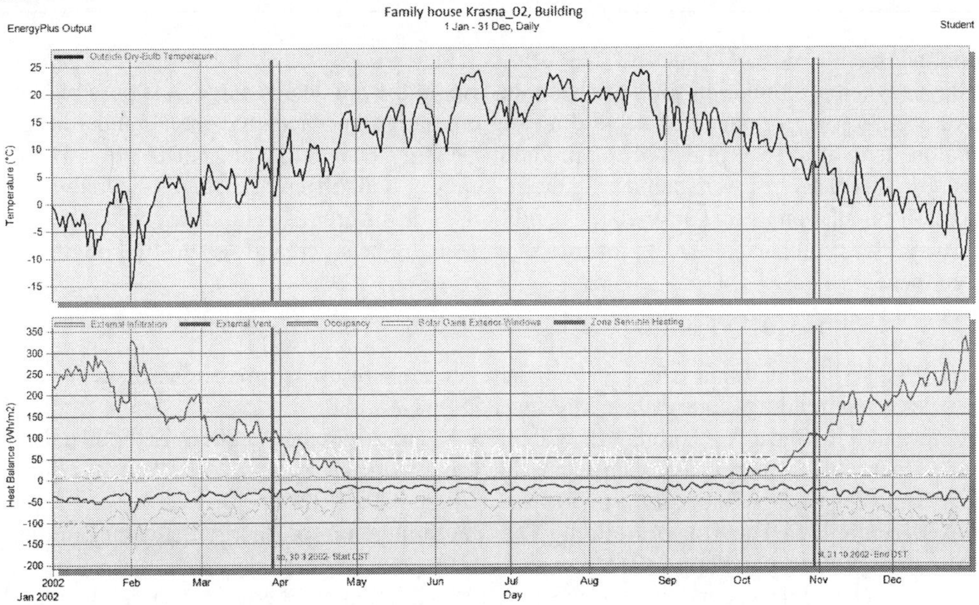

Figure 5. Heating demand /red line/ of the family house "B" – daily method.

As it can be seen (Figure 4, Figure 5), heating demand is directly to weather conditions, occupancy, solar gains through exterior glazing, external infiltration and external ventilation.

Table 3. Family house "B" results.

Simulation	Calculation
Heat demand for heating according to standard valid until 31.12.2020	
Qh,nd,1 ≤ Qh,nd,r1	Qh,nd,1 ≤ Qh,nd,r1
34,01 kWh/m2.a ≤ 41,90 kWh/m2.a	44,93 kWh/m2.a ≤ 41,90 kWh/m2.a
Heat demand for heating according to standard valid from 01.01.2021	
Qh,nd,2 ≤ Qh,nd,r2	Qh,nd,2 ≤ Qh,nd,r2
34,01 kWh/m2.a ≤ 21,15 kWh/m2.a	44,93 kWh/m2.a ≤ 21,15 kWh/m2.a
Difference between simulation and calculation 19,8 %	

Table 4. Family house "A".

Simulation	Calculation
Heat demand for heating according to standard valid until 31.12.2020	
Qh,nd,1 ≤ Qh,nd,r1	Qh,nd,1 ≤ Qh,nd,r1
22.43 kWh/m2.a ≤ 29.75 kWh/m2.a	28.30 kWh/m2.a ≤ 29.75 kWh/m2.a
Heat demand for heating according to standard valid from 01.01.2021	
Qh,nd,2 ≤ Qh,nd,r2	Qh,nd,2 ≤ Qh,nd,r2

IOP Conf. Series: Materials Science and Engineering **566** (2019) 012035 doi:10.1088/1757-899X/566/1/012035

22.43 kWh/m2.a ≤ 29.75 kWh/m2.a	28.30 kWh/m2.a ≤ 29.75 kWh/m2.a
Difference between simulation and calculation 19,8 %	

4. Conclusion

Nowadays, dynamic simulation software is on the rise. In the pre design stage, it is possible to optimize the future energy performance of a building more detailed and accurate than standardized national calculation. It is a very helpful tool for predicting energy behaviour, and all in all – to save future expenses. In this paper, we were comparing the accuracy of both different methods – dynamic simulation vs. standard calculation using non-variable conditions. The difference in both cases is close to 20%. The reason why I was comparing two approaches of calculating heat demand was to find out the difference between them.

5. References

[1] H Wang S, Yan Ch and Xio F 2012 Quantitative energy performance assessment methods for existing buildings *Energy and Buildings 55* pp 873–888

[2] Giordano R, Serra V, Tortalla E, Valentini V and Aghemo C 2015 Embodied Energy and Operational Energy assessment in the framework of Nearly Zero Energy Building and Building Energy Rating *Energy Procedia* 78 pp 3204 – 3209

[3] Andarini R 2014 The Role of Building Thermal Simulation for Energy Efficient Building Design *Energy Procedia* 47 pp 217-226

[4] Mateus N M, Pinto A and da Graça G C 2014 Validation of EnergyPlus thermal simulation of a double skin naturally and mechanically ventilated test cell *Energy Build 75* pp 511-522

Acknowledgements

This paper has been prepared within the Grant No. 2/0142/19 of the Slovak Grant Agency for Science.

Assessment of shear capacity of concrete bridge deck slabs using theoretical formulations and FEM analysis

A Vidaković[1], J Halvonik[1]

[1] Slovak University of Technology in Bratislava, Faculty of Civil Engineering, Department of Concrete Structures and Bridges, Slovakia

Abstract. Reinforced concrete (RC) slabs without shear reinforcement are typical cases of bridge deck slabs. For such structures, shear has been a challenging problem in the assessment based on the current standards. This paper deals with the comparison of the methods for assessment of bridge deck slabs subjected to a concentrated load. Two experimental campaigns were selected as case studies and analysed by the simplified theoretical formulations and linear finite element analysis (LFEA). The differences between analysis methods were discussed regarding one-way shear behaviour of the bridge deck slabs.

1. Introduction

Shear in RC slabs is a challenging problem in the assessment based on the current standards. Verifications used in the past provided higher shear resistance than current codes of practice and this often leads to the requirement to provide shear reinforcement in the areas where it was not necessary in the past. This also raises the question of reliability of existing structures and therefore it has become the major motivation for the experimental campaigns following with theoretical and numerical assessments of the shear resistance. The greatest differences between outcomes of the past and current models for prediction of shear capacity are in the case of slabs subjected to a concentrated load. The typical example are RC bridge deck slabs loaded by wheel pressure located close to the support. In this case two models have to be merged, the first one is model for one-way shear and the second one is for the distribution of a concentrated load. The obtained product, method for the assessment of shear capacity has to be verified by the test results from experiments. An article is focused on the verification of two methods; the first method is based on the design equations from relevant codes of practice and the second method is based on the linear analysis using FEM models. Test results from two experimental campaigns are used for the verification of the methods.

2. Experimental campaigns

2.1. *Campaign by R Vida and J Halvonik*

The geometry of the tested specimens by R Vida and J Halvonik [1] are introduced in figure 1(a). The slabs were loaded on both sides of cantilevers with a concentrated load on a 250 x 250 mm steel plates, generated by a two synchronized hydraulic jacks supported by a strong steel frame. The position of the loaded area was constant, at the distance of 335 mm (twice the effective depth of the slab) from the face of the central beam under the slab. The measured concrete strength before tests varied between 30 and 38 MPa. Three different reinforcement ratios of 0.73%, 1.27% and 1.69% with standard B500B grade of steel were used.

2.2. Campaign by G Rombach and L Henze

The geometry of the cantilever slabs tested by G Rombach and L Henze [2] can be seen in figure 1(b). The load was applied in the symmetry axis through a 400 x 400 mm steel plate and the cantilever on the backside was fixed by means of four vertically prestressed bars against the laboratory strong floor. Grade B500S of steel was used and a reinforcement ratio of 1.17% was the same for all slabs. The measured concrete strength varied between 38 and 56 MPa. Five different loading locations were investigated with a_v/d ratio from 2 to 6, where a_v is the distance between the edge of a support and the inner edge of the loading plate and d is the effective depth of slab.

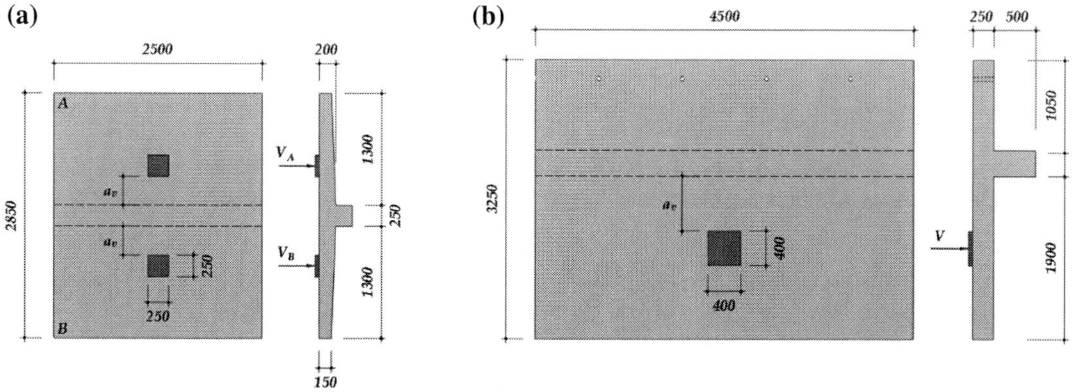

Figure 1. Investigated slab specimens tested by: (a) R Vida and J Halvonik [1]; and (b) G Rombach and L Henze [2].

2.3. Material properties

Material properties for all specimens from both experimental campaigns can be seen in table 1.

Table 1. Material properties of the tested specimens [1] and [2].

Authors	Specimen	d	a_v/d	ρ_l	f_c	E_c	d_g
R Vida and J Halvonik (2018) [1]	**SL0.1**	168	2	0.73	36.05	35.20	16
	SL0.2				38.18	38.20	
	SL0.3				34.49	40.50	
	SL1.1			1.69	34.80	38.30	
	SL1.2				32.15	36.80	
	SL2.1			1.27	30.47	34.50	
	SL2.2				29.41	36.30	
G Rombach and L Henze (2017) [2]	**P-2d-1**	215	2	1.17	47.30	34.45	16
	P-2d-2				55.50	36.45	
	P-3d-1		3		37.80	28.32	
	P-3d-2				49.50	34.29	
	P-4d-1		4		40.00	32.75	
	P-5d		5		46.40	30.18	
	P-6d		6		43.00	29.87	

3. Models for assessment of shear resistance

3.1. Eurocode 2

The shear strength formulation according to current EN 1992-1-1 (Eurocode 2) [3], for a member without shear reinforcement was empirically calibrated with experimental data through a statistical approach. The shear strength is given by:

IOP Conf. Series: Materials Science and Engineering **566** (2019) 012036 doi:10.1088/1757-899X/566/1/012036

$$V_{R,c} = \left[C_{R,c} \cdot k \cdot (100 \cdot \rho_l \cdot f_c)^{\frac{1}{3}} \right] \cdot b_w \cdot d \geq 0.035 \cdot k^{\frac{3}{2}} \cdot \sqrt{f_c} \cdot b_w \cdot d \tag{1}$$

where $C_{R,c}$ is an empirical factor (0.18 MPa), f_c is the concrete compressive strength (in MPa), d is the effective depth (in mm), ρ_l is the longitudinal reinforcement ratio and b_w is the effective shear width (in mm). The size-effect parameter k can be determined as:

$$k = 1 + \sqrt{\frac{200}{d}} \leq 2.0 \quad d \text{ in [mm]} \tag{2}$$

The effective shear width b_w usually depends on the national practices [4]. French and Dutch models are introduced in figure 2(a) and 2(b), respectively. The distribution under 45° is measured from the outer edges of the loaded area in the case of French and from the center of the loaded area in the case of Dutch model.

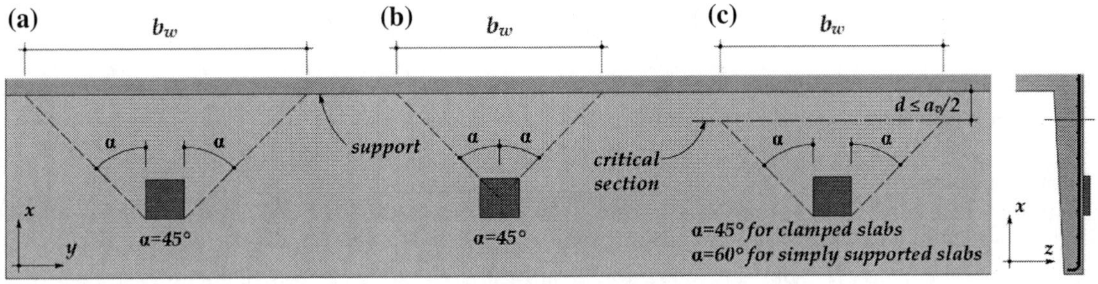

Figure 2. Effective shear width according to the: (a) French practice; (b) Dutch practice; and (c) Model Code 2010.

3.2. Model Code 2010

The *fib* Model Code 2010 (MC2010) [5] proposes three methods for calculating one-way shear resistance of concrete slabs: Level I, Level II and Level IV approximations, which complexity and accuracy increasing as the level arises. The shear strength formulation is based on The Simplified Modified Compression Field Theory [6] and is given by:

$$V_{R,c} = k_v \cdot \sqrt{f_c} \cdot z \cdot b_w \tag{3}$$

where the lever arm z can be taken as $0.9d$ and the effective shear width b_w is determined by the load distribution angle of 45° for clamped and 60° for simply supported slabs (figure 2(c)). The critical section is assumed at the smaller distance of d and $a_v/2$ from the face of the support.

For a Level I approximation, k_v is determined as:

$$k_v = \frac{180}{1000 + 1.25z} \tag{4}$$

For a Level II approximation, k_v is determined as:

$$k_v = \frac{0.4}{1 + 1500\varepsilon_x} \cdot \frac{1300}{1000 + k_{dg} \cdot z} \tag{5}$$

$$\varepsilon_x = \frac{1}{2E_s \cdot a_s} \cdot \left(\frac{m}{z} + v \right) \tag{6}$$

where ε_x is the longitudinal strain at the mid-depth of the effective depth in the control section, k_{dg} is a parameter depending on the maximum aggregate size which is 1.0 in this case, since $d_{g,}$ = 16 mm.

3.3. Critical Shear Crack Theory

The Critical Shear Crack Theory (CSCT) [7] assumes that the critical crack width w_{cr} is proportional to the product of the longitudinal strain in the control depth ε times the effective flexural depth d:

$$w_{cr} \propto \varepsilon \cdot d \tag{7}$$

The following CSCT failure criterion has been proposed [8][9]:

$$v_c(\varepsilon) = \frac{d \cdot \sqrt{f_c}}{3} \cdot \frac{1}{1 + \dfrac{120 \cdot \varepsilon \cdot d}{16 + d_g}} \tag{8}$$

The longitudinal strain ε and the depth of the compression zone c_{flex} are defined by (9) and (10):

$$\varepsilon = \frac{m}{d \cdot \rho \cdot E_s \cdot \left(d - c_{flex}/3\right)} \cdot \frac{0.6d - c_{flex}}{d - c_{flex}} \tag{9}$$

$$c_{flex} = d \cdot \rho \cdot \frac{E_s}{E_c} \cdot \left(\sqrt{1 + \frac{2E_c}{\rho \cdot E_s}} - 1 \right) \tag{10}$$

where m is the maximum unitary bending moment in the control section and E_c and E_s is the modulus of elasticity of concrete and steel, respectively.

4. Numerical analysis

Numerical analysis based on the linear elastic finite element method (LFEM) is becoming a standard tool for modelling reinforced concrete members like bridge deck slabs. However, the results from this analysis are not always reliable, due to the nonlinear behaviour of the reinforced concrete caused by the cracking and yielding of the reinforcement. Experimental results show that the cantilever slabs subjected to concentrated load have the ability to distribute the load in transverse direction. In order to take into account for this distribution of internal forces when using LFEA, Natário et al. [8] proposed a new method based on the CSCT for the assessment of bridge deck slabs. An average shear force $v_{avg,4d}$ is calculated along a distance $4d$ assuming unitary shear forces obtained from LFEA. The control section of the proposed method is located at the distance of $d/2$ from the support as can be seen in figure 3(a). The reference longitudinal strain (see equation (9)) is calculated at the maximum unitary bending moment m at the control section through an iterative procedure. Finally, the ultimate shear failure force V_R is calculated by equation (11) at the intersection with the failure criterion defined previously in equation (8):

$$V_R = \frac{v_c(\varepsilon)}{\beta} \cdot b \tag{11}$$

where b is calculated from LFEA results as the ratio between the applied load and $v_{avg,4d}$ and in order to take into account the arching action, a factor β is applied as [8]:

$$\beta = \frac{a_v}{2.75d} \leq 1.0 \tag{12}$$

(a)

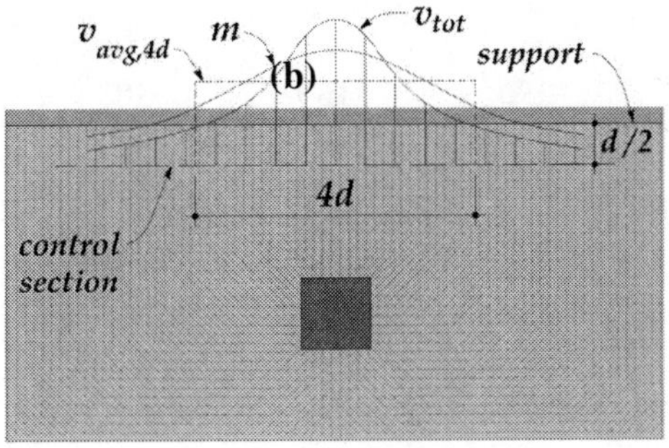

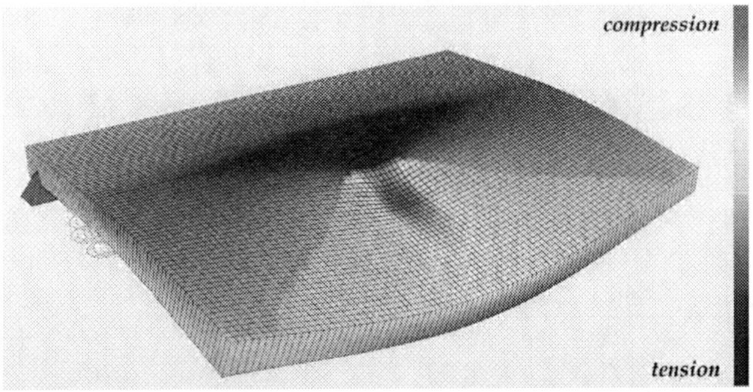

Figure 3. (a) Definition of reduced shear force $v_{avg,4d}$ according to Natário et al. [8]; and (b) Deformed numerical model under concentrated load.

Numerical analysis was carried out in SOFiSTiK [10] software, figure 3(b). Quadrilateral 4-node shell elements with shear deformations based on the Mindlin-Reissner theory were used in this study. Self-weight was neglected and the concentrated loads were modelled as surface loads. The central beam of the tested specimens was modelled with linear elastic compression-only springs, representing the stiffness of the steel base in compression and having zero stiffness in tension. Fixed line support was used on the opposite side of the cantilever slabs, allowing rotations only around its axis.

5. Analysis of the test results

Comparisons of the test results with calculated shear resistances according to the relevant models and selected methods are shown in table 2. The best results provided method introduced by Natário et al. [8]. This method is based on the LFEA combined with one-way shear model based on the CSCT. High quality of the method is given by a very low value of CoV = 0.09 and average value of the ratio V_{exp}/V_{model} equals nearly to one. The model works perfectly for any given a_v/d ratio.

Good results are recorded in the case of the current EC2 model in connection with the model for the distribution of shear forces according to the French recommendation. However, the safety level of the model decreases with increasing of a_v/d ratio. It works well for $a_v/d \leq 3.0$, then the model safety becomes unacceptable and probably the model for two-way shear (punching) shall be applied or the control section should not be assumed further than $2d$ from the edge of the loading plate. The similar results were observed in the case of the EC2 model with Dutch recommendation. Shorter effective width increases model safety in comparison with the French model but for $a_v/d \geq 4.0$ the model becomes unsafe too.

Table 2. Comparison between tests and shear strength predictions.

Authors	Specimen	V_{exp}	$V_{EC2,French}$		$V_{EC2,Dutch}$		$V_{MC2010,I}$		$V_{MC2010,II}$		$V_{LFEA+CSCT}$	
			kN	*[-]	kN	*[-]	kN	*[-]	kN	*[-]	kN	*[-]
R Vida and	**SL0.1A**	347	256	1.36	166	2.09	146	2.38	194	1.79	377	0.92
J Halvonik	**SL0.1B**	362		1.42		2.18		2.48		1.87		0.96
(2018) [1]	**SL0.2A**	393	261	1.51	169	2.33	150	2.62	197	1.99	385	1.02

	SL0.2B	360		1.38		2.13		2.40		1.83		0.94
	SL0.3A	390	252	1.55	163	2.39	143	2.73	191	2.04	370	1.06
	SL0.3B	383		1.52		2.35		2.68		2.01		1.04
	SL1.1A	373	334	1.12	216	1.72	143	2.60	251	1.48	457	0.82
	SL1.1B	451		1.35		2.08		3.15		1.80		0.99
	SL1.2A	382	325	1.17	211	1.81	138	2.77	244	1.56	442	0.86
	SL1.2B	419		1.29		1.99		3.04		1.72		0.95
	SL2.1A	380	291	1.31	188	2.02	134	2.83	220	1.73	408	0.93
	SL2.1B	350		1.20		1.86		2.61		1.59		0.86
	SL2.2A	430	287	1.50	186	2.31	132	3.26	217	1.98	401	1.07
	SL2.2B	413		1.44		2.22		3.14		1.90		1.03
G Rombach and L Henze (2017) [2]	**P-2d-1**	984	597	1.65	365	2.69	314	3.13	412	2.39	805	1.22
	P-2d-2	897	629	1.42	385	2.33	341	2.63	433	2.07	852	1.05
	P-3d-1	630	669	0.94	454	1.39	355	1.77	445	1.42	590	1.07
	P-3d-2	680	732	0.93	497	1.37	406	1.67	484	1.40	643	1.06
	P-4d-1	677	800	0.85	581	1.17	442	1.53	510	1.33	636	1.06
	P-5d	699	964	0.73	734	0.95	558	1.25	588	1.19	708	0.99
	P-6d	656	1061	0.62	836	0.78	616	1.06	624	1.05	693	0.95
$*V_{exp}/V_{model}$			Average	1.25		1.91		2.46		1.72		0.99
			CoV	0.23		0.27		0.26		0.19		0.09

The similar tendency as for the EC2 model is observed in the case of the Model Code 2010 model, which means greater a_v/d ratio leads to the lower model safety. For $a_v/d = 2.0$ the model safety is very high, the average value is 1.86. It might be due to the direct transfer of shear forces to the support by concrete struts. For greater a_v/d ratio the safety level is still quite high.

6. Summary and future work

The presented analyses of the models for prediction of the shear capacity of bridge deck slabs subjected to a concentrated load allow us to draw following conclusions:

- The safety level of the design models that are based on the calculation of the effective width with distribution of shear forces under 45° in the case of clamped slabs is significantly influenced by a_v/d ratio. With increasing a_v/d the safety level is decreasing.
- All models provide safe solution if $a_v/d \leq 2.0$. The EC2 model in connection with the French recommendation is on the unsafe side for $a_v/d \geq 3.0$ and with the Dutch recommendation for $a_v/d \geq 5.0$. The Model Code 2010 model is very conservative for $a_v/d \leq 2.0$ and for $a_v/d > 3.0$ it provides reasonable solution.
- The best results were obtained for Natário et al.'s CSCT model. The safety of the model is independent on the a_v/d ratio. The average value of the ratio V_{exp}/V_{model} which is close to one, can be increased by simple update of the failure criterion in one-way shear CSCT model.
- Future research should be focused particularly on the cases where a_v/d ratio is higher than two. It is necessary to extend database of the test results, where concentrated load is placed at the distance higher than $2d$ measured from the edge of a support.
- An amendment of the current Eurocode 2 is needed in the case of the design of concrete bridge deck slabs for shear. The minimum and maximum distance of the wheel pressure from the edge of a slab support should be introduced in the standard as well as position of the control section if a_v/d ratio is higher than two.

7. References

[1] Vida R and Halvonik J 2018 Tests of shear capacity of deck slabs under concentrated load *Proceedings of the 12th fib International PhD Symposium in Civil Engineering* (Czech Technical University in Prague) pp 773–9

IOP Conf. Series: Materials Science and Engineering **566** (2019) 012036 doi:10.1088/1757-899X/566/1/012036

[2] Rombach G and Henze L 2017 Shear capacity of concrete slabs under concentrated loads close to support *Proceedings of the 2017 PhD Symposium* (Maastricht: Hordijk D and Luković M (eds.) *High Tech Concrete: Where Technology and Engineering Meet*) pp 719–26
[3] CEN 2005 Eurocode 2: *Design of concrete structures – Part 1-1: General rules and rules for buildings* p 198
[4] Lantsoght E , van der Veen C and Walraven J 2013 Shear in one-way slabs under concentrated load close to support *ACI Structural Journal* vol **110** no 2 pp 275–84
[5] fib Model Code 2010 – final draft 2013 *fib Bulletin 65* vols **1** and **2** 350p, *fib Bulletin 66* p 370
[6] Bentz E C, Vecchio F J and Collins M P 2006 Simplified modified compression field theory for calculating shear strength of reinforced concrete elements *ACI Structural Journal* vol **103** no 4 pp 614–24
[7] Muttoni A and Fernández Ruiz M 2008 Shear strength of members without transverse reinforcement as function of critical shear crack width *ACI Structural Journal* vol **105** no 2 pp 163–72
[8] Natário F, Muttoni A and Fernández Ruiz M 2014 Shear strength of RC slabs under concentrated loads near clamped linear supports *Engineering Structures* vol **76** pp 10–23
[9] Belletti B, Scolari M, Muttoni A and Cantone R 2015 Shear strength evaluation of RC bridge deck slabs according to CSCT with multi–layered shell elements and PARC_CL Crack Model *IABSE Conference* (Geneva) pp 1158–65
[10] SOFiSTiK AG 2018 *SOFiSTiK Analysis Programs* version 2018-7 (Oberschleissheim)

Acknowledgments
This work was supported by the Slovak Research and Development Agency under the contract No. APVV-17-0204.

IOP Conf. Series: Materials Science and Engineering **566** (2019) 012037 doi:10.1088/1757-899X/566/1/012037

Zero house by 3D printer technology

I Wawrek[1]

[1] Technical university of Košice, Faculty of Civil Engineering, Department of architectural engineering, Slovakia

Abstract. Implementation of 3D printing technologies into building process is recently one of the most developing sectors. Research teams come every day with new ways of use of this technology. Research is focused on testing thermal – technical properties of certain wall structure built by 3D printing additive technology. Based on results of calculation and measurements will be done optimization of design of mentioned wall structure. The goal of the research is to design wall structure with performance of "zero" house. Another goal is to extend knowledge in area of 3D printing and create new possibilities to better adaptation autonomics 3D printing to building process of the future.

1. Introduction

3D printer technology implementation to building process is currently one of the most dynamically developing sectors. Research groups from all over the world are continuously coming with new ways how to bring this method into actual building practice. This article is mapping current situation of knowledge in this area. It is creating new demands for implementation of mentioned technology and is searching for the answers with help of 3D computer analysis. It then verifies the accuracy of the analysis results by measuring the prototype solution under laboratory conditions. Research results would be optimized wall construction, suitable for direct application on building site with use of autonomous device. This wall would meet the "zero" heat transfer trough the construction parameters (according to STN 40 2020).

FDM (Fused Deposition Modelling) this technology of 3D printing was patented in 1992. FDM is printing method realised by printing head (extruder), which is moving in three dimensional space (axe x, y, z) and prints horizontal layers of materials on each other and the whole object high is growing. By this principle is possible to create device with whom it would be possible to layer wall constructions directly in place of construction and thus to create the building object as if growing from the ground without the need for additional personnel [1].

2. Building process

Currently the 3D printing is used in constructional process in two ways, wall construction prefabrication and creation of monolithic buildings directly on site.

3. Prefabrication

Wall constructions are printed in the production hall and then transported on site where they are completed to building objects. This method is used in WINSUN company (Shanghai - China). WINSUN built their first building with use of 3D printing tools in 2014. There were years of printing material development behind this first attempt.

2002 – GRG – special plasterboard reinforced with glass fibre

2006 – SRC - special cement reinforced with glass fibre

IOP Conf. Series: Materials Science and Engineering **566** (2019) 012037 doi:10.1088/1757-899X/566/1/012037

2006 – FRP – special composite material with glass fibre
2007 – 3D printing of natural stone [3]

Figure 1. Interior of WINSUNs printed house. **Figure 2.** WINSUNs wall *construction.*

4. Construction on site

Building is constructed with help of robotic mechanism directly on chosen site. Such a process greatly reduces the construction costs as well as the construction process time. APIS COR company from Boston in USA is the first company which designed and assembled tool for 3D printing of entire building as monolithic unit directly at the destination. Functionality of their design was verified by printing family house which was finished with additional finishes in 24 hours.

Building was additionally insulated with blown insulation into the outer wall construction envelope. Wall construction is open and made by two closed air gaps, which are continuous in the vertical direction. A narrower gap (closer to the outer surface) is filled with blown insulation after printing is done, thus creating a thermal insulation layer of the wall structure. The gap at the inner edge is reinforced in the horizontal direction by glass-fibre bars (the method of placing these reinforcement elements during the printing process is not known) [4].

Figure 3. House printed in 24 hours by APIS COR.

IOP Conf. Series: Materials Science and Engineering **566** (2019) 012037 doi:10.1088/1757-899X/566/1/012037

Figure 4. Cooperation of NASA and APIS COR.

5. Evaluation

Table 1. Advantages and disadvantages

ADVANTAGES OF BULDING CONSTRUCTION 3D PRINTING	DISADVANTAGES OF BUILDING CONSTRUCTION 3D PRINTING
Minimal construction waste	High initial printer equipment cost
Lower technology costs	Litle efficient construction
Less construction workers	Little known properties of bulding materials for printing
Safer construction site	Need of continuous control during the building process
Higher precision	Need for additional equipment (fitting of the fillings of construction openings)
Possibility of shape changes during construction process without additional costs.	Need to wider and deepen the level of knowledge on the subject.
Possibility of estimating exact finish time	
Less dust production impact on surrounding environment	
Shorter building time	
Easier logistic	
Little site space requirements	
Little material storage requirements	
Less CO_2 emissions	
Less energy consumption for construction	

IOP Conf. Series: Materials Science and Engineering **566** (2019) 012037 doi:10.1088/1757-899X/566/1/012037

6. Problem determination

As it is visible from table 1 one of the main disadvantages of 3D printed constructions shows to be little effectiveness of construction. This construction weakness is caused by incorrect shape design of inner wall structure, which is in most examples adapted to create ideal traces for printing head. This way it is shortening building time and it helps to promote the technology of construction.

7. Aim

The aim of research is to design fitting optimization of shape, thickness and inner wall structure with mind on new possibilities, which are offered with use of layer construction printing method. Adjust suitable construction parameters to achieve a heat transfer coefficient U for this construction at the zero-level house level.

8. Zero house

It's a term which describes very energy-saving efficient buildings, mostly using energy from renewable sources in such amount that during a year they get almost equal amount of energy from environment as they consume. This means that their energy balance is equal to zero or near to zero. [2]

9. Requirements for envelope constructions

According to energy efficiency, buildings are divided into energy classes. By the end of 2015 for new buildings, it was enough to reach the upper limit of class B (global primary energy indicator) for their successful certification. From 1 January 2016 until the end of 2020, it is necessary to enter A1 and after 2020 into A0. [5]

Table 2. Standardized requirement for building envelope construction.

Building enevelope structure	Heat transfer coefficient U (W/(m².K))		
	Till 31.12.20015	From 1.1.2016	From 1.1.2021
	NED ...B	UNB ...A1	TNB (PD) ...A0
Outher wall, pitched roof with angle above 45^0	0,32	0,22	0,15
Flat and sloping roof (up to 45^0), ceiling above the exterior	0,20	0,10	0,10
Ceiling above unheated space	0,25	0,15	0,15
Windows and glazed walls, roof windows	1,40	1,00	0,60
Doors without backrest / with backrests	3,00/4,00	2,50 / 3,00	2,00
Average U Building (shape factor 0.3 to 1.0)	0,39 - 0,58	0,27 - 0,38	0,20 - 0,25

10. Current construction

The currently used cross-section of the structure has only minimal thermal insulating properties which greatly reduces its use in the construction of residential buildings.

IOP Conf. Series: Materials Science and Engineering **566** (2019) 012037 doi:10.1088/1757-899X/566/1/012037

Figure 5. Example of current solution.

11. Design

The optimization design of the perimeter structure is inspired by the organic structure of the bone section. Such a spatial structure creates small closed air gaps which greatly reduce the coefficient of heat transfer in the structure. The dense inner mesh forms, similarly to the bone, a support grid on the outer surface, which improves the overall static properties of the element. For better application possibilities of this structure to mathematical models, it is necessary to reduce it to a spatial structure that consists of repeating the same geometric element (Figure 8). The correct design of this structure, the size of the holes in it, and the spacing of the cavities in horizontal and vertical planes will be checked by computer analysis.

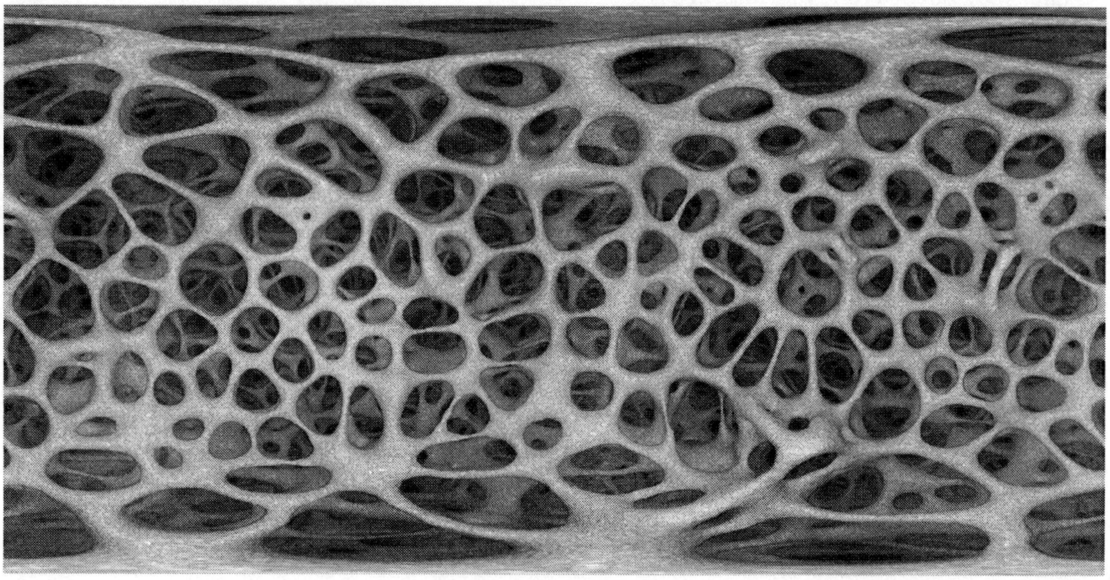

Figure 6. Bone section.

IOP Conf. Series: Materials Science and Engineering **566** (2019) 012037 doi:10.1088/1757-899X/566/1/012037

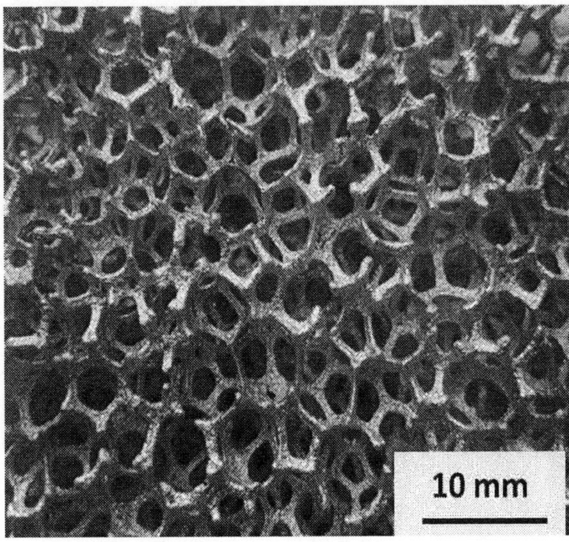

Figure 7.

Figure 8. Example of current solution

Final sample, which will came out of computer analysis as the most appropriate solution, will be made and subjected to laboratory tests for strength and heat resistance. Verification of the design sample confirms or invalidates the accuracy of the computer analysis.

12. Vizualization of possible solutions

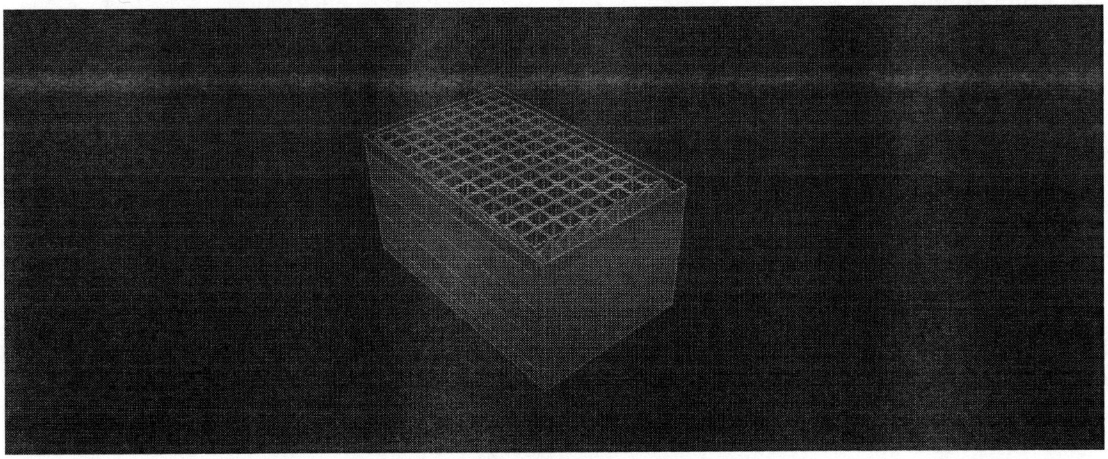

Figure 9. Horizontal section

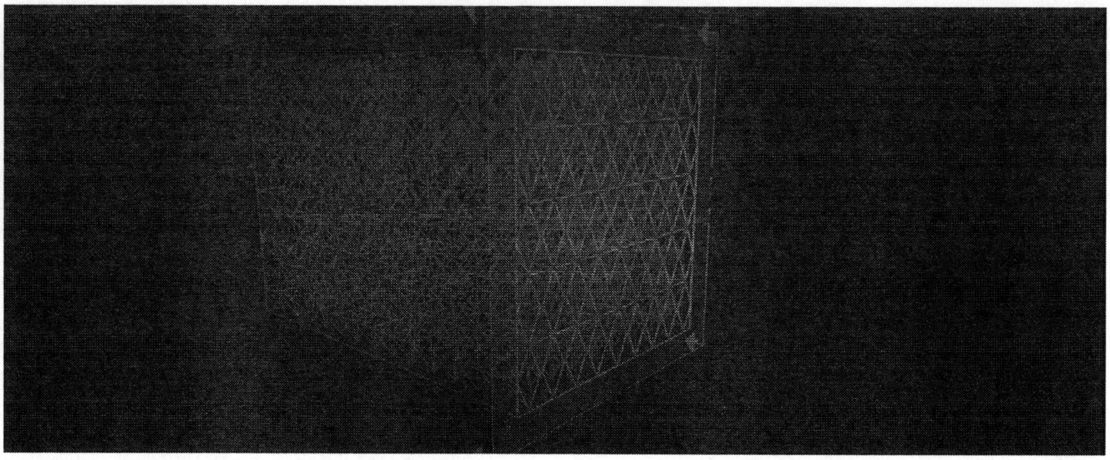

Figure 10. Vertical section

13. Use

The automated construction system, using the right construction parameters, has wide range of uses. It is possible to apply it on humanitarian missions in areas affected by natural disasters, where it can autonomously create objects from temporary to permanent use. It would also be good to apply this system of construction as one of the possible housing solutions for immigrants, homeless people and unappealing citizens. It is not excluded even its adaptation to the system of individual housing construction where it would be able to cover the needs of so called starting flats.

14. Conclusion

Research of the optimal setting and shape of a wall construction that meets the prerequisites for direct print technology at the building site is the answer to one of the major drawbacks of this construction method. By verifying possible solutions to the problem, it moves the level of knowledge on the subject and creates favourable conditions for the introduction of an automated construction method into building practice.

References

[1] Sobotka A and Pacewicz K Building Site Organization with3D Technology in Use *World Multidisciplinary Civil Engeneering-Architecture-Urban Planning Symposium*

[2] Pifko H 2017 *NEED Navrhovanie energeticky efektívnych domov* 14

[3] Winsun 2018 Winsun-Global Harbor Online: http://www.winsun3d.com/En/About/

[4] Apis Cor 2018 Apis Cor https://www.apis-cor.com/

[5] STN 73 0540-2:2012/71: 216

Acknowledgments

This article was created with the support of a grant project 1/0307/16 "Research on the quality of the indoor environment of buildings for education and education, aimed at protecting the health of children and youth". (1/0307/16 „Výskum kvality vnútorného prostredia budov pre školstvo a vzdelávanie zameraný na ochranu zdravia detí a mládeže“.)